C000127598

8421748

UNIVERSITY OF SURREY LIBRARY

DYNAMIC MODELING AND CONTROL OF ENGINEERING SYSTEMS

THIRD EDITION

This textbook is ideal for a course in Engineering System Dynamics and Controls. The work is a comprehensive treatment of the analysis of lumped-parameter physical systems. Starting with a discussion of mathematical models in general, and ordinary differential equations, the book covers input–output and state-space models, computer simulation, and modeling methods and techniques in mechanical, electrical, thermal, and fluid domains. Frequency-domain methods, transfer functions, and frequency response are covered in detail. The book concludes with a treatment of stability, feedback control (PID, lag–lead, root locus), and an introduction to discrete-time systems. This new edition features many new and expanded sections on such topics as Solving Stiff Systems, Operational Amplifiers, Electrohydraulic Servovalves, Using MATLAB® with Transfer Functions, Using MATLAB with Frequency Response, MATLAB Tutorial, and an expanded Simulink® Tutorial. The work has 40 percent more end-of-chapter exercises and 30 percent more examples.

Bohdan T. Kulakowski, Ph.D. (1942–2006) was Professor of Mechanical Engineering at Pennsylvania State University. He was an internationally recognized expert in automatic control systems, computer simulations and control of industrial processes, systems dynamics, vehicle–road dynamic interaction, and transportation systems. His fuzzy-logic algorithm for avoiding skidding accidents was recognized in 2000 by *Discover* magazine as one of its top 10 technological innovations of the year.

John F. Gardner is Chair of the Mechanical and Biomedical Engineering Department at Boise State University, where he has been a faculty member since 2000. Before his appointment at Boise State, he was on the faculty of Pennsylvania State University in University Park, where his research in dynamic systems and controls led to publications in diverse fields from railroad freight car dynamics to adaptive control of artificial hearts. He pursues research in modeling and control of engineering and biological systems.

J. Lowen Shearer (1921–1992) received his Sc.D. from the Massachusetts Institute of Technology. At MIT, between 1950 and 1963, he served as the group leader in the Dynamic Analysis & Control Laboratory, and as a member of the mechanical engineering faculty. From 1963 until his retirement in 1985, he was on the faculty of Mechanical Engineering at Pennsylvania State University. Professor Shearer was a member of ASME's Dynamic Systems and Control Division and received that group's Rufus Oldenberger Award in 1983. In addition, he received the Donald P. Eckman Award (ISA, 1965), and the Richards Memorial Award (ASME, 1966).

DYNAMIC MODELING AND CONTROL OF ENGINEERING SYSTEMS

THIRD EDITION

Bohdan T. Kulakowski

Deceased, formerly Pennsylvania State University

John F. Gardner

Boise State University

J. Lowen Shearer

Deceased, formerly Pennsylvania State University

SHELF 620· 0011 KUL
SEQ   TYPE
CN

2 8 APR 2008

FUND J59S   DDC 22
BARCODE 8421 748
UNIVERSITY OF SURREY LIBRARY

CAMBRIDGE
UNIVERSITY PRESS

CAMBRIDGE UNIVERSITY PRESS
Cambridge, New York, Melbourne, Madrid, Cape Town, Singapore, São Paulo

Cambridge University Press
32 Avenue of the Americas, New York, NY 10013-2473, USA

www.cambridge.org
Information on this title: www.cambridge.org/9780521864350

© John F. Gardner 2007

This publication is in copyright. Subject to statutory exception
and to the provisions of relevant collective licensing agreements,
no reproduction of any part may take place without
the written permission of Cambridge University Press.

First published 2007

Printed in the United States of America

A catalog record for this publication is available from the British Library.

Library of Congress Cataloging in Publication Data

Kulakowski, Bohdan T.
Dynamic modeling and control of engineering systems / Bohdan T. Kulakowski, John F.
Gardner, J. Lowen Shearer. – 3rd ed.
    p. cm.
Includes bibliographical references and index.
ISBN-13: 978-0-521-86435-0 (hardback)
ISBN-10: 0-521-86435-6 (hardback)
1. Engineering – Mathematical models.   2. System engineering – Mathematical models.
I. Gardner, John F. (John Francis), 1958–   II. Shearer, J. Lowen.   III. Title.
TA342.S54   2007
620.001′1 – dc22   2006031544

Cambridge University Press has no responsibility for
the persistence or accuracy of URLs for external or
third-party Internet Web sites referred to in this publication
and does not guarantee that any content on such
Web sites is, or will remain, accurate or appropriate.

MATLAB® and Simulink® are trademarks of The MathWorks, Inc. and are used with
permission. The MathWorks does not warrant the accuracy of the text or exercises in this
book. This book's use or discussion of MATLAB® and Simulink® software or related
products does not constitute endorsement or sponsorship by The MathWorks of a particular
pedagogical approach or particular use of the MATLAB® and Simulink® software.

Dedicated to the memories of Professor Bohdan T. Kulakowski (1942–2006), the victims of the April 16, 2007 shootings at Virginia Tech, and all who are touched by senseless violence. May we never forget and always strive to learn form history.

Contents

Preface   *page* xi

1 INTRODUCTION   1

  1.1 Systems and System Models   1
  1.2 System Elements, Their Characteristics, and the Role of Integration   4
      Problems   9

2 MECHANICAL SYSTEMS   14

  2.1 Introduction   14
  2.2 Translational Mechanical Systems   16
  2.3 Rotational–Mechanical Systems   30
  2.4 Linearization   34
  2.5 Synopsis   44
      Problems   45

3 MATHEMATICAL MODELS   54

  3.1 Introduction   54
  3.2 Input–Output Models   55
  3.3 State Models   61
  3.4 Transition Between Input–Output and State Models   68
  3.5 Nonlinearities in Input–Output and State Models   71
  3.6 Synopsis   76
      Problems   76

4 ANALYTICAL SOLUTIONS OF SYSTEM INPUT–OUTPUT EQUATIONS   81

  4.1 Introduction   81
  4.2 Analytical Solutions of Linear Differential Equations   82
  4.3 First-Order Models   84
  4.4 Second-Order Models   92
  4.5 Third- and Higher-Order Models   106
  4.6 Synopsis   109
      Problems   111

5 NUMERICAL SOLUTIONS OF ORDINARY DIFFERENTIAL EQUATIONS   120

  5.1 Introduction   120
  5.2 Euler's Method   121
  5.3 More Accurate Methods   124
  5.4 Integration Step Size   129

5.5   Systems of Differential Equations   133
5.6   Stiff Systems of Differential Equations   133
5.7   Synopsis   138
       Problems   139

6   SIMULATION OF DYNAMIC SYSTEMS   141

6.1   Introduction   141
6.2   Simulation Block Diagrams   143
6.3   Building a Simulation   147
6.4   Studying a System with a Simulation   150
6.5   Simulation Case Study: Mechanical Snubber   157
6.6   Synopsis   164
       Problems   165

7   ELECTRICAL SYSTEMS   168

7.1   Introduction   168
7.2   Diagrams, Symbols, and Circuit Laws   169
7.3   Elemental Diagrams, Equations, and Energy Storage   170
7.4   Analysis of Systems of Interacting Electrical Elements   175
7.5   Operational Amplifiers   179
7.6   Linear Time-Varying Electrical Elements   186
7.7   Synopsis   188
       Problems   189

8   THERMAL SYSTEMS   198

8.1   Introduction   198
8.2   Basic Mechanisms of Heat Transfer   199
8.3   Lumped Models of Thermal Systems   202
8.4   Synopsis   212
       Problems   213

9   FLUID SYSTEMS   219

9.1   Introduction   219
9.2   Fluid System Elements   220
9.3   Analysis of Fluid Systems   225
9.4   Electrohydraulic Servoactuator   228
9.5   Pneumatic Systems   235
9.6   Synopsis   243
       Problems   244

10   MIXED SYSTEMS   249

10.1   Introduction   249
10.2   Energy-Converting Transducers and Devices   249
10.3   Signal-Converting Transducers   254
10.4   Application Examples   255
10.5   Synopsis   261
       Problems   261

11	**SYSTEM TRANSFER FUNCTIONS**	**273**
	11.1 Introduction	273
	11.2 Approach Based on System Response to Exponential Inputs	274
	11.3 Approach Based on Laplace Transformation	276
	11.4 Properties of System Transfer Functions	277
	11.5 Transfer Functions of Multi-Input, Multi-Output Systems	283
	11.6 Transfer Function Block-Diagram Algebra	286
	11.7 MATLAB Representation of Transfer Function	293
	11.8 Synopsis	298
	Problems	299

12	**FREQUENCY ANALYSIS**	**302**
	12.1 Introduction	302
	12.2 Frequency-Response Transfer Functions	302
	12.3 Bode Diagrams	307
	12.4 Relationship between Time Response and Frequency Response	314
	12.5 Polar Plot Diagrams	317
	12.6 Frequency-Domain Analysis with MATLAB	319
	12.7 Synopsis	323
	Problems	323

13	**CLOSED-LOOP SYSTEMS AND SYSTEM STABILITY**	**329**
	13.1 Introduction	329
	13.2 Basic Definitions and Terminology	332
	13.3 Algebraic Stability Criteria	333
	13.4 Nyquist Stability Criterion	338
	13.5 Quantitative Measures of Stability	341
	13.6 Root-Locus Method	344
	13.7 MATLAB Tools for System Stability Analysis	349
	13.8 Synopsis	351
	Problems	352

14	**CONTROL SYSTEMS**	**356**
	14.1 Introduction	356
	14.2 Steady-State Control Error	357
	14.3 Steady-State Disturbance Sensitivity	361
	14.4 Interrelation of Steady-State and Transient Considerations	364
	14.5 Industrial Controllers	365
	14.6 System Compensation	378
	14.7 Synopsis	383
	Problems	383

15	**ANALYSIS OF DISCRETE-TIME SYSTEMS**	**389**
	15.1 Introduction	389
	15.2 Mathematical Modeling	390
	15.3 Sampling and Holding Devices	396
	15.4 The z Transform	400

15.5 Pulse Transfer Function   405
15.6 Synopsis   407
      Problems   408

16 DIGITAL CONTROL SYSTEMS   **410**

16.1 Introduction   410
16.2 Single-Loop Control Systems   410
16.3 Transient Performance   412
16.4 Steady-State Performance   418
16.5 Digital Controllers   421
16.6 Synopsis   423
      Problems   424

APPENDIX 1. Fourier Series and the Fourier Transform   **427**

APPENDIX 2. Laplace Transforms   **432**

APPENDIX 3. MATLAB Tutorial   **438**

APPENDIX 4. Simulink Tutorial   **463**

Index   481

Preface

From its beginnings in the middle of the 20th century, the field of systems dynamics and feedback control has rapidly become both a core science for mathematicians and engineers and a remarkably mature field of study. As early as 20 years ago, textbooks (and professors) could be found that purported astoundingly different and widely varying approaches and tools for this field. From block diagrams to signal flow graphs and bond graphs, the diversity of approaches, and the passion with which they were defended (or attacked), made any meeting of systems and control professionals a lively event.

Although the various tools of the field still exist, there appears to be a consensus forming that the tools are secondary to the insight they provide. The field of system dynamics is nothing short of a unique, useful, and utterly different way of looking at natural and manmade systems. With this in mind, this text takes a rather neutral approach to the tools of the field, instead emphasizing insight into the underlying physics and the similarity of those physical effects across the various domains.

This book has its roots as lecture notes from Lowen Shearer's senior-level mechanical engineering course at Penn State in the 1970s with additions from Bohdan Kulakowski's and John Gardner's experiences since the 1980s. As such, it reveals those roots by beginning with lumped-parameter mechanical systems, engaging the student on familiar ground. The following chapters, dealing with types of models (Chapter 3) and analytical solutions (Chapter 4), have seen only minimal revisions from the original version of this text, with the exception of modest changes in order of presentation and clarification of notation. Chapters 5 and 6, dealing with numerical solutions (simulations), were extensively rewritten for the second edition and further updated for this edition. Although we made a decision to feature the industry-standard software package (MATLAB®) in this book (Appendices 3 and 4 are tutorials on MATLAB and Simulink®), the presentation was specifically designed to allow other software tools to be used.

Chapters 7, 8, and 9 are domain-specific presentations of electric, thermal, and fluid systems, respectively. For the third edition, these chapters have been extensively expanded, including operational amplifiers in Chapter 7, an example of lumped approximation of a cooling fin in Chapter 8, and an electrohydraulic servovalve in Chapter 9. Those using this text in a multidisciplinary setting, or for nonmechanical engineering students, may wish to delay the use of Chapter 2 (mechanical systems) to this point, thus presenting the four physical domains sequentially. Chapter 10 presents some important issues in dealing with multidomain systems and how they interact.

Chapters 11 and 12 introduce the important concept of a transfer function and frequency-domain analysis. These two chapters are the most revised and (hopefully) improved parts of the text. In previous editions of this text, we derived the complex transfer function by using complex exponentials as input. For the third edition, we retain this approach, but have added a section showing how to achieve the same ends using the Laplace transform. It is hoped that this dual approach will enrich student understanding of this material. In approaching these, and other, revisions, we listened carefully to our colleagues throughout the world who helped us see where the presentation could be improved. We are particularly grateful to Sean Brennan (of Penn State) and Giorgio Rizzoni (of Ohio State) for their insightful comments.

This text, and the course that gave rise to it, is intended to be a prerequisite to a semester-long course in control systems. However, Chapters 13 and 14 present a very brief discussion of the fundamental concepts in feedback control, stability (and algebraic and numerical stability techniques), closed-loop performance, and PID and simple cascade controllers. Similarly, the preponderance of digitally implemented control schemes necessitates a discussion of discrete-time control and the dynamic effects inherent in sampling in the final chapters (15 and 16). It is hoped that these four chapters will be useful both for students who are continuing their studies in electives or graduate school and for those for which this is a terminal course of study.

Supplementary materials, including MATLAB and Simulink files for examples throughout the text, are available through the Cambridge University Press web site (http://www.cambridge.org/us/engineering) and readers are encouraged to check back often as updates and additional case studies are made available.

Outcomes assessment, at the program and course level, has now become a fixture of engineering programs. Although necessitated by accreditation criteria, many have discovered that an educational approach based on clearly stated learning objectives and well-designed assessment methods can lead to a better educational experience for both the student and the instructor. In the third edition, we open each chapter with the learning objectives that underlie each chapter. Also in this edition, the examples and end-of-chapter problems, many of which are based on real-world systems encountered by the authors, were expanded.

This preface closes on a sad note. In March of 2006, just as the final touches were being put on this edition, Bohdan Kulakowski was suddenly and tragically taken from us while riding his bicycle home from the Penn State campus, as was his daily habit. His family, friends, and the entire engineering community suffered a great loss, but Bohdan's legacy lives on in these pages, as does Lowen's. As the steward of this legacy, I find myself "standing on the shoulders of giants" and can take credit only for its shortcomings.

JFG
Boise, ID
May, 2007

DYNAMIC MODELING AND CONTROL OF ENGINEERING SYSTEMS

1

Introduction

LEARNING OBJECTIVES FOR THIS CHAPTER

1–1 To work comfortably with the engineering concept of a "system" and its interaction with the environment through inputs and outputs.

1–2 To distinguish among various types of mathematical models used to represent and predict the behavior of systems.

1–3 To recognize through (T-type) variables and across (A-type) variables when examining energy transfer within a system.

1–4 To recognize analogs between corresponding energy-storage and energy-dissipation elements in different types of dynamic systems.

1–5 To understand the key role of energy-storage processes in system dynamics.

1.1   SYSTEMS AND SYSTEM MODELS

The word "system" has become very popular in recent years. It is used not only in engineering but also in science, economics, sociology, and even in politics. In spite of its common use (or perhaps because of it), the exact meaning of the term is not always fully understood. A system is defined as a combination of components that act together to perform a certain objective. A little more philosophically, a system can be understood as a conceptually isolated part of the universe that is of interest to us. Other parts of the universe that interact with the system comprise the system environment, or neighboring systems.

All existing systems change with time, and when the rates of change are significant, the systems are referred to as dynamic systems. A car riding over a road can be considered as a dynamic system (especially on a crooked or bumpy road). The limits of the conceptual isolation determining a system are entirely arbitrary. Therefore any part of the car given as an example of a system – its engine, brakes, suspension, etc. – can also be considered a system (i.e., a subsystem). Similarly, two cars in a passing maneuver or even all vehicles within a specified area can be considered as a major traffic system.

The isolation of a system from the environment is purely conceptual. Every system interacts with its environment through two groups of variables. The variables in the first group originate outside the system and are not directly dependent on what happens in the system. These variables are called input variables, or simply inputs. The other group comprises variables generated by the system as it interacts with its

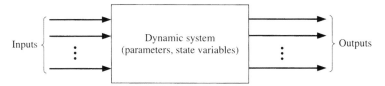

Figure 1.1. A dynamic system.

environment. Those dependent variables in this group that are of primary interest to us are called output variables, or simply outputs.

In describing the system itself, one needs a complete set of variables, called state variables. The state variables constitute the minimum set of system variables necessary to describe completely the state of the system at any given instant of time; and they are of great importance in the modeling and analysis of dynamic systems. Provided the initial state and the input variables have all been specified, the state variables then describe from instant to instant the behavior, or response, of the system. The concept of the state of a dynamic system is discussed in more detail in Chap. 3. In most cases, the state-variable equations used in this text represent only simplified models of the systems, and their use leads to only approximate predictions of system behavior.

Figure 1.1 shows a graphical presentation of a dynamic system. In addition to the state variables, parameters also characterize the system. In the example of the moving car, the input variables would include throttle position, position of the steering wheel, and road conditions such as slope and roughness. In the simplest model, the state variables would be the position and velocity of the vehicle as it travels along a straight path. The choice of the output variables is arbitrary, determined by the objectives of the analysis. The position, velocity, or acceleration of the car, or perhaps the average fuel flow rate or the engine temperature, can be selected as the output(s). Some of the system parameters would be the mass of the vehicle and the size of its engine. Note that the system parameters may change with time. For instance, the mass of the car will change as the amount of fuel in its tank increases or decreases or when passengers embark or disembark. Changes in mass may or may not be negligible for the performance of a car but would certainly be of critical importance in the analysis of the dynamics of a ballistic missile.

The main objective of system analysis is to predict the manner in which a system will respond to various inputs and how that response changes with different system parameter values. In the absence of the tools introduced in this book, engineers are often forced to build prototype systems to test them. Whereas the data obtained from the testing of physical prototypes are very valuable, the costs, in time and money, of obtaining these data can be prohibitive. Moreover, mathematical models are inherently more flexible than physical prototypes and allow for rapid refinement of system designs to optimize various performance measures. Therefore one of the early major tasks in system analysis is to establish an adequate mathematical model that can be used to gain the equivalent information that would come from several different physical prototypes. In this way, even if a final prototype is built to verify the mathematical model, the modeler has still saved significant time and expense.

A mathematical model is a set of equations that completely describes the relationships among the system variables. It is used as a tool in developing designs or control algorithms, and the major task for which it is to be used has basic implications for the choice of a particular form of the system model.

In other words, if a model can be considered a tool, it is a specialized tool, developed specifically for a particular application. Constructing universal mathematical models, even for systems of moderate complexity, is impractical and uneconomical. Let us use the moving automobile as an example once again. The task of developing a model general enough to allow for studies of ride quality, fuel economy, traction characteristics, passenger safety, and forces exerted on the road pavement (to name just a few problems typical for transportation systems) could be compared to the task of designing one vehicle to be used as a truck, for daily commuting to work in New York City, and as a racing car to compete in the Indianapolis 500. Moreover, even if such a supermodel were developed and made available to researchers (free), it is very likely that the cost of using it for most applications would be prohibitive.

Thus, system models should be as simple as possible, and each model should be developed with a specific application in mind. Of course, this approach may lead to different models being built for different uses of the same system. In the case of mathematical models, different types of equations may be used in describing the system in various applications.

Mathematical models can be grouped according to several different criteria. Table 1.1 classifies system models according to the four most common criteria: applicability of the principle of superposition, dependence on spatial coordinates as well

Table 1.1. Classification of system models

Type of model	Classification criterion	Type of model equation
Nonlinear	Principle of superposition does not apply	Nonlinear differential equations
Linear	Principle of superposition applies	Linear differential equations
Distributed	Dependent variables are functions of spatial coordinates and time	Partial differential equations
Lumped	Dependent variables are independent of spatial coordinates	Ordinary differential equations
Time-varying	Model parameters vary in time	Differential equations with time-varying coefficients
Stationary	Model parameters are constant in time	Differential equations with constant coefficients
Continuous	Dependent variables defined over continuous range of independent variables	Differential equations
Discrete	Dependent variables defined only for distinct values of independent variables	Time-difference equations

as on time, variability of parameters in time, and continuity of independent variables. Based on these criteria, models of dynamic systems are classified as linear or nonlinear, lumped or distributed, stationary time invariant or time varying, continuous or discrete, respectively. Each class of models is also characterized by the type of mathematical equations employed in describing the system. All types of system models listed in Table 1.1 are discussed in this book, although distributed models are given only limited attention.

1.2 SYSTEM ELEMENTS, THEIR CHARACTERISTICS, AND THE ROLE OF INTEGRATION

The modeling techniques developed in this text focus initially on the use of a set of simple ideal system elements found in four main types of systems: mechanical, electrical, fluid, and thermal. Transducers, which enable the coupling of these types of system to create mixed-system models, will be introduced later.

   This set of ideal linear elements is shown in Table 1.2, which also provides their elemental equations and, in the case of energy-storing elements, their energy-storage equations in simplified form. The variables, such as force F and velocity v used in mechanical systems, current i and voltage e in electrical systems, fluid flow rate Q_f and pressure P in fluid systems, and heat flow rate Q_h and temperature T in thermal systems, have also been classified as either T-type (through) variables, which act through the elements, or A-type (across) variables, which act across the elements. Thus force, current, fluid flow rate, and heat flow rate are called T variables, and velocity, voltage, pressure, and temperature are called A variables. Note that these designations also correspond to the manner in which each variable is measured in a physical system. An instrument measuring a T variable is used in series to measure what goes *through* the element. On the other hand, an instrument measuring an A variable is connected in parallel to measure the difference *across* the element. Furthermore, the energy-storing elements are also classified as T-type or A-type elements, designated by the nature of their respective energy-storage equations: for example, mass stores kinetic energy, which is a function of its velocity, an A variable; hence mass is an A-type element. Note that although T and A variables have been identified for each type of system in Table 1.2, both T-type and A-type energy-storing elements are identified in mechanical, electrical, and fluid systems only. In thermal systems, the A-type element is the thermal capacitor but there is no T-type element that would be capable of storing energy by virtue of a heat flow through the element.

   In developing mathematical models of dynamic systems, it is very important not only to identify all energy-storing elements in the system but also to determine how many energy-storing elements are independent or, in other words, in how many elements the process of energy storage is independent. The energy storage in an element is considered to be independent if it can be given any arbitrary value without changing any previously established energy storage in other system elements. To put it simply, two energy-storing elements are not independent if the amount of energy stored in one element completely determines the amount of energy stored in the other element. Examples of energy-storing elements that are not independent are rack-and-pinion gears, and series and parallel combinations of springs, capacitors, inductors,

Table 1.2. Ideal system elements (linear)

System type	Mechanical translational	Mechanical rotational	Electrical	Fluid	Thermal
A-type variable	Velocity, v	Velocity, Ω	Voltage, e	Pressure, P	Temperature, T
A-type element	Mass, m	Mass moment of inertia, J	Capacitor, C	Fluid Capacitor, C_f	Thermal capacitor, C_h
Elemental equations	$F = m\dfrac{dv}{dt}$	$T = J\dfrac{d\Omega}{dt}$	$i = C\dfrac{de}{dt}$	$Q_f = C_f\dfrac{dP}{dt}$	$Q_h = C_h\dfrac{dT}{dt}$
Energy stored	Kinetic	Kinetic	Electric field	Potential	Thermal
Energy equations	$\mathscr{E}_k = \dfrac{1}{2}mv^2$	$\mathscr{E}_k = \dfrac{1}{2}J\Omega^2$	$\mathscr{E}_e = \dfrac{1}{2}Ce^2$	$\mathscr{E}_p = \dfrac{1}{2}C_f P^2$	$\mathscr{E}_t = \dfrac{1}{2}C_h T^2$
T-type variable	Force, F	Torque, T	Current, i	Fluid flow rate, Q_f	Heat flow rate, Q_h
T-type element	Compliance, $1/k$	Compliance, $1/K$	Inductor, L	Inertor, I	None
Elemental equations	$v = \dfrac{1}{k}\dfrac{dF}{dt}$	$\Omega = \dfrac{1}{K}\dfrac{dT}{dt}$	$e = L\dfrac{di}{dt}$	$P = I\dfrac{dQ_f}{dt}$	
Energy stored	Potential	Potential	Magnetic field	Kinetic	
Energy equations	$\mathscr{E}_P = \dfrac{1}{2k}F^2$	$\mathscr{E}_P = \dfrac{1}{2K}T^2$	$\mathscr{E}_m = \dfrac{1}{2}Li^2$	$\mathscr{E}_k = \dfrac{1}{2}IQ_f^2$	
D-type element	Damper, b	Rotational damper, B	Resistor, R	Fluid resistor, R_f	Thermal resistor, R_h
Elemental equations	$F = bv$	$T = B\Omega$	$i = \dfrac{1}{R}e$	$Q_f = \dfrac{1}{R_f}P$	$Q_h = \dfrac{1}{R_h}T$
Rate of energy dissipated	$\begin{aligned}\dfrac{dE_D}{dt} &= Fv \\ &= \dfrac{1}{b}F^2 \\ &= bv^2\end{aligned}$	$\begin{aligned}\dfrac{dE_D}{dt} &= T\Omega \\ &= \dfrac{1}{B}T^2 \\ &= B\Omega^2\end{aligned}$	$\begin{aligned}\dfrac{dE_D}{dt} &= ie \\ &= Ri^2 \\ &= \dfrac{1}{R}e^2\end{aligned}$	$\begin{aligned}\dfrac{dE_D}{dt} &= Q_f P \\ &= R_f Q_f^2 \\ &= \dfrac{1}{R_f}P^2\end{aligned}$	$\dfrac{dE_D}{dt} = Q_h$

Note: A-type variable represents a spatial difference across the element.

etc. As demonstrated in the following chapters, the number of independent energy storing elements in a system is equal to the order of the system and to the number of state variables in the system model.

The A-type elements are said to be analogous to each other; T-type elements are also analogs of each other. This physical analogy is also demonstrated mathematically by the same form of the elemental equations for each type of element. The general form of the elemental equations for an A-type element in mechanical, electrical, fluid, and thermal systems is

$$V_T = E_A \frac{dV_A}{dt}, \tag{1.1}$$

where V_T is a T variable, V_A is an A variable, and E_A is the parameter associated with an A-type element. The general form of the elemental equations for a T-type element in mechanical, electrical, and fluid systems is

$$V_A = E_T \frac{dV_T}{dt}. \tag{1.2}$$

Equation (1.2) does not apply to thermal systems because of lack of a T-type element in those systems.

Because differentiation is seldom, if ever, encountered in nature, whereas integration is very commonly encountered, the essential dynamic character of each energy-storage element is better expressed when its elemental equation is converted from differential form to integral form. Thus general elemental equations (1.1) and (1.2) in integral form are

$$V_A(t) = V_A(0) + \frac{1}{E_A} \int_0^t V_T dt, \tag{1.3}$$

$$V_T(t) = V_T(0) + \frac{1}{E_T} \int_0^t V_A dt. \tag{1.4}$$

To better understand the physical significance of integral equations (1.3) and (1.4), consider a mechanical system. The A-type element in a mechanical system is mass, and the equation corresponding to Eq. (1.3) is

$$v(t) = v(0) + \frac{1}{m} \int_0^t F dt. \tag{1.5}$$

This equation states that the velocity of a given mass m increases as the integral (with respect to time) of the net force applied to it. This concept is formally known as Newton's second law of motion. It also implicitly says that, lacking a very, very large (infinite) force F, the velocity of mass m cannot change instantaneously. Thus the kinetic energy $\mathscr{E}_k = (m/2)v^2$ of the mass m is also accumulated over time when the force F is finite and cannot be changed in zero time.

The integral equation for a T-type element in mechanical systems, compliance $(1/k)$, corresponding to Eq. (1.4) is

$$F_k(t) = F_k(0) + k \int_0^t v_{21} dt, \tag{1.6}$$

where F_k is the force transmitted by the spring k and v_{21} is the velocity of one end of the spring relative to the velocity at the other end. This equation states that the spring force F_k cannot change instantaneously and thus the amount of potential energy stored in the spring $\mathcal{E}_p = (1/2k)F_k^2$ is accumulated over time and cannot be changed in zero time in a real system. Although Eq. (1.6) might seem to be a particularly clumsy statement of Hooke's law for springs ($F = kx$), it is essential for the purposes of system dynamic analysis that the process of storing energy in the spring as one of a cumulative process (integration) over time.

Similar elemental equations in integral form may be written for all the other energy-storage elements, and similar conclusions can be drawn concerning the role of integration with respect to time and how it affects the accumulation of energy with respect to time. These two phenomena, integration and energy storage, are very important aspects of dynamic system analysis, especially when energy-storage elements interact and exchange energy with each other.

The energy-dissipation elements, or D elements, store no useful energy and have elemental equations that express instantaneous relationships between their A variables and their T variables, with no need to wait for time integration to take effect. For example, the force in a damper is instantaneously related to the velocity difference across it (i.e., no integration with respect to time is involved).

Furthermore, these energy dissipators absorb energy from the system and exert a "negative-feedback" effect (to be discussed in detail later), which provides damping and helps ensure system stability.

EXAMPLE 1.1

Consider a simplified diagram of one-fourth of an automobile, often referred to as a "quarter-car" model, shown schematically in Fig. 1.2. Such a model of vehicle dynamics is useful when only bounce (vertical) motion of the car is of interest, whereas both pitch and roll motions can be neglected.

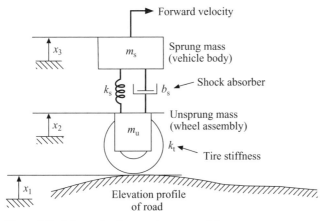

Figure 1.2. Schematic of a quarter-car model.

Table 1.3. Elements of the quarter-car model

Element	Element type	Type of energy stored	Energy equation
m_s	A-type energy storing	Kinetic	$\mathcal{E}_k = \dfrac{1}{2} m_s v_3^2$
m_u	A-type energy storing	Kinetic	$\mathcal{E}_k = \dfrac{1}{2} m_u v_2^2$
k_s	T-type energy storing	Potential	$\mathcal{E}_p = \dfrac{1}{2} k_s (x_2 - x_3)^2$
k_t	T-type energy storing	Potential	$\mathcal{E}_p = \dfrac{1}{2} k_t (x_1 - x_2)^2$
b_s	D-type energy dissipating	None	$\dfrac{d\mathcal{E}_D}{dt} = b_s (v_2 - v_3)^2$

List all system elements, indicate their type, and write their respective energy equations. Draw input–output block diagrams, such as that shown in Fig. 1.1, showing what you consider to be the input variables and output variables for two cases:

(a) in a study of passenger ride comfort, and
(b) in a study of dynamic loads applied by vehicle tires to road pavement.

SOLUTION

There are four independent energy-storing elements, m_s, m_u, k_s, and k_t. There is also one energy-dissipating element, damper b_s, representing the shock absorber. The system elements, their respective types, and energy-storage or -dissipation equations are given in Table 1.3.

The input variable to the model is the history of the elevation profile, $x_1(t)$, of the road surface over which the vehicle is traveling. In most cases, the elevation profile is measured as a function of distance traveled, and it is then combined with vehicle forward velocity data to obtain $x_1(t)$.

In studies of ride comfort, the main variable of interest is usually acceleration of the vehicle body,

$$a_3 = \frac{dv_3}{dt}.$$

In studies of dynamic tire loads, on the other hand, the variable of interest is the vertical force applied by the tire to the road surface:

$$F_t = k_t (x_1 - x_2).$$

Simple block diagrams for the two cases are shown in Fig. 1.3. There is an important observation to make in the context of this example. When a given physical system is modeled, different output variables can be selected as needed for the modeling task at hand.

(a)

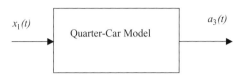

Figure 1.3. Block diagrams of the quarter-car models used in (a) ride comfort and (b) dynamic tire load studies.

(b)

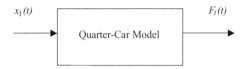

PROBLEMS

1.1 Using an input–output block diagram, such as that shown in Fig. 1.1. show what you consider to be the input variables and the output variables for an automobile engine, shown schematically in Fig. P1.1.

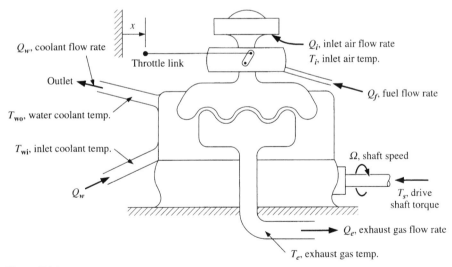

Figure P1.1.

1.2 For the automotive alternator shown in Fig. P1.2, prepare an input–output diagram showing what you consider to be inputs and what you consider to be outputs.

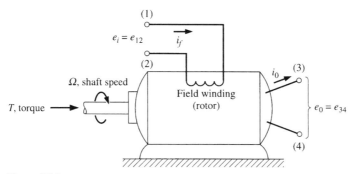

Figure P1.2.

1.3 Prepare an input–output block diagram showing what you consider to be the inputs and the outputs for the domestic hot water furnace shown schematically in Fig. P1.3.

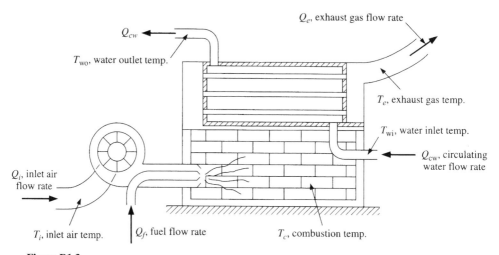

Figure P1.3.

1.4 A simple drawing of a hand-powered egg beater is shown in Fig. P1.4. The handle driven by torque T_i turns a large double-sided crown wheel, which in turn drives two bevel pinions to spin the beaters. The diameter of the crown wheel is much larger than the diameter of the bevel pinions, so the rotational velocity of the pinions (and the beaters), Ω_b, is much greater the rotational velocity of the crown wheel, Ω_c. The shafts that connect the bevel pinions to the beaters are slightly compliant, and the beaters experience a frictional resistance as they spin while beating the eggs. Make reasonable simplifying assumptions and list all elements that you would include in a mathematical model of the egg beater, indicate their types, and write their corresponding energy equations. How many independent energy-storing elements are in the system?

1.5 Figure P1.5 is a schematic representation of a wind turbine used for irrigation. The turbine is located on the rim of a canyon where the wind speed (V_w) is highest. The velocity of the rotor (the blade assembly) is Ω_r, and the electrical generator runs at Ω_g because of a gearbox in the nacelle of the wind turbine. Electrical power (V_e and I_e) is supplied to the motor and pump, located on the riverbank at the bottom of the canyon.

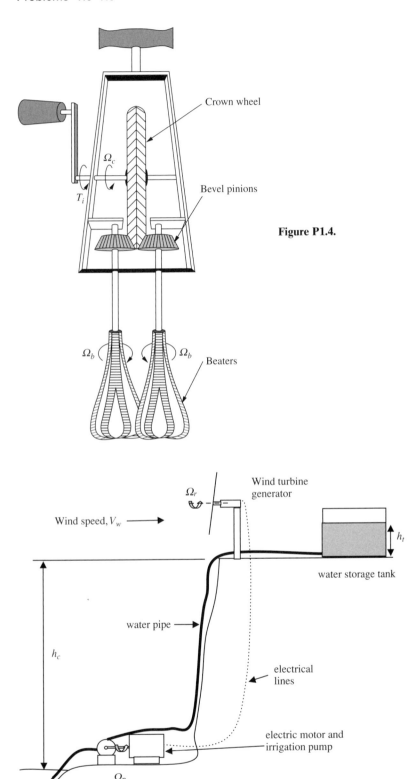

Figure P1.4.

Crown wheel

Bevel pinions

Ω_c

T_i

Ω_b Ω_b Beaters

Wind speed, V_w

Ω_r

Wind turbine
generator

h_t

water storage tank

water pipe

h_c

electrical
lines

electric motor and
irrigation pump

Ω_p

Figure P1.5. Wind turbine system to power a canyon irrigation pump.

The motor spins the pump at a fixed velocity, Ω_p, and the water is pumped to a holding pond above the canyon rim at a height of h_c above the river.

(a) Prepare an input–output block diagram of the wind turbine–generator as a system. Identify the energy-storing elements.

(b) Prepare an input–output block diagram of the motor–pump as a system.

(c) Prepare an input–output block diagram of this system and specify the energy-storing elements of the system. For each energy-storing element, chose the appropriate A- or T-type element from Table 1.2 that would be appropriate for each component of the system. Briefly explain your choices.

1.6 A schematic representation of an artificial human heart is shown in Fig. P1.6. The heart consists of two flexible chambers (blood sacs) enclosed in a rigid case. The chambers are alternately squeezed by flat plates that are, in turn, moved back and forth by a dc motor and rollerscrew arrangement. Two one-way valves (prosthetic heart valves) are used in each blood sac to ensure directional flow. The motor is powered by a battery pack through a controller circuit. The blood sac shown on the left-hand side in the figure (the right heart) takes blood returning from the body (through the vena cavae) and pumps it to the lungs (through the pulmonary artery). Oxygenated blood returns from the lungs (via the pulmonary veins) to the left heart where it is then pumped to the body through the aorta.

Identify the energy-storing elements of the system. If the intent of the model is to design a control system, what would be appropriate inputs and outputs of this system?

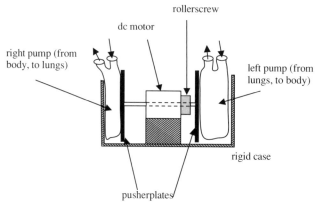

Figure P1.6. Schematic of a total artificial heart.

1.7 As mobile electronic systems become more efficient and require less energy, it becomes more attractive to generate power by scavenging energy from the motion inherent in moving these systems around. Analogous to the self-winding wristwatches that were popular in the 1970s, this concept makes use of a mechanical oscillator coupled with an electromagnetic generator to keep a battery charged.

Figure P1.7 shows a schematic of just such a system, proposed for a cell phone. The motion the phone experiences while in your backpack as you walk across campus is given as y_b in the figure. The mass (m) and spring (k) make up the mechanical oscillator. The mass is actually a permanent magnet that generates a time-varying (because of its motion) magnetic field in the vicinity of wire coil. The coil has both resistance (R) and inductance (L) and a current (i) is induced in the coil because of the motion of the magnet. This current supplies a charging circuit that maintains a set voltage (e) on the battery.

Identify the energy-storing elements of this system and draw an input–output block diagram. From Table 1.2, identify the A- and T-type elements appropriate for each energy-storing element and briefly discuss your choices.

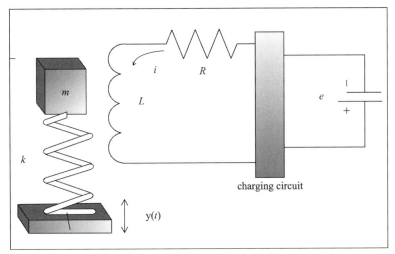

Figure P1.7. Schematic of energy-generating oscillator.

1.8 A motorized wheelchair uses a battery pack to supply two dc motors that, in turn, drive the left and right wheels through a belt transmission. The wheelchair is controlled through a joystick that allows the user to select forward and backward rotation of the wheels. The user accomplishes turning by running the two motors in opposite directions, thus rotating the chair about a vertical axis. The battery voltage is E_b, the internal moving parts of the motors have rotational inertia (J_a), and the electrical coils of the motors have both resistance (R_a) and inductance (L_a). The belt transmissions drive the wheels, which have rotational inertia (J_w) through a drive ratio given by the parameter N. The mass of the chair and rider is given by m_t and the grade is given by the angle γ.

Is the energy stored in the motor armatures, wheels, and chair (because of kinetic energy of the inertias) independent? Why or why not?

2

Mechanical Systems

LEARNING OBJECTIVES FOR THIS CHAPTER

2–1 To apply constitutive equations for the fundamental translational and rotational mechanical components: lumped mass, springs, and dampers.

2–2 To derive correct equations of motion for systems involving multiple instances of these fundamental components, some of which may be nonlinear.

2–3 To apply a linearization procedure based on Taylor series expansion to approximate nonlinear systems with simplified linear models.

2.1 INTRODUCTION

As indicated in Chap. 1, three basic ideal elements are available for modeling elementary mechanical systems: masses, springs, and dampers. Although each of these elements is itself a system with all the attributes of a system (inputs, parameters, state variables, and outputs), the use of the term "system" usually implies a combination of interacting elements. In this chapter, systems composed of only mechanical elements are discussed. In addition to the translational elements (moving along a single axis) introduced in Chap. 1, a corresponding set of rotational elements (rotating about a single axis) is introduced to deal with rotational–mechanical systems and mixed (translational and rotational) systems.

Also, this chapter deals with only so-called lumped-parameter models of real mechanical systems. In certain situations, such as modeling a real spring having both mass and stiffness uniformly distributed from one end to the other, suitable lumped-parameter models can be conceived that will adequately describe the system under at least limited conditions of operation. For example, if a real spring is compressed very slowly, the acceleration of the distributed mass is very small so that all the force acting on one end is transmitted through it to the other end; under these conditions the spring may be modeled as an ideal spring. On the other hand, if the real spring is being driven in such a manner that the forces acting on it cause negligible deflection of its coils, it may be modeled as an ideal mass. When a real spring is driven so as to cause both significant acceleration of its mass and significant deflection of its coils, combinations of lumped ideal mass(es) and lumped ideal spring(s) are available to model the real spring adequately, depending on the type of vibration induced in it. And, if the real spring is being driven at frequencies well below the lowest-frequency mode of vibration for that type of forcing, simple combinations of ideal masses and

(a) Real spring (distributed m and k)

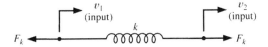

(b) Ideal spring (negligible acceleration; velocity driven)

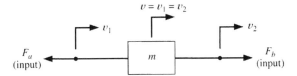

(c) Ideal mass (negligible deflection; force driven)

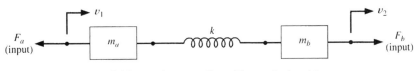

(d) Mass–spring–mass (force driven, at both ends)

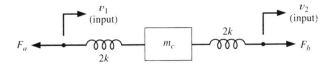

(e) Spring–mass–spring (velocity driven at both ends)

(f) Mass–spring (force driven at mass end; velocity driven at spring end)

Figure 2.1. Several possible lumped-parameter models of a real spring. In (d), (e), and (f), the choice of values for m_a or m_b and m_c depends on relative amplitudes of the forcing functions, i.e., the inputs.

ideal springs will usually suffice to model it adequately. These combinations are shown for the corresponding forcing conditions in Fig. 2.1.

As the frequencies of the forcing functions approach the lowest natural frequency for that type of forcing, the use of lumped-parameter models becomes questionable; unless a many-element model is used (i.e., a finite-element model), the formulation of the partial differential equations for a distributed-parameter model is advisable.

Additionally, it is important to note that essentially every existing engineering system is nonlinear when considered over the entire possible (even if sometimes not

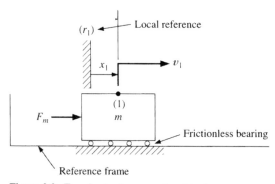

Figure 2.2. Free-body diagram of an ideal mass.

practical) range of its input variables. However, developing and solving mathematical models for nonlinear systems is usually much more difficult than it is for systems that can reasonably be considered to be linear. In Section 2.4, a systematic linearization procedure is introduced whereby the nonlinear system characteristics are replaced with approximate linear formulas over a relatively small range of variations of the input variables.

2.2   TRANSLATIONAL MECHANICAL SYSTEMS

2.2.1 Translational Masses

Analysis of mechanical systems is based on the principles embodied in Newton's laws of motion and the principle of compatibility (no gaps between connected elements). An ideal mass, depicted schematically in free-body diagram form in Fig. 2.2, moves in relation to a nonaccelerating frame of reference, which is usually taken to be a fixed point on the earth (ground) – however, the frame of a nonaccelerating vehicle could be used instead.

The elemental equation for an ideal mass m, based on Newton's second law,[1] is

$$m\left(\frac{dv_1}{dt}\right) = F_m,\tag{2.1}$$

where v_1 is the velocity of the mass m relative to the ground reference point and F_m is the net force (i.e., the sum of all the applied forces) acting on the mass in the x direction.

Because $v_1 = dx_1/dt$, the variation of the distance x_1 of the mass from the reference point is related to F_m by

$$m\left(\frac{d^2 x_1}{dt^2}\right) = F_m.\tag{2.2}$$

---

[1] Newton's second law expressed in more general form is

$$m\left(\frac{d^2 x}{dt^2}\right) = \sum_{i=1}^{n} F_i.$$

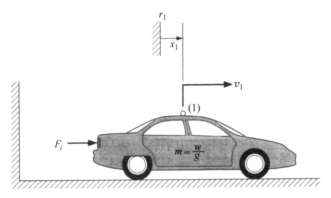

Figure 2.3. Schematic diagram of automobile with applied force.

EXAMPLE 2.1

Find the response (in terms of its acceleration, velocity, and position versus time) of a 3000-lb automobile to a force F_i of 500 lb, which is suddenly applied by three members of the football squad (i.e., a step change in force occurring at $t = 0$), ignoring friction effects (see Fig. 2.3). Assume that the football players are able to maintain the applied force of 500 lb regardless of how fast the automobile moves. Reference r_1 is the local ground reference – in other words, the starting point for vehicle motion.

SOLUTION

Using Eq. (2.1) – that is, ignoring friction effects – we have

$$\frac{dv_1}{dt} = \left(\frac{1}{m}\right) F_i(t). \tag{2.3}$$

Because the vehicle acceleration $a_1(t) = dv_1/dt$, we see that it undergoes a step change from 0 to $(32.2)(500)/3000$ at $t = 0$ and remains at that value until the applied force is removed. Next, we may separate variables in Eq. (2.3) and integrate with respect to time to solve for $v_1 = (t)$,

$$\int_{v_1(0)}^{v_1(t)} dv_1 = \left(\frac{1}{m}\right) \int_0^t F_i(t)dt, \tag{2.4}$$

which yields

$$v_1(t) - v_1(0) = \frac{(32.2)}{6}t - 0$$

or

$$v_1(t) = v_1(0) + 5.37t \text{ ft s; i.e., a ramp starting at } t = 0 \text{ having a slope of}$$
$$5.37 \text{ ft/s}^2. \text{ Similarly, a second integration (assuming the initial}$$
$$\text{velocity is zero) with respect to time yields} \tag{2.5}$$

$$\int_{x_1(0)}^{x_1(t)} dx = 5.37 \int_0^t t dt$$

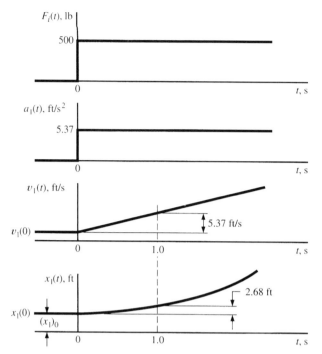

Figure 2.4. Response of an automobile to a suddenly applied force.

or

$$x_1(t) = x_1(0) + 2.68t^2 \text{ ft}; \quad \text{i.e., a parabola starting at } t = 0. \tag{2.6}$$

The results are shown as functions of time in Fig. 2.4 along with the input force F_i.

Note that it takes time to build up changes in velocity and displacement because of the integrations involved. Example 2.1 displays one of the first indications of the role played by integration in determining the dynamic response of a system. From another point of view, the action of the applied force represents work being done on the mass as it accelerates, increasing the kinetic energy stored in it as time goes by. The rate at which energy is stored in the system is equal to the rate at which work is expended on it by the members of the football squad (the first law of thermodynamics):

$$\frac{d\mathscr{E}_K}{dt} = F_i v_1. \tag{2.7}$$

Separating variables and integrating with respect to time, we find

$$\int_{\mathscr{E}_k(0)}^{\mathscr{E}_k(t)} d\mathscr{E}_K = F_i v_1 dt = \int_0^t m v_1 \left(\frac{dv_1}{dt} \right) dt = m \int_{v_1(0)}^{v_1(t)} v_1(t) dv_1,$$

so that

$$\mathscr{E}_K(t) = \mathscr{E}_K(0) + \left(\frac{m}{2} \right) v_1(t)^2. \tag{2.8}$$

Thus the stored energy accumulates over time, proportional to the square of the velocity, as the work is being done on the system; and the mass is an A-type element storing energy that is a function of the square of its A variable v_1.

As this text proceeds to the analysis of more complex systems, the central role played by integration in shaping dynamic system response will become more and more evident.

It should be noted in passing that it would *not* be reasonable to try to impose on the automobile a step change in velocity, because this would be an impossible feat for three football players – or even for 10 million football players! Such a feat would require a very, very great (infinite) force as well as a very, very great (infinite) source of power – infinite sources! Considering that one definition of infinity is that it is a number greater than the greatest possible imaginable number, would it be likely to find or devise an infinite force and an infinite power source?

2.2.2 Translational Springs

An ideal translational spring that stores potential energy as it is deflected along its axis may also be depicted within the same frame of reference used for a mass. Figure 2.5 shows such a spring in two ways: in mechanical drawing format in relaxed state with $F_k = 0$ [Fig. 2.5(a)] and in stylized schematic form [Fig. 2.5(b)] with the left end displaced relative to the right end as a result of the action of F_k shown acting at both ends in free-body diagram fashion. The references r_1 and r_2 are local ground references.

Note that, because an ideal spring contains no mass, the force transmitted by it is undiminished during acceleration; therefore the forces acting on its ends must always be equal and opposite (Newton's third law of motion). The elemental equation for such a spring derives from Hooke's law, namely,

$$F_k = k[x_{21} - (x_{21})_0],$$ (2.9)

where $(x_{21})_0$ is the free length of the spring, i.e., its length when $F_k = 0$. Differentiating Eq. (2.9) with respect to time yields

$$\frac{dF_k}{dt} = kv_{21},$$ (2.10)

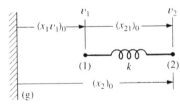

(a) Initial relaxed state, no force acting, no displacement

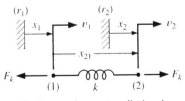

(b) Force acting, system displaced

Figure 2.5. Free-body diagram representation of an ideal spring.

where $v_{21} = dx_{21}/dt$ is the velocity of the right-hand end of the spring, point (2), relative to the velocity of the left-hand end, point (1). Because $[x_{21} - (x_{21})_0] = x_2 - x_1$, Eq. (2.9) is simplified to

$$F_k = k(x_2 - x_1), \tag{2.11}$$

where $(x_2 - x_1)$ is the deflection of the spring is from its initial free length.

When the spring is nonlinear, it does not have truly constant stiffness k, and it is advisable to employ a nonlinear symbol designation and a nonlinear function to describe it, as follows:

$$F_{NLS} = f_{NL}(x_2 - x_1).$$

Linearization of this equation for small perturbations will yield an incremental stiffness k_{inc} that is often adequate for use in a small region near the normal operating point for the spring, as discussed in Section 2.4.

EXAMPLE 2.2

Find the response (in terms of force F_k and deflection x_1) of the spring shown in Fig. 2.6, having $k = 8000$ lb/in., when it is subjected to a 20-in./s step change in input velocity from zero, starting from its free length at $t = 0$.

SOLUTION

Separating variables in Eq. (2.10) and integrating with respect to time, we have

$$\int_{F_k(0)}^{F_k(t)} dF_k = k \int_0^t v_1 dt = (8000)(20) \int_0^t dt,$$

$$F_k(t) - 0 = (8000)(20)t - 0,$$

or

$$F_k(t) = 160{,}000t \text{ lb.} \tag{2.12}$$

We may use the definition

$$v_1 \equiv \frac{dx_1}{dt}. \tag{2.13}$$

Alternatively

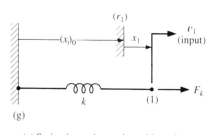

(a) Spring in tension, x_1 is positive when F_k is positive

(b) Spring in compression, x_1 is positive when F_k is positive

Figure 2.6. An ideal spring subjected to a step change in velocity.

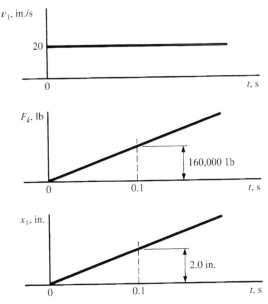

Figure 2.7. Responses for an ideal spring subjected to a step change in velocity.

Separating variables in Eq. (2.13) and integrating again with respect to time yields

$$\int_{x_1(0)}^{x_1(t)} dx_1 = \int_0^t v_1 dt = \int_0^t 20\, dt,$$

or

$$x_1(t) - 0 = 20t - 0,$$

or

$$x_1(t) = 20t \text{ in.} \tag{2.14}$$

The results are shown in Fig. 2.7.

Again, the essential role played by integration in finding the dynamic response of a system is evident. From an energy point of view, the rate at which potential energy is stored in the spring is equal to the rate at which the steadily increasing force does work on the spring as it deflects the spring:

$$\frac{d\mathscr{E}_P}{dt} = F_k v_1.$$

Separating variables and integrating yields

$$\int_{\mathscr{E}_p(0)}^{\mathscr{E}_p(t)} d\mathscr{E}_p = \int_0^t F_k v_1 dt = \left(\frac{1}{k}\right) \int_0^t F_k \left(\frac{dF_k}{dt}\right) dt = \left(\frac{1}{k}\right) \int_{F_k(0)}^{F_k(t)} F_k dF_k,$$

so that

$$\mathscr{E}_p(t) - 0 = \frac{1}{2k}F_k^2 - 0$$

Figure 2.8. Free-body diagram of an ideal translational damper.

or

$$\mathscr{E}_p(t) = \left(\frac{1}{2k}\right) F_k^2. \tag{2.15}$$

Thus it takes time for the work input to add to the accumulated energy stored in the spring. A spring is a T-type element, storing energy as a function of the square of its T variable F_k.

Note again, as for changing velocity in Example 2.1, it would *not* be realistic here to try to apply a step change in *force* to a spring; such a force source would have to move at a very, very great velocity to deflect the spring suddenly, which would require a very, very great power source. *In general it can be stated that inputs that would suddenly add to the stored energy in a system are not realistic and cannot be achieved in the natural world.*

2.2.3 Translational Dampers

An ideal damper is shown in free-body diagram form in Fig. 2.8. Because an ideal damper contains no mass, the force transmitted through it is undiminished during acceleration; therefore the forces acting at its ends must always be equal and opposite. The elemental equation for an ideal damper is

$$F_b = b(v_2 - v_1) = bv_{21}. \tag{2.16}$$

A damper is a D-type element that dissipates energy. With this element there is no storage of retrievable mechanical work (work being done by an applied force becomes dissipated as thermal internal energy), and the relationship between force and velocity is instantaneous. Thus it is realistic to apply step changes of either force or velocity to such an element. Damping plays a key role in influencing speed of response and stability of many systems.

Although ideal damping does exist, arising from viscous friction between well-lubricated moving mechanical parts of a system, nonideal forms of damping are very often present. Nonideal damping is usually characterized by nonlinearities that can be severe, especially where poorly lubricated parts move with metal-to-metal contact. In other cases the nonlinearity is the result of hydrodynamic flow effects, in which internal inertia forces predominate, such as in the fluid coupling discussed briefly in Section 2.4 under the topic of linearization and in hydraulic shock absorbers that use orifice-type energy dissipation.

For the case in which the damping is nonlinear, there is no damping constant b, and the elemental equation is expressed as a nonlinear function of velocity:

$$F_{\text{NLD}} = f_{\text{NL}}(v_{12}). \tag{2.17}$$

Linearization for variations around an operating point is often feasible, especially when no discontinuities exist in the F versus v characteristic. Otherwise, computer simulation of the damper characteristic is required.

2.2.4 Elementary Systems – Combinations of Translational Elements

The equations used to describe a combination of interacting elements constitute a mathematical model for the system. Other types of models are discussed in later chapters.

Newton's second law was used to model mathematically the motion of an ideal mass in Subsection 2.2.1. Now Newton's third law is used to sum forces at interconnection points between the elements, ensuring *continuity* of force in the system. A choice of common variables for common motions (position, velocity) at connection points ensures *compatibility*[2] (i.e., no gaps between connected elements).

EXAMPLE 2.3

Consider the spring–damper system shown in Fig. 2.9. This combination of elements is useful for absorbing the impulsive interaction with an impinging system, i.e., a kind of shock absorber. As developed here, it is intended that an input force F_i be the forcing function, and the resulting motion, $x_1 - x_2$ (or $v_1 - v_2$), is then to be considered the resulting output. The relationship between input and output is to be modeled mathematically. Qualitatively speaking, the system responds to the force F_i, storing energy in the spring and dissipating energy in the damper until the force is reduced to zero, whereupon the spring gives up its stored energy and the damper continues to dissipate energy until the system returns to its original state. The net result, after the force has been removed, is that energy that has been delivered to the system by the action of the force F_i has been dissipated by the damper and the system has returned to its original relaxed state.

The object here is to develop a mathematical model relating the output motion to the input force. The use of this mathematical model in solving for the output motion as a function of time is left to a later chapter.

SOLUTION

As an introductory aid in visualizing the action of each member of the system and defining variables, Fig. 2.9 shows diagrams of three different kinds for this system: a cross-sectioned mechanical drawing showing the system in its initial relaxed state with $F_i = 0$ [Fig. 2.9(a)], a stylized diagram showing the system in an active, displaced state when the force F_i is acting [Fig. 2.9(b)], and a free-body diagram of the system "broken open" to show the free-body diagram for each member of the system [Fig. 2.9(c)]. Applying Newton's third law at point (1) yields

$$F_i = F_k + F_b. \tag{2.18}$$

For the elemental equations,

$$F_k = k(x_1 - x_2), \tag{2.19}$$

$$F_b = b(v_1 - v_2). \tag{2.20}$$

---

[2] The principles of continuity and compatibility are discussed in detail in J. L. Shearer, A. T. Murphy, and H. H. Richardson, *Introduction to System Dynamics* (Addison-Wesley, Reading, MA, 1967).

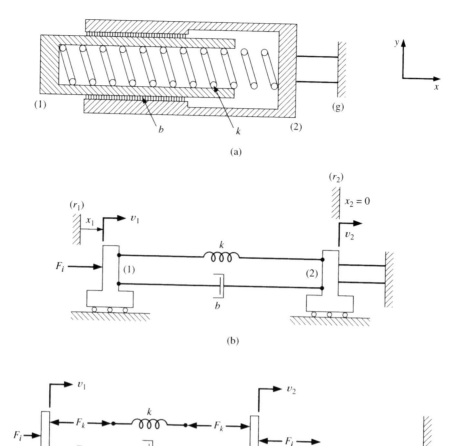

Figure 2.9. Diagrams for spring–damper system.

Definitions:

$$v_1 \equiv \frac{dx_1}{dt}, \tag{2.21}$$

$$v_2 \equiv \frac{dx_2}{dt}. \tag{2.22}$$

The system is now described completely by a *necessary and sufficient set* of five equations containing the five unknown variables x_1, F_k, F_b, v_1, and v_2. Note: *The number of independent describing equations must equal the number of unknown variables before one can proceed to eliminate the unwanted unknown variables.*

Combining Eqs. (2.18)–(2.22) to eliminate F_k, F_b, v_1, and v_2 yields

$$b\left(\frac{dx_1}{dt} - \frac{dx_2}{dt}\right) + k(x_1 - x_2) = F_i. \tag{2.23}$$

Note that x_2 and dx_2/dt have been left in Eq. (2.23) for the sake of generality to cover the situation in which the right-hand side of the system might be in motion, as it could be in some systems. Because the right-hand side here is rigidly connected to the frame of reference g, these variables are zero, leaving

$$b\frac{dx_1}{dt} + kx_1 = F_i. \tag{2.24}$$

This first-order differential equation is the desired mathematical model, describing in a very concise way the events described earlier in verbal form. It may be noted, in the context of state variables to be discussed in Chap. 3, that a first-order system such as this requires only one state variable, in this case x_1, for describing its state from instant to instant as time passes.

Given the initial state of the system in Example 2.3 and the nature of F_i as a function of time, it is possible to solve for the response of the system as a function of time. The procedure for doing this is discussed in later chapters.

The steps involved in producing mathematical models of simple mechanical systems are also illustrated in the following additional examples.

EXAMPLE 2.4

A mass m, supported only by a bearing having a pressurized film of viscous fluid, undergoes translation (i.e., motion along a straight line in the x direction) as the result of having a time-varying input force F_i applied to it, as shown in Fig. 2.10. The object is to develop a mathematical model that relates the velocity v_1 of the mass to the input force F_i. Expressed verbally, the system responds to the input force as follows. Initially the input

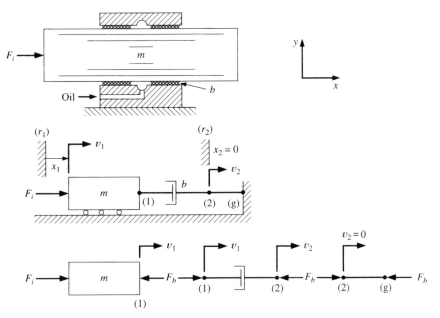

Figure 2.10. Diagrams for mass–damper system.

force accelerates the mass so that its velocity increases, accompanied by an increase in its kinetic energy; however, as the velocity increases, the damper force increases, opposing the action of the input force and dissipating energy at an increasing rate. Thus the action of the damper is to reduce the acceleration of the mass resulting from the input force. If the input force is then removed, the damper force continues to oppose the motion of the mass until it comes to rest, having lost all of its kinetic energy by dissipation in the damper.

SOLUTION

Newton's second law applied to the mass m yields

$$F_i - F_b = m\frac{dv_1}{dt}. \tag{2.25}$$

The elemental equation for the damper is

$$F_b = b(v_1 - v_2). \tag{2.26}$$

Equations (2.25) and (2.26) constitute a necessary and sufficient set of two equations containing the two unknowns F_b and v_1 (v_2 is zero here).

Combining Eqs. (2.25) and (2.26) to eliminate F_b gives

$$m\frac{dv_1}{dt} + b(v_1 - v_2) = F_i. \tag{2.27}$$

Because the bearing block is rigidly connected to ground, $v_2 = 0$, leaving

$$m\frac{dv_1}{dt} + bv_1 = F_i. \tag{2.28}$$

Again, this is a simple first-order system, requiring only one state variable, v_1, for describing its state as a function of time.

EXAMPLE 2.5

This system, typical of spring-suspended mass systems, is shown in Fig. 2.11. The object is to develop the mathematical model relating displacement of the mass x_1 to the input force F_i. Expressed verbally, the system responds to the input in the following fashion. First, the velocity of the mass increases, accompanied by an increase in its stored kinetic energy, while the rate of energy dissipated in the damper increases. Meanwhile, the motion results in a displacement x_1, which is the time integral of its velocity, so that the spring is compressed, accompanied by an increase in its stored potential energy. If the damping coefficient is small enough, the spring will cause a rebounding action, transferring some of its potential energy back into kinetic energy of the mass on a cyclic basis; in this case the decaying oscillation associated with a lightly damped second-order system will occur, even after the input force has been removed.

SOLUTION

Newton's second law applied to m – i.e., the elemental equation for m – yields

$$F_i + mg - F_k - F_b = m\frac{dv_1}{dt}. \tag{2.29}$$

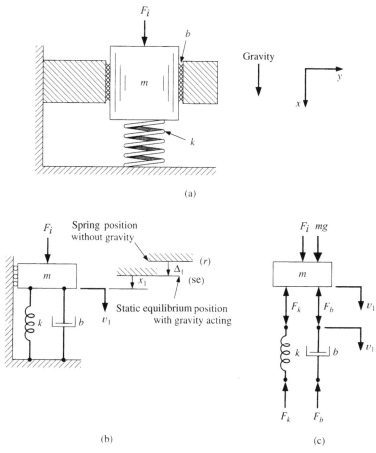

Figure 2.11. Diagrams for mass-spring-damper system in a gravity field.

The elemental equations for the damper and the spring are

$$F_b = bv_1, \tag{2.30}$$

$$F_k = k(x_1 + \Delta_1), \tag{2.31}$$

where Δ_1 is the displacement of the spring that is due to gravity force mg, and x_1 is the displacement that is due to force F_i. In other words, as illustrated in Fig. 2.11(b), Δ_1 is the spring's displacement caused by gravity and measured from the position where the spring is fully relaxed (r), whereas x_1 is the displacement caused by the external input force and measured from the static equilibrium position (se), at which the spring is deflected by the gravity force but no external force is acting, $F_i = 0$. It is therefore implied that when $F_i = 0$, then $x_1 = 0$ and $v_1 = 0$, and under these conditions Eq. (2.29) yields

$$mg - k\Delta_1 = 0. \tag{2.32}$$

Substituting back into Eq. (2.29) eliminates the gravity term:

$$F_i - kx_1 - bv_1 = m\frac{dv_1}{dt}. \tag{2.33}$$

Hence the resulting mathematical model for the system is

$$m\frac{d^2x_1}{dt^2} + b\frac{dx_1}{dt} + kx_1 = F_i. \tag{2.34}$$

Note that, in the final model equation, displacement x_1 is measured from the static equilibrium (se) position and the gravity term does not appear in the equation. The gravity term, mg, has been canceled by a portion of the spring force, $k\Delta_1$. The cancellation of the gravity term by a portion of the spring force can be performed only if the spring is linear. If the spring is nonlinear, $F_k = f_{NL}(\Delta_1 + x_1)$, the displacement must be measured from the spring-relaxed (r) position and the gravity term cannot be eliminated from the equation.

The mathematical solution of Eq. (2.34) for specific forcing functions is carried out in Chap. 4.

This second-order system contains two independent energy-storage elements, and it requires a set of two state variables (e.g., x_1 and dx_1/dt, or v_1 and F_k, or some other pair) to describe its state as a function of time.

Equation (2.34), a second-order differential equation, expresses in a very succinct way the action described earlier in a paragraph of many words. Moreover, it is a more precise description, capable of providing a detailed picture of the system responses to various inputs.

EXAMPLE 2.6

The six-element system shown in Fig. 2.12 is a simplified representation of a vibrating spring–mass assembly (k_1, m_1, b_1) with an attached vibration absorber, subjected to a displacement input x_1, as shown. The object is to develop a mathematical model capable of relating the motions x_2 and x_3 to the input displacement x_1.

Also presented in Fig. 2.12 are diagrams showing the system in an active displaced state and "broken open" for free-body representation.

SOLUTION

The elemental equation for the spring k_1 in derivative form is

$$\frac{dF_{k1}}{dt} = k_1(v_1 - v_2). \tag{2.35}$$

Integration of Equation (2.35) with respect to time with x_1 and x_2, both zero in the relaxed state, yields

$$F_{k1} = k_1(x_1 - x_2). \tag{2.36}$$

For the mass m_1,

$$F_{k1} - F_{k2} - F_{b1} = m_1\frac{d^2x_2}{dt^2}. \tag{2.37}$$

For the damper b_1,

$$F_{b1} = b_1 v_2. \tag{2.38}$$

For the spring k_2,

$$F_{k2} = k_2(x_2 - x_3). \tag{2.39}$$

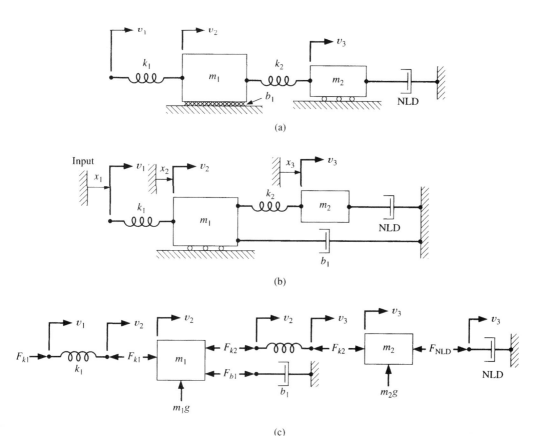

Figure 2.12. Six-element system responding to a displacement input. NLD, nonlinear damper.

For the mass m_2,

$$F_{k2} - F_{\text{NLD}} = m_2 \frac{d^2 x_3}{dt^2}, \tag{2.40}$$

and for the nonlinear damper (NLD),

$$F_{\text{NLD}} = f_{\text{NL}}(v_3) = f_{\text{NL}}\left(\frac{dx_3}{dt}\right). \tag{2.41}$$

Equations (2.36)–(2.39) may now be combined, yielding

$$m_1 \frac{d^2 x_2}{dt^2} + b_1 \frac{dx_2}{dt} + (k_1 + k_2)x_2 = k_1 x_1 + k_2 x_3, \tag{2.42}$$

and Eqs. (2.39)–(2.41) are combined to yield

$$m_2 \frac{d^2 x_3}{dt^2} + f_{\text{NL}}\left(\frac{dx_3}{dt}\right) + k_2 x_3 = k_2 x_2. \tag{2.43}$$

It can be seen that two second-order differential equations are needed to model this fourth-order system (four independent energy-storage elements), one of which is non-linear. The nonlinear damping term in Eq. (2.43) complicates the algebraic combination of Eqs. (2.42) and (2.43) into a single fourth-order differential equation model. In some cases the NLD characteristic may be linearized, making it possible to combine Eqs. (2.42) and (2.43) into a single fourth-order differential equation for x_2 or x_3. Because this system has four independent energy-storage elements, a set of four state variables is required for describing the state of this system (e.g., x_2, v_2, x_3, and v_3, or F_{k1}, v_2, F_{k2}, and v_3). The exchange of energy among the input source and the two springs and two masses, together with the energy dissipated by the dampers, would require a very long and complicated verbal description. Thus the mathematical model is a very compact, concise description of the system. Further discussion of the manipulation and solution of this mathematical model is deferred to later chapters.

2.3   ROTATIONAL–MECHANICAL SYSTEMS

Corresponding respectively to the translational elements mass, spring, and damping are rotational inertia, rotational spring, and rotational damping. These rotational elements are used in the modeling of systems in which each element rotates about a single nonaccelerating axis.

2.3.1 Rotational Inertias

An ideal inertia, depicted schematically in free-body diagram form in Fig. 2.13, moves in relation to a nonaccelerating rotational frame of reference, which is usually taken to be the earth (ground). However, the frame of a steadily rotating space vehicle, for instance, could be used as a reference.

The elemental equation for an ideal inertia J, based on Newton's second law for rotational motion, is

$$T_J = J \frac{d\Omega_1}{dt}, \tag{2.44}$$

where Ω_1 is the angular velocity of the inertia relative to the ground reference and T_J is the sum of all the external torques (twisting moments) applied to the inertia. Because $\Omega_1 = d\theta_1/dt$, the variation of θ is related to T_J by

$$J \frac{d^2\theta_1}{dt^2} = T_J. \tag{2.45}$$

The kinetic energy stored in an ideal rotational inertia is

$$\mathscr{E}_k = \left(\frac{J}{2}\right) \Omega_1^2.$$

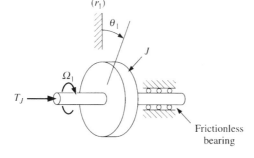

Figure 2.13. Free-body diagram of am ideal rotational inertia.

Hence it is designated as an A-type element, storing energy as a function of the square of its A variable Ω_1.

The response of an inertia to an applied torque T_J is analogous to the response of a mass to an applied force F_m. It takes time for angular velocity, kinetic energy, and angular displacement to accumulate after the application of a finite torque, and it would not be realistic to try to impose a sudden change in angular velocity on a rotational inertia.

2.3.2 Rotational Springs

A rotating shaft may be modeled as an ideal spring if the torque required for accelerating its rotational inertia is negligible compared with the torque that it transmits. Sometimes, however, the torque transmitted by the shaft is small compared with that required for accelerating its own inertia so that it should be modeled as an inertia; and sometimes a combination of springs and inertias may be required for modeling a real shaft, as illustrated in Section 2.1 for a translational spring.

An ideal rotational spring stores potential energy as it is twisted (i.e., wound up) by the action of equal but opposite torques, as shown in Fig. 2.14. Here the rotational spring is shown in a relaxed state [Fig. 2.14(a)], with no torques acting, and while transmitting torque T_K [Fig. 2.14(b)], with both ends displaced rotationally from their local references r_1 and r_2, with T_K shown acting at both ends in free-body diagram form.

The elemental equation for a rotational spring is similar to Hooke's law for a translational spring, given by

$$T_K = K(\theta_1 - \theta_2), \tag{2.46}$$

where θ_1 and θ_1 are the angular displacements of the ends from their local references r_1 and r_2 In derivative form, this equation becomes

$$\frac{dT_K}{dt} = K(\Omega_1 - \Omega_2), \tag{2.47}$$

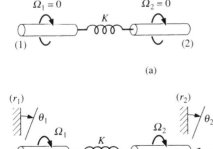

(a)

(b)

Figure 2.14. Schematic and free-body diagrams of an ideal spring.

where Ω_1 and Ω_2 are the angular velocities of the ends. In each case, the sign convention used for motion is clockwise positive when viewed from the left, and the sign convention for torque is clockwise positive when acting from the left (with the head of the torque vector toward the left end of the spring). The potential energy stored in a rotational spring is given by

$$\mathcal{E}_p = \frac{1}{2K} T_K^2. \tag{2.48}$$

Hence it is designated as a T-type element, storing energy as a function of the square of its T variable, T_K.

The comments about the response of a translational spring to a step change in velocity difference between its ends apply equally well to the response of a rotational spring to a step change in angular velocity difference between its ends. Thus it would be unreasonable to try to impose a step change of torque in a rotational spring because that would represent an attempt to change suddenly the energy stored in it in a real world that does not contain sources of infinite power.

When a rotational spring is nonlinear, it does not have truly constant stiffness K, and it is advisable to use a nonlinear symbol designation and a nonlinear function to describe it:

$$T_{\text{NLS}} = f_{\text{NL}}(\theta_1 - \theta_2).$$

Linearization to achieve a local incremental stiffness K_{inc} is often feasible, as in the previously discussed case of a nonlinear translational spring.

2.3.3 Rotational Dampers

Just as friction between moving parts of a translational system gives rise to translational damping, friction between rotating parts in a rotational system is the source of rotational damping. When the interfaces are well lubricated, the friction is a result of the shearing of a thin film of viscous fluid, yielding a constant damping coefficient B, as shown in Fig. 2.15, which uses a cross-sectioned diagram with no torque being transmitted so that $\Omega_1 = \Omega_2 = \Omega$ [Fig. 2.15(a)] and a free-body diagram with torque being transmitted [Fig. 2.15(b)].

The elemental equation of an ideal damper is given by

$$T_B = B(\Omega_1 - \Omega_2), \tag{2.49}$$

where T_B is the torque transmitted by the damper.

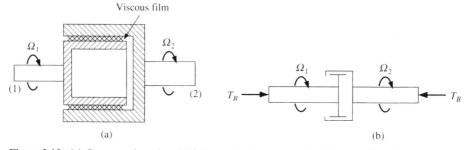

Figure 2.15. (a) Cross-sectioned and (b) free-body diagrams of an ideal rotational damper.

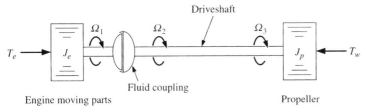

Figure 2.16. Schematic diagram of a simplified model of a ship propulsion system.

A rotational damper is designated as a D-type element because it dissipates energy.

When the lubrication is imperfect, so that direct contact occurs between two parts of the damper, dry friction becomes evident and the damping effect cannot be described by means of a simple damping constant. Also, in some cases, the interaction between the two parts of the damper involves hydrodynamic fluid motion, as in the fluid coupling considered in the following example, in which a square-law nonlinear function is involved the elemental equation for a NLD. A NLD is expressed in nonlinear form as

$$T_{\text{NLD}} = f_{\text{NL}}(\Omega_1, \Omega_2).$$ (2.50)

EXAMPLE 2.7

The power transmission system from a diesel engine to a propeller for a ship is shown in simplified form in Fig. 2.16. The role of the fluid coupling is to transmit the main flow of power from the engine to the propeller shaft without allowing excessive vibration, which would otherwise be caused by the pulsations of engine torque resulting from the cyclic firing of its cylinders. The object here is to develop a mathematical model for this system in order to relate the shaft torque T_K to the inputs T_e and T_w.

SOLUTION

The complete set of free-body diagrams for this system is shown in Fig. 2.17. We develop the system analysis by beginning at the left-hand end and writing the describing equation for each element and any necessary connecting point equations until the system has been completely described.

For the engine (moving parts and flywheel lumped together into an ideal inertia in which friction is ignored),

$$\frac{d\Omega_1}{dt} = \left(\frac{1}{J_e}\right)(T_e - T_c).$$ (2.51)

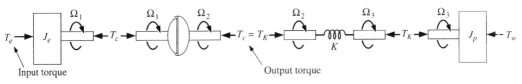

Figure 2.17. Complete set of free-body diagrams for a ship propulsion system.

For the fluid coupling (negligible inertia),

$$T_c = C_c \left(\Omega_1^2 - \Omega_2^2 \right). \tag{2.52}$$

At the junction between the fluid coupling and the driveshaft,

$$T_c = T_K. \tag{2.53}$$

For the driveshaft (ideal rotational spring – negligible friction and inertia),

$$\frac{dT_K}{dt} = K(\Omega_2 - \Omega_3). \tag{2.54}$$

For the propeller (ideal inertia – negligible friction),

$$\frac{d\Omega_3}{dt} = \left(\frac{1}{J_P} \right) (T_k - T_w). \tag{2.55}$$

Equations (2.51)–(2.55) constitute a necessary and sufficient set of five equations for this system containing five unknowns: Ω_1, Ω_2, T_c, T_K, and Ω_3.

Note that additional dampers to ground at points (1), (2), and (3) would be required if bearing friction at these points were not negligible. Rearranging Eq. (2.52) into the form $\Omega_2 = f_2(T_c, \Omega_1)$ yields

$$\Omega_2^2 = \Omega_1^2 - \frac{T_c}{C_c} \tag{2.52a}$$

or

$$\Omega_2 = \text{SSR} \left(\Omega_1^2 - \frac{T_c}{C_c} \right), \tag{2.52b}$$

where SSR denotes "signed square root," e.g.,

$$\text{SSR}(X) = \frac{X}{\sqrt{|X|}}.$$

Combining Eqs. (2.51) and (2.53), we have

$$\frac{d\Omega_1}{dt} = \left(\frac{1}{J_e} \right) (T_e - T_k); \tag{2.56}$$

and combining Eqs. (2.52b), (2.53), and (2.54) yields

$$\frac{dT_K}{dt} = K \left[\text{SSR} \left(\Omega_1^2 - \frac{T_K}{C_c} \right) - \Omega_3 \right]. \tag{2.57}$$

Equations (2.55), (2.56), and (2.57) constitute a necessary and sufficient set of equations for this system, containing the three unknown variables Ω_3, T_K, and Ω_1. Because of the nonlinearity in the fluid coupling, it is not possible to combine these equations algebraically into a single input–output differential equation. However, in their present form they are a complete set of state-variable equations ready to be integrated numerically on a computer (see Chap. 5).

2.4 LINEARIZATION

It should be emphasized that the arbitrary classifications presented in Table 1.1 and the ideal elements listed in Table 1.2 are used only as aids in system modeling. Real

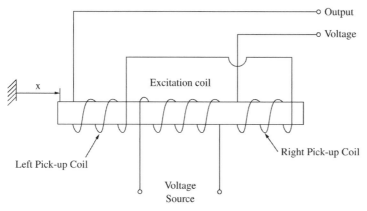

Figure 2.18. Schematic of a LVDT.

systems usually exhibit nonideal characteristics and/or combinations of characteristics that depart somewhat from the ideal models used here. However, analysis of
a system represented by nonlinear, partial differential equations with time-varying
coefficients is extremely difficult and requires extraordinary computational resources
to perform complex, iterative solution procedures, an effort that can rarely be justified by the purpose of the analysis. Naturally, therefore, simplified descriptions of the
actual system behavior are used whenever possible. In particular, nonlinear system
characteristics are approximated by linear models for which many powerful methods
of analysis and design, involving noniterative solution procedures, are available.

    Consider for example a linear variable differential transformer (LVDT), which
is commonly used to measure displacement in high-performance applications. A
simplified schematic diagram of a LVDT is show in Fig. 2.18. In this simple device,
a ferromagnetic core is placed inside three coils. The center coil is energized by
an external ac power source, and the two end coils pick up the voltage induced by a
magnetic field established by the center excitation coil. When the core is in the center
of the coils, the amplitudes of the voltages induced in the end coils are the same and
because the end coils are connected in opposite phase, the output voltage is zero.
When the core is displaced left or right from the center position, the voltage induced
in one of the end coils becomes greater and the resulting output voltage is not zero.
Figure 2.19 shows the nonlinear relationship between the amplitude of the output
voltage and the displacement of the core relative to its center position. It is somewhat
ironic that the word "linear" appears in the name of this inherently nonlinear system,
but it is justified by the fact that the range of measured displacements is usually limited
to a relatively small vicinity of the core's center position, where the relationship
between output voltage and the core's displacement is approximately linear (the
smaller the displacement, the better the approximation).

    The nonlinear performance characteristic of a LVDT requires that the opera-
tional range of the device be restricted and is thus an unwelcome feature. Although
nonlinearities are often unwelcome because they make the system analysis and design
more difficult or because they limit the system's applications, there are also many
systems in which nonlinear characteristics are desired and sometimes even necessary.

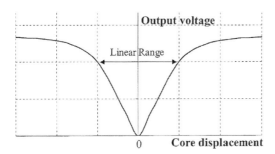

Figure 2.19. Output voltage vs. core displacement characteristic for a LVDT.

One example of a device that is designed to be nonlinear is a fuel gauge in an automobile. A typical characteristic of the gauge's needle position versus the amount of fuel remaining in a 16-gal tank is shown in Fig. 2.20. This characteristic is distinctly nonlinear near the needle positions indicating that the fuel tank is empty (E) and full (F). Both nonlinearities are deliberately introduced to benefit both the driver and the manufacturer of the vehicle. The needle reaches the "E" position when there is still a considerable amount (usually around 2 gal) of fuel in the tank. The main reason is to protect the driver from running out of fuel. Another reason is to protect a fuel-injected engine that requires certain minimum fuel pressure to keep running. On the other end of the scale, after the tank is filled, the needle stays at the "F" position while the vehicle is driven for a while. The fuel gauge is designed to behave that way because it makes an impression of better fuel economy and satisfies most drivers' expectations of being able to drive a while on a full tank. In summary, a fuel gauge is designed to be nonlinear, and most drivers prefer this.

In general, one of the following three options is usually taken in analysis of a nonlinear system:

(a) replacing nonlinear elements with "roughly equivalent" linear elements,
(b) developing and solving a nonlinear model,
(c) linearizing the system equations for small perturbations.

Use of option (a) often leads to invalid models, to say the least. Approach (b) leads to the most accurate results, but the cost involved in nonlinear model analysis may be excessively high, often not justified by the benefits of a very accurate solution.

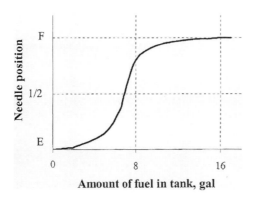

Figure 2.20. Nonlinear characteristic of a typical automotive fuel gauge.

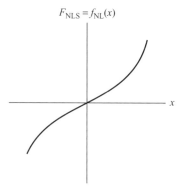

$F_{NLS} = f_{NL}(x)$

x

Figure 2.21. NLS characteristic.

Finally, option (c) represents the most rational approach, especially in the preliminary stages of system analysis. Thus option (c) is now discussed in considerable detail.

Consider a nonlinear spring (NLS) characterized by a relationship, $F_{NLS} = f_{NL}(x)$, describing a force exerted by the spring, F_{NLS}, when subjected to a change in length, x, as shown in Fig. 2.21. The value of x is considered to be the extension of the spring from its relaxed length x_r when no external forces are applied, as illustrated in Fig. 2.22. The purpose of linearization is to replace a nonlinear characteristic with a linear approximation. In other words, linearizing the nonlinear function $f_{NL}(x)$ means replacing it locally with an approximating straight line. Such a formulation of the linearization process is not precise and may yield inaccurate results unless restrictions are placed on its use. In particular, a limit must be somehow established for the small variation $\hat{x}$ of the whole variable $x = \bar{x} + \hat{x}$ from its normal operating-point value $\bar{x}$. This limit on the range of acceptable variation of the independent variable x is influenced by the shape of the nonlinear function curve and the location of the normal operating point on the curve. (In the case of two independent variables, of course, a surface would assume the role of the curve.)

The term "normal operating point" here refers to the condition of a system when it is in a state of equilibrium with the input variables constant and equal to their mean values averaged over time. The variations of the inputs to the system containing the nonlinear element from these mean values must be small enough collectively for the linearization error to be acceptable. In this case, the maximum variation of x resulting from all the system inputs acting simultaneously must be small enough so that the maximum resulting variation of the force F_{NLS} is adequately described by the straight-line approximation. At the normal operating point for the NLS, the normal

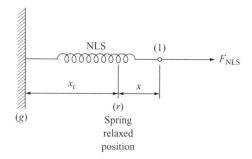

NLS   (1)

F_{NLS}

x_r x

(r)

(g) Spring
             relaxed
             position

Figure 2.22. Elongation of spring subjected to force F_{NLS}.

operating-point force $\overline{F}_{NLS}$ is related to the normal operating-point displacement $\overline{x}$ by

$$\overline{F}_{NLS} = f_{NL}(\overline{x}). \tag{2.58}$$

As variations from these normal operating-point values occur,

$$x = \overline{x} + \hat{x}, \quad F_{NLS} = \overline{F}_{NLS} + \hat{F}_{NLS},$$
$$F_{NLS} = f_{NL}(\overline{x} + \hat{x}) = \overline{f}_{NL} + \hat{f}_{NL}, \tag{2.59}$$

where $\overline{f}_{NL}$, the normal operating value of f_{NL}, is the first term of the Taylor's series expansion of a function near its operating point:

$$f_{NL}(\overline{x} + \hat{x}) = f_{NL}(\overline{x}) + \hat{x} \frac{d f_{NL}(x)}{dx}\bigg|_{\overline{x}}$$
$$+ \frac{\hat{x}^2}{2!} \frac{d^2 f_{NL}(x)}{dx^2}\bigg|_{\overline{x}}$$
$$+ \frac{\hat{x}^3}{3!} \frac{d^3 f_{NL}(x)}{dx^3}\bigg|_{\overline{x}} + \cdots + .$$

Hence, based on Eq. (2.59) and the preceding Taylor's series expansion,

$$\overline{f}_{NL} = f_{NL}(\overline{x}), \tag{2.60}$$

and the incremental or "hat" value $\hat{f}_{NL}$ would represent the remaining terms of the Taylor's series expansion:

$$\hat{f}_{NL} = \hat{x} \frac{d f_{NL}}{dx}\bigg|_{\overline{x}}$$
$$+ \frac{\hat{x}^2}{2!} \frac{d^2 f_{NL}}{dx^2}\bigg|_{\overline{x}} + \cdots + . \tag{2.61}$$

However, imposition of the conditions discussed earlier, which justify linearization, makes it possible to neglect the terms involving higher powers of $\hat{x}$ in Eq. (2.61). Thus linearization for small perturbations of x about the normal operating point uses the following approximation of $\hat{f}_{NL}$:

$$\hat{x} \frac{d f_{NL}}{dx}\bigg|_{\overline{x}} \approx \hat{f}_{NL} \tag{2.62}$$

Hence the NLS force F_{NLS} may be approximated adequately in a small vicinity of the normal operating point where $x = \overline{x}$ by

$$F_{NLS} \approx f_{NL}(\overline{x}) + \hat{x} \frac{d f_{NL}}{dx}\bigg|_{\overline{x}}, \tag{2.63}$$

where

$$f_{NL}(\overline{x}) = \overline{F}_{NLS}, \quad \hat{x} \frac{d f_{NL}}{dx}\bigg|_{\overline{x}} \cong \hat{F}_{NLS}.$$

The linearized equation for the NLS is represented by a new incremental coordinate system having its origin at the normal operating point $(\overline{x}, \overline{F}_{NLS})$, as shown in

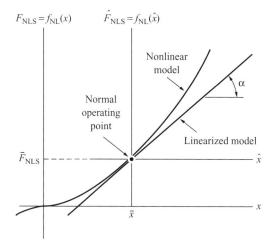

$F_{\text{NLS}} = f_{\text{NL}}(x)$ $\hat{F}_{\text{NLS}} = f_{\text{NL}}(\hat{x})$

Nonlinear model

α

Normal operating point

Linearized model

$\overline{F}_{\text{NLS}}$ $\hat{x}$

$\overline{x}$ x

Figure 2.23. Incremental coordinate system.

Fig. 2.23. The linearized spring characteristic is represented by the diagonal straight line passing through the new origin with a slope having angle α. The slope itself is $\tan \alpha$, which is often referred to as the local incremental spring constant k_{inc}, where

$$k_{\text{inc}} = \tan \ \alpha = \left. \frac{d F_{\text{NLS}}}{dx} \right|_{\overline{x}}$$

$$= \left. \frac{d f_{\text{NL}}}{dx} \right|_{\overline{x}}. \tag{2.64}$$

The approximating linear formula can now be written in the simple form

$$F_{\text{NLS}} = \overline{F}_{\text{NLS}} + \hat{F}_{\text{NLS}} \approx \hat{F}_{\text{NLS}} + k_{\text{inc}}\hat{x}. \tag{2.65}$$

A similar procedure may be applied to a nonlinear element described by a function of two variables. Consider the example of a hydrokinetic type of fluid coupling (rotational damper) such as that shown in Fig. 2.24.

In this case the fluid coupling torque T_{fc} is a nonlinear function of two variables $f_{\text{NL}}(\Omega_1, \Omega_2)$ that have a Taylor's series expansion of the following form:

$$f_{\text{NL}}[(\overline{\Omega}_1 + \hat{\Omega}_1), (\overline{\Omega}_2 + \hat{\Omega}_2)] = f_{\text{NL}}(\overline{\Omega}_1, \overline{\Omega}_2)$$

$$+ \hat{\Omega}_1 \left. \frac{\partial f_{\text{NL}}}{\partial \Omega_1} \right|_{\overline{\Omega}_1, \overline{\Omega}_2} + \hat{\Omega}_2 \left. \frac{\partial f_{\text{NL}}}{\partial \Omega_2} \right|_{\overline{\Omega}_1, \overline{\Omega}_2}$$

$$+ \frac{\hat{\Omega}_1^2}{2!} \left. \frac{\partial^2 f_{\text{NL}}}{\partial \Omega_1^2} \right|_{\overline{\Omega}_1, \overline{\Omega}_2} + \frac{\hat{\Omega}_2^2}{2!} \left. \frac{\partial^2 f_{\text{NL}}}{\partial \Omega_2^2} \right|_{\overline{\Omega}_1, \overline{\Omega}_2} + \cdots + . \tag{2.66}$$

Neglecting Taylor's series terms involving higher powers of $\hat{\Omega}_1$ and $\hat{\Omega}_2$ yields

$$f_{\text{NL}} = C_T(\Omega_1^2 - \Omega_2^2) \approx \overline{f}_{\text{NL}} + \hat{f}_{\text{NL}},$$

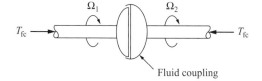

Ω_1 Ω_2

T_{fc} T_{fc}

Fluid coupling

Figure 2.24. Fluid coupling described by a function of two shaft speeds Ω_1 and Ω_2.

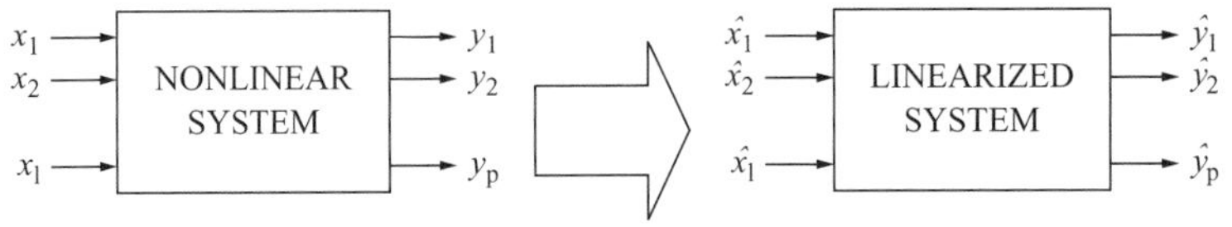

Figure 2.25. Graphical illustration of the linearization process.

where[3]

$$\overline{f}_{NL} = C_T(\overline{\Omega}_1^2 - \overline{\Omega}_2^2),$$

$$\hat{f}_{NL} = 2C_T(\overline{\Omega}_1\hat{\Omega}_1 - \overline{\Omega}_2\hat{\Omega}_2).$$

Thus we may write

$$T_{fc} \approx \overline{f}_{NL} + \hat{f}_{NL}$$

$$\approx C_T(\overline{\Omega}_1^2 - \overline{\Omega}_2^2) + 2C_T(\overline{\Omega}_1\hat{\Omega}_1 - \overline{\Omega}_2\hat{\Omega}_2). \tag{2.67}$$

The process of linearization can also be thought of as a transformation of a nonlinear system defined in terms of original input and output variables, $x_1, x_2, \ldots, x_l$ and $y_1, y_2, \ldots, y_p$ into a linearized model defined in terms of incremental input and output variables, $\hat{x}_1, \hat{x}_2, \ldots, \hat{x}_l$ and $\hat{y}_1, \hat{y}_2, \ldots, \hat{y}_p$ representing small deviations of the original variables from their respective normal operating point values. This transformation is illustrated graphically in Fig. 2.25.

In general, the incremental variables are defined as

$$\hat{x}_i = x_i - \overline{x}_i \ \text{ for i } = 1, 2, \ldots, l,$$
$$\hat{y}_j = y_j - \overline{y}_j \ \text{ for j } = 1, 2, \ldots, p.$$

Simplification of a system model obtained as a result of linearization is certainly a benefit, but – as is usually the case with benefits – it does not come free. The price that must be paid in this case represents the error involved in approximating the actual nonlinear characteristics by linear models, which can be called a linearization error. The magnitude of this error depends primarily on the particular type of nonlinearity being linearized and on the amplitudes of deviations from a normal operating point experienced by the system. The effects of these two factors are illustrated in Fig. 2.26. In both parts of the figure, the same nonlinear function, representing combined viscous and dry friction forces, is linearized. However, because different normal operating points are selected, different types of nonlinearities are locally approximated by the dashed lines. In Fig. 2.26(a), the normal operating value of the velocity v is very close to zero, where a very large discontinuity in the nonlinear function occurs.

---

[3] Instead of using the Taylor's series expansion, another approach here would be simply to multiply the whole variables and drop terms that are second-order small. In other words,

$$f_{NL} = C_T[(\overline{\Omega}_1 + \hat{\Omega}_1)^2 - (\overline{\Omega}_2 + \hat{\Omega}_2)^2]$$

$$= C_T\left[\left(\overline{\Omega}_1^2 + 2\overline{\Omega}_1\hat{\Omega}_1 + \hat{\Omega}_2^2\right) - \left(\overline{\Omega}_2^2 + 2\overline{\Omega}_2\hat{\Omega}_2 + \hat{\Omega}_2^2\right)\right]$$

$$= C_T\left(\overline{\Omega}_1^2 - \overline{\Omega}_2^2\right) + 2C_T(\overline{\Omega}_1\hat{\Omega}_1 - \overline{\Omega}_2\hat{\Omega}_2).$$

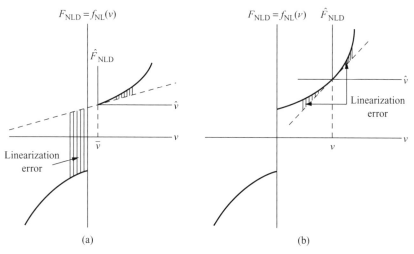

Figure 2.26. Effect of the type of nonlinearity on linearization error.

In this case, a sudden change in the force F_{NLD} occurs at the dry-friction discontinuity when the system deviates only a small amount from its normal operating state. On the other hand, if the normal operating point is far away from the discontinuity at zero velocity, as in the case shown in Fig. 2.26(b), the error resulting from linearization is reasonably small if the magnitude of the change in v is small.

A complete linearization procedure can be performed in the following five steps:

Step 1. Derive the nonlinear model.
Step 2. Determine the normal operating point.
Step 3. Introduce incremental variables.
Step 4. Linearize all nonlinear terms by use of Taylor's series expansion.
Step 5. Arrange the linearized equation into a final form.

This five-step procedure is applied in the following example.

EXAMPLE 2.8

In the mechanical system shown in Fig. 2.27, the relationship between the force exerted by the spring and the change of the spring's length measured from the spring's relaxed position has been approximated mathematically by the following nonlinear equation:

$$F_{NLS} = f_{NL}(x) = 2.5\sqrt{x}. \tag{2.68}$$

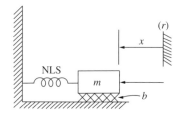

Figure 2.27. Schematic diagram of the system considered in Example 2.8.

Obtain a linearized mathematical model of the system that approximates the system dynamics in a small vicinity of the normal operating point determined by the average value of the input force, $\overline{F}_i = 0.1\,\text{N}$.

SOLUTION

This problem is solved following the five-step linearization process introduced earler in this section.

Step 1. The mathematical model of the system is a second-order nonlinear differential equation:

$$m\ddot{x} + b\dot{x} + f_{\text{NL}}(x) = F_i, \tag{2.69}$$

$$m\ddot{x} + b\dot{x} + 2.5\sqrt{x} = F_i. \tag{2.70}$$

Step 2. The normal operating point is defined by the given constant input force $\overline{F}_i = 0.1\,\text{N}$ and by the corresponding displacement $\overline{x}$. One can find the unknown value of $\overline{x}$ from model equation (2.70) by setting $F_i = \overline{F}_i$ and $x = \overline{x}$, which yields

$$m\ddot{\overline{x}} + b\dot{\overline{x}} + 2.5\sqrt{\overline{x}} = \overline{F}_i. \tag{2.71}$$

The first two terms of Eq. (2.71) drop out because $\overline{x}$ is, by definition, a constant, and does not vary with time. Hence

$$2.5\sqrt{\overline{x}} = \overline{F}_i, \tag{2.72}$$

$$\overline{x} = \left(\frac{0.1}{2.5}\right)^2 = 0.0016 \quad \text{m}. \tag{2.73}$$

Thus the normal operating point corresponding to the constant portion of the input force of 0.1 N is a deflection of 0.0016 m.

Step 3. Introduce the incremental variables by substituting $x = \overline{x} + \hat{x}$ and $F_i = \overline{F}_i + \hat{F}_i$ into Eq. (2.71):

$$m\ddot{\hat{x}} + b\dot{\hat{x}} + f_{\text{NL}}(\overline{x} + \hat{x}) = \overline{F}_i + \hat{F}_i. \tag{2.74}$$

Step 4. The nonlinear term in Eq. (2.74) is approximated by the first two terms of the Taylor's series expansion:

$$f_{\text{NL}}(\overline{x} + \hat{x}) \approx 2.5\sqrt{\overline{x}} + \left(\frac{2.5}{2\sqrt{\overline{x}}}\right)\hat{x} = 0.1 + 31.25\hat{x}. \tag{2.75}$$

Step 5. Substitute the linear approximation in Eq. (2.75) into (2.74) to get the following result:

$$m\ddot{\hat{x}} + b\dot{\hat{x}} + 0.1 + 31.25\hat{x} = \overline{F}_i + \hat{F}_i. \tag{2.76}$$

Note that the constant portion of the input force appears on both sides of Eq. (2.76) and therefore cancels out, leaving a linear second-order ordinary differential equation (ODE):

$$m\ddot{\hat{x}} + b\dot{\hat{x}} + 31.25\hat{x} = \hat{F}_i. \tag{2.77}$$

Note that the coefficient of the incremental displacement term occupies the position where one normally finds a spring constant. In point of fact, this is the incremental stiffness k_{inc} of the NLS in the vicinity near the normal operating position established by the force $\overline{F}_i = 0.1$ N:

$$m\ddot{\hat{x}} + b\dot{\hat{x}} + k_{inc}\,\hat{x} = \hat{F}_i. \tag{2.78}$$

where

$$k_{inc} = 31.25 \text{ N/m}. \tag{2.79}$$

To further illustrate this important concept, both the linearized and nonlinear spring characteristics can be plotted by use of MATLAB. Appendix 3 is a tutorial on the MATLAB environment, and readers not well versed in its usage are encouraged to review the tutorial before proceeding with the book material.

The nonlinear function to be plotted is given by Eq. (2.68). The linearized model of the spring is represented by a straight line tangent to the nonlinear function at the normal operating point ($\overline{x} = 0.0016$ m). The slope of the straight line is the incremental stiffness, 31.25 N/m and its y intercept is 0.05 N, which can be found by substitution of the coordinates of the normal operating point into a general equation for the tangent line. Thus the approximated linearized spring force equation is

$$F_{NLS} \approx 31.25x + 0.05. \tag{2.80}$$

The following MATLAB commands can be used to generate the plots:

```
>> x = [0.0:0.0001:0.01];        % set up a vector of x values
>> Fnls = 2.5*x^0.5;             % compute NLS force
>> Flins = 31.25*x+0.05;         % compute linearized spring force
>> plot (x, Fnls, x, Flins)      % plot both force against deflection
>> xlabel('Deflection (m)')      % add x and y labels and grid to plot
>> ylabel('Force (N)')
>> grid
```

The plot is shown in Fig. (2.28). The main outcome of the linearization process performed in Example 2.8 was the incremental stiffness of the spring (k_{inc}), which is equal to the slope of the line that is tangent to the nonlinear function at the normal operating point of interest. Some interesting observations can be made from the plot. First, it can be seen that the slope of the tangent line (and the incremental stiffness) will vary a great deal, depending on the location of the normal operating point, from a very large positive number near the point of zero deflection (approaching infinity) to a more modest value of 31.25 N/m at a deflection of 0.0016 m. Second, it is quite clear that the linearization error, or the difference in the vertical direction between the nonlinear function and the tangent line, will increase as the spring deflection deviates from the normal operating point. Finally, it can also be observed that the

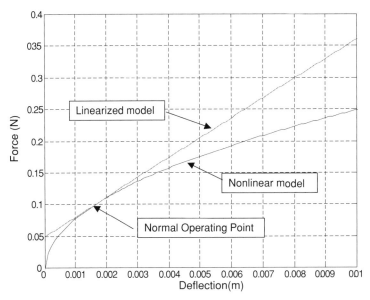

Figure 2.28. MATLAB-generated plot of NLS and linearized spring characteristics considered in Example 2.8.

linear approximation is better as the normal operating point moves farther away from the spring-relaxed position, $x = 0$.

In summary, although linearized models should be used whenever possible (because they lend themselves to powerful analytical methods), it should be kept in mind that they constitute local (usually for small deviations from a nominal operating point) approximations of the nonlinear system and that the results obtained with linearized models provide a simplified picture of the actual system behavior.

2.5 SYNOPSIS

This chapter demonstrated the principles involved in developing simplified lumped-parameter mathematical models of mechanical systems of two basic types: (a) translational systems and (b) rotational systems. In each case, the system model was developed through the use of Newton's laws dealing with summation of forces at a massless point (or torques at an inertialess point) and acceleration of a lumped mass (or lumped inertia), together with elemental equations for springs, dampers, or both. When carried out properly, this results in a set of n equations containing n unknown variables. Subsequent mathematical manipulation of these equations was carried out to eliminate unwanted variables, producing the desired model involving the variables of greatest interest.

Usually, this desired model consisted of a single input–output differential equation relating a desired output to one or more given inputs. In some cases, a reduced set of first-order equations, called state-variable equations, was developed as part of the process of eliminating unwanted variables. This was done because, in some instances in the future, this is all the reduction needed to proceed with a computer simulation or analysis of the system. The definition of state variables is left to

Chap. 3, which covers the topic of this aspect of system modeling in considerable detail.

The energy converters required for coupling translational with rotational systems are discussed in Chap. 10, in which the general topic of energy converters in "mixed" systems is covered in some detail.

During the development of this chapter, the basic system elements were classified as to their energy-storage or energy-dissipation traits. Furthermore, in the case of energy-storage elements, precautions were emphasized relative to the impossibility of storing a finite amount of energy in an energy-storage element with finite sources of force (or torque) and velocity in zero time. Knowledge of these limitations on energy storage establishes limits on the kinds of sudden changes of input that are physically realizable when only finite sources of force (or torque) and velocity are available. Knowledge of these limitations on energy storage is also essential to the determination of the initial conditions of a system (in other words, the values of the system's variables) after a sudden change of a system input occurs.

Finally, examples of nonlinear elements were introduced. Methods of linearization for small perturbations about a given operating point were demonstrated, and limitations on the use of this linearization were discussed. The underlying motivation of simplification to augment mathematical analysis was balanced against the limitations imposed by linear approximation for a sufficiently small region near the selected normal operating point.

PROBLEMS

2.1 A rather heavy compression spring weighing 1.0 lb has a stiffness k_s of 2000 lb/in. To the casual observer it looks like a spring but "feels" like a mass. This problem deals with the choice of a suitable lumped-parameter model for such a spring.

According to vibration theory, this spring, containing both mass and stiffness itself, will respond with different kinds of oscillations, depending on how it is forced (i.e., its boundary conditions).[4] This means that the choice of an approximate lumped-parameter model for this element will depend partly on the elements with which it interacts and partly on the range of frequencies, or rate of variation, of inputs applied to the system containing it. Obviously, in a simple system containing this heavy spring and an attached mass, the spring may be modeled to a good degree of approximation as a pure spring if the attached mass is at least an order of magnitude greater than the self-mass of the spring and if the portion of the force applied to the spring required for accelerating its self-mass is small compared with the force required for deflecting the spring. Likewise, the spring might be approximately modeled as a pure mass if it interacts with another spring having a stiffness at least an order of magnitude smaller than its own stiffness and if its self-deflection is small compared with the deflection of the other spring as a result of the acceleration force of its mass.

You are asked to propose approximate lumped-parameter models for such a spring in the following situations.

(a) The heavy spring supports a mass m weighing 5.0 lb with a force source F_s acting on the mass, as shown in Fig. P2(a). The maximum frequency of a possible sinusoidal

---

[4] See D. J. Inman, *Engineering Vibration*, 2nd ed. (Prentice-Hall, Englewood Cliffs, NJ, 2001), pp. 431–75.

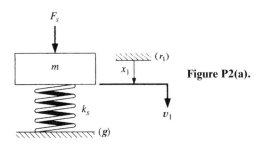

Figure P2(a).

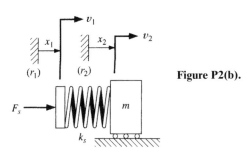

Figure P2(b).

variation of F_s is about half the lowest natural frequency of the heavy spring itself, operating in free–free or clamped–clamped mode.[5]

(b) The spring acts between a mass weighing 3.0 lb and a force source, as shown in Figure P2(b), with a maximum possible frequency of sinusoidal variation that is about one-half the lowest natural frequency of the heavy spring itself, operating in free–clamped mode.[6]

2.2 An automobile weighing 3000 lb is put in motion on a level highway and then allowed to coast to rest. Its speed is measured at successive increments of time, as recorded in the following table:

Time (s)	Speed (ft/s, approx.)
0	15.2
10	10.1
20	6.6
30	4.5
40	2.9
50	2.1
60	1.25

(a) Using a mass–damper model for this system, draw the system diagram and set up the differential equation for the velocity v_1 of the vehicle.

(b) Evaluate m and then estimate b by using a graph of the data given in the preceding table. (*Hint:* Use the slope dv_1/dt and v_1 itself at any time t.)

2.3 An automobile weighing 2200 lb is released on a 5.8° slope (i.e., 1/10 rad), and its speed at successive increments of time is recorded in the following table. Estimate the

5 *Ibid.*
6 *Ibid.*

effective damping constant b for this system, expressed in Newton seconds per meter (N s/m). (See *hint* in Problem 2.2. Also check first for Coulomb friction.)

Time (s)	Speed (m/s)
0	0
10	2.05
20	3.30
30	4.15
40	4.85
50	5.20
60	5.55

2.4 An electric motor has been disconnected from its electrical driving circuit and set up to be driven mechanically by a variable-speed electric hand drill mounted in a simple dynamometer arrangement. The torque versus speed data given in the following table were obtained. Then, starting at a high speed, the motor was allowed to coast so that the speed versus time data given in the table could be taken.

Driven		Coasting	
Shaft speed (rpm)	Torque (N m)	Time (s)	Shaft speed (rpm)
100	0.85	0	600
200	1.35	10	395
300	2.10	20	270
400	2.70	30	180
500	3.70	40	110
600	4.50	50	70
–	–	60	40

(a) Evaluate the Coulomb friction torque T_c and the linear rotational damping coefficient B for the electric motor.

(b) Draw a system model for the electric motor and set up the differential equation for the shaft speed Ω_1 during the coasting interval.

(c) Estimate the rotational inertia J for the rotating parts of the electric motor.

2.5 (a) Draw a complete free-body diagram for all the elements of the system shown in Fig. P2.5.

(b) When the input force F_i is zero, the displacement x_1 is zero, the force in each spring is F_s, and the system is motionless. Find the spring forces F_{k1} and F_{k2} and the displacement x_1 when the input force is 5.0 lb and the system is again motionless.

(c) Derive the differential equation relating the displacement x_1 to the input force F_1.

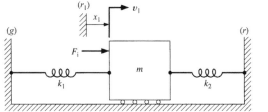

Figure P2.5.

(d) Find the initial values of u_1 and x_1 at $= 0^+$, i.e., $u_1 (0+)$ and $x_1(0+)$, after the input force F_1 is suddenly reduced to zero from 5.0 lb.

2.6 A metal plate weighing 5.0 kg and measuring 10 cm × 30 cm is suspended by means of a 30-cm-long, 5.0-mm-diameter steel rod, as shown in Fig. P2.6.

(a) Draw a complete free-body diagram for all the elements of this system and write the differential equation for the angular displacement θ_1 of the plate, using a simple inertia-spring model for the system.

(b) Using your knowledge about vibrations, find the natural frequency of oscillation for this system model in radians per second.

(c) Comment on the suitability of using this simple model for the system. Take into consideration that the input torque T_1 will never change with sinusoidal variations that have frequencies exceeding about 1/10 of the lowest clamped–clamped natural frequency of the suspension rod itself.

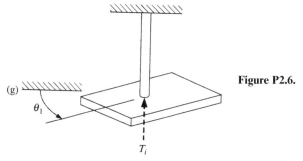

Figure P2.6.

2.7 A simplified schematic diagram of a quarter-car model is shown in Fig. 1.2.

(a) Derive the differential equation relating the motion x_3 to the road profile motion x_1.

(b) Derive the differential equation relating the spring force F_k to the road profile motion x_1.

2.8 The rotational system shown schematically in Fig. P2.8 is an idealized model of a machine-tool drive system. Although friction in the bearings is considered negligible in this case, the nonlinear friction effect NLD between the inertia J_2 and ground is significant. It is described by

$$T_{\mathrm{NLD}} = \Omega_2 \left(\frac{T_0}{|\Omega_2|} \right) + C\,|\Omega_2|\,\Omega_2.$$

(a) Draw the complete free-body diagram for this system, showing each element separately and clearly delineating all variables.

(b) Write the necessary and sufficient set of describing equations for this complete system.

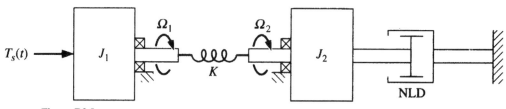

Figure P2.8.

(c) Write corresponding equations for small perturbations of all variables for the case in which Ω_2 is always positive and varies about a mean value Ω_2. Find corresponding mean values of T_s and Ω_2 in terms of $\overline{\Omega}_2$

(d) Combine the small-perturbation equations to eliminate unwanted variables and develop the system differential equation, relating the output $\hat{\Omega}_2$ to the input $\hat{T}_s(t)$.

2.9 The system shown schematically in Fig. P2.9 is an idealized model of a cable lift system for which the springs k_1 and k_2 represent the compliances of connecting flexible cables.

(a) Draw a complete free-body diagram for the system, showing each element separately, including the massless pulley, and clearly delineating all variables.

(b) Develop expressions for Δ_2 and Δ_3, the displaced references resulting from the action of gravity on the mass m_2 when $x_1 = 0$.

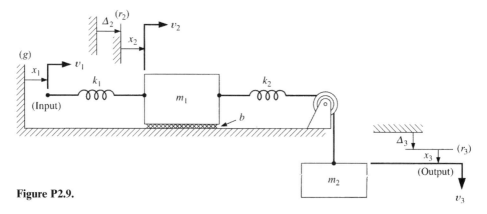

Figure P2.9.

(c) Write the necessary and sufficient set of describing equations for this system.

(d) Combine these equations to remove unwanted variables and develop the system differential equation relating x_3 to x_1.

2.10 Two masses are connected by springs to a rotating lever, as is shown schematically in Fig. P2.10.

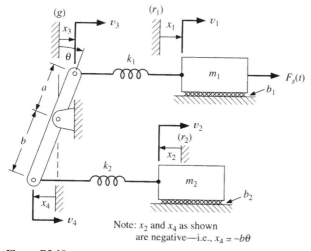

Figure P2.10.

(a) Draw a complete free-body diagram for this system, showing each element, including the massless lever, and delineating clearly all variables.

(b) Write the necessary and sufficient set of equations describing this system, assuming that the angle θ never changes by more than a few degrees.

(c) Combine equations to remove unwanted variables and obtain the system differential equation relating x_3 to $F_3(t)$.

2.11 A nonlinear hydraulic shock absorber using symmetrical orifice-type flow resistances, shown schematically in Fig. P2.11, has the nonlinear force versus velocity characteristic shown in the accompanying graph.

(a) Find the operating-point value $\overline{F}_{NLD}$ of the force F_{NLD} in terms of v_1.

(b) Linearize this characteristic, showing how $\hat{F}_{NLD}$ is related to $\hat{v}_1$ for small variations of v_1 around an operating-point value of $\bar{v}$. In other words, find the incremental damping coefficient $b_{inc} = \hat{F}_{NLD}/\hat{v}_1$.

(c) Suggest what should be used as the approximate equivalent damping coefficient b_{eq} when large periodic variations occur through the large range v_{R1} about zero velocity, as shown in the graph.

(d) Suggest what should be used for b_{eq} when a large periodic "one-sided" variation, v_{R2}, of v_1 occurs.

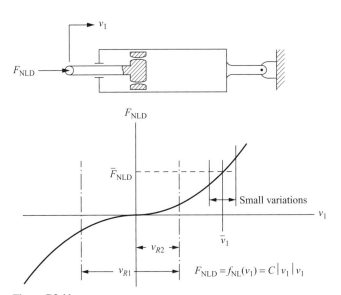

Figure P2.11.

2.12 The height y of the liquid in a leaking tank, shown in Fig. P2.12, is mathematically modeled by the differential equation

$$\frac{dy}{dt} = -\frac{by}{A_0 + ay} + \frac{Q_i}{A_0 + ay},$$

where Q_i is the input flow rate of liquid and a, b, and A_0 are constant parameters.

(a) Find the normal operating-point value y of the liquid height for the input flow rate $Q_1 = Q_1 = \text{const.}$

(b) Linearize the liquid height equation in a small vicinity of the normal operating point.

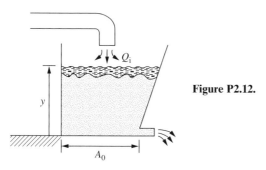

Figure P2.12.

2.13 An approximate model of a hydraulic turbine has been derived in the following form:

$$T_t = T_0 - \frac{T_0}{6}\left(1 - \frac{\Omega_1}{\Omega_0}\right)^2 - T_0\left(1 - \frac{\Delta P}{P_0}\right),$$

where T_t is the torque generated by the turbine, Ω_1 is the shaft velocity, ΔP is the hydraulic pressure, and $T_0, \Omega_0,$ and P_0 are constants. This model is shown graphically in Fig. P10.2(b). Linearize the torque model for small deviations from a normal operating point determined by $\Omega_1 = \overline{\Omega}_1$ and $\Delta P = \Delta \overline{P}$.

2.14 In an electric circuit shown in Fig. P2.14, resistor R and inductor L are assumed to be linear but the semiconductor diode behaves like an exponential resistor, whose current–voltage relationship is approximated by the following equation:

$$i = I_0 e^{\alpha e_D}.$$

The mathematical model of the circuit is given by

$$e_i = Ri + L\frac{di}{dt} + e_D.$$

Follow the five-step procedure used in Example 2.8 to linearize the circuit equation in the vicinity of the normal operating point established by a constant input voltage of $\overline{e}_i$.

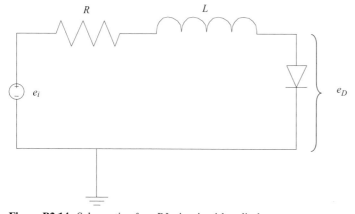

Figure P2.14. Schematic of an RL circuit with a diode.

2.15 Figure P2.15 shows the free-body diagram of a steel ball in a magnetic levitation apparatus. A large electromagnet is used to suspend the 1-in.-diameter steel ball. The magnetic force applied to the ball can be expressed as a function of the current through the coil i and the distance of the center of the ball to the base of the magnet x, as subsequently shown:

$$F_m = -K_i \frac{i^2}{x^2}.$$

(a) At equilibrium, the force generated by the coil is equal to the weight of the ball. For a given steady coil current, $\bar{i}$, find the corresponding equilibrium position $\bar{x}$.

(b) Linearize this equation for small perturbations about the equilibrium position.

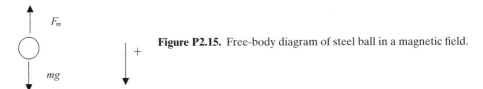

Figure P2.15. Free-body diagram of steel ball in a magnetic field.

2.16 Figure P2.16 shows a schematic of a mechanical drive system in which a servomotor (inertia J_1) drives a pinion gear (inertia J_2, radius R) that is meshed with a rack (mass m). The rack moves laterally against a spring element (k). The linear bearing that guides the rack is well lubricated and produces a viscous friction effect (b). Assume that the shaft between the motor (J_1) and the gear (J_2) is rigid. Develop the input–output differential equation of motion for this system with T_i as the input and x as the output.

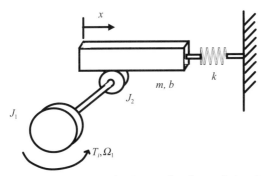

Figure P2.16. Schematic of a rotational–translational mechanical drive system.

2.17 Consider the mechanical system sketched in Fig. P2.17. It represents a motor driving a heavy inertial load through a belt drive. The belt is flexible and is modeled as a torsional spring. Also, the friction inherent in the motor is modeled as a linear damping factor between the motor inertia and the ground. Develop the equation relating the input torque T_i to the speed of the large inertia Ω_2.

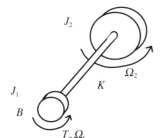

Figure P2.17. Schematic of a mechanical drive system with two rotary inertias and a torsional spring.

2.18 Figure P2.18 shows a simple model of a mass, two springs, and a damper. Derive the differential equation that relates the input force F to the motion of the mass x. The final differential equation should have only x and F as variables; there should be no extraneous coordinates in the final equation.

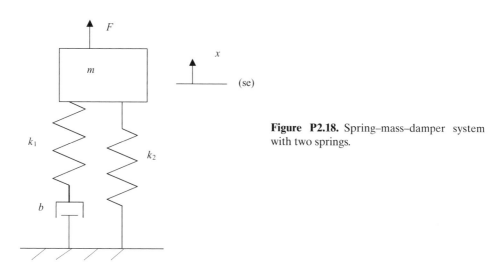

Figure P2.18. Spring–mass–damper system with two springs.

2.19 For the magnetic levitation problem introduced in Problem 2.15, derive the equation of motion for the steel ball. Use MATLAB to graph the nonlinear characteristic between the equilibrium position and the nominal current in the coil. For a nominal position of 15 mm, find the range of motion for which the linearization error is less than 10%. Use $m = 0.068$ kg and $K_i = 3.2654 \times 10^{-5}$.

2.20 The mechanical system shown in Fig. P2.20 depicts a model of a packaging system to protect delicate electronic items during shipping. Derive a differential equation relating the force F_i to the motion of m_2. Assume that all displacements are referenced to the static equilibrium position of the masses.

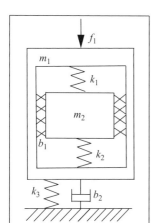

Figure P2.20. Schematic representation of a packaging system.

3

Mathematical Models

LEARNING OBJECTIVES FOR THIS CHAPTER

3–1 To derive input–output models of linear mechanical systems.

3–2 To distinguish between input–output and state-space forms of mathematical models of engineering systems.

3–3 To discuss the concept of a system state and the related concepts of state variable, state vector, and state space.

3–4 To derive state-space models of linear and nonlinear mechanical systems.

3–5 To recognize advantages and disadvantages of input–output models and state models and transform input–output models to state-space models and vice versa.

3.1 INTRODUCTION

In almost all areas of engineering, and certainly in all those areas for which new processes or devices are being developed, considerable efforts are directed toward acquiring information on various aspects of system performance. This process is generally referred to as a system analysis. Traditionally, system analysis was carried out by investigation of the performance of an existing physical object subjected to selected test input signals. Although there is no doubt that such an experimental approach provides extremely valuable and most reliable information about system characteristics, experimenting with an actual full-scale system is not always feasible and is very often practically impossible, especially in the early stages of the system analysis. There are several reasons for this. First, an actual system must be available for testing or a new test object must be constructed, which may involve high cost in terms of time and money. Second, the extent to which the parameters of an existing engineering system can be varied in order to observe their effects on system performance is usually very limited. Third, the experimental results are always "object specific," because they represent only a particular system under investigation, and may be difficult to generalize.

To overcome these problems, researchers develop simplified representations of actual systems, called system models. The system models may be physical in nature (downscaled actual systems), or they may be developed in the form of abstract descriptions of the relationships existing among the system variables. In the latter case, a dynamic system can be described by verbal text, plots and graphs, tables of relevant numerical data, or mathematical equations. The language of mathematics

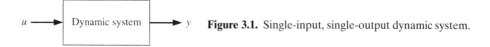

Figure 3.1. Single-input, single-output dynamic system.

is preferred in modeling engineering systems because of its superior precision and generality of expression. Mathematical methods of system modeling lead to better ability to generalize the results and to apply them to solving control and design problems. A representation of the relationships existing among the system variables in the form of mathematical equations is a mathematical model of the system.

Two types of mathematical models are introduced in this chapter, input–output models and state models. Both types of models carry essentially the same information about the system dynamics, but the sets of model differential equations are different from each other in several respects. These differences carry serious implications for practical usefulness and applicability of the types of models in various engineering problems.

3.2 INPUT–OUTPUT MODELS

A basic concept of a dynamic system interacting with its surroundings by means of input variables and output variables was introduced in Chap. 1. Recall that the input variables, or simply *inputs*, originate outside the system and are not directly dependent on what happens in the system. The output variables, or simply *outputs*, are chosen from the set of variables generated by the system as it is subjected to the input variables. The choice of the outputs is arbitrary, determined by the objectives of the system analysis.

Consider the single-input, single-output dynamic system shown in Fig. 3.1. In general, the relationship between the input and output signals of the system can be represented by an nth-order differential equation of the following form:

$$f\left(y, \frac{dy}{dt}, \ldots, \frac{d^n y}{dt^n}, u, \frac{du}{dt}, \ldots, \frac{d^m u}{dt^m}, t\right) = 0, \tag{3.1}$$

where $m \leq n$ for existing and realizable engineering systems because of the inherent inertia of those systems. Also, having $m > n$ is physically impossible because it would imply the ability to "predict the future" of the system input. A set of n initial conditions, $y(0^+)$, $y(0^+)$, $\ldots$, $y^{(n-1)}(0^+)$, must be known in order to solve the equation for a given input $u(t)$, $t \geq 0$.[1]

If the system is nonlinear, f is a nonlinear function of its arguments. If the system is stationary, time t is not an explicit argument of function f and the model differential equation can then be written as

$$f\left(y, \frac{dy}{dt}, \ldots, \frac{d^n y}{dt^n}, u, \frac{du}{dt}, \ldots, \frac{d^m u}{dt^m}\right) = 0, \tag{3.2}$$

where f may be a nonlinear function.

---

[1] To solve a differential equation and find the system response to an input beginning at time $t = 0$, it is necessary to have the initial conditions just after the input starts, i.e., at $t = 0^+$. All initial conditions throughout this book are therefore meant to be taken at $t = 0^+$, unless otherwise stated.

For a stationary linear model, the function f is a sum of terms that are linear with respect to the arguments of f, and the model input–output equation can then be presented in the following form:

$$a_n \frac{d^n y}{dt^n} + a_{n-1} \frac{d^{n-1} y}{dt^{n-1}} + \cdots + a_1 \frac{dy}{dt} + a_0 y$$

$$= b_m \frac{d^m u}{dt^m} + b_{m-1} \frac{d^{m-1} u}{dt^{m-1}} + \cdots + b_1 \frac{du}{dt} + b_0 u, \tag{3.3}$$

where $a_0, a_1, \ldots, a_n$ and $b_0, b_1, \ldots, b_m$ are all constants. As before, the order of the highest derivative of the input variable cannot be greater than the order of the highest derivative of the output variable, $m \leq n$.

Equations (3.1), (3.2), and (3.3) represent general forms of input–output models for single-input, single-output dynamic systems.

EXAMPLE 3.1

Derive an input–output equation for the system described in Example 2.4 by using force $F_i(t)$ as the input and velocity $v(t)$ as the output. In this system mass m is supported by an oil film bearing. The bearing produces a resisting force proportional to the velocity of the mass. The system is presented schematically in Fig. 3.2.

SOLUTION

The system equation of motion, derived in Example 2.4, is

$$m \frac{dv_1}{dt} + bv_1 = F_i$$

or, rearranged slightly,

$$m \frac{dv_1}{dt} + bv_1 - F_i = 0.$$

By comparing the preceding input–output equations with the general forms of Eqs. (3.1)–(3.3), it can be observed that this model is stationary (because time does not appear explicitly as a variable) and linear (the terms involving the first derivative of the output variable, the output variable itself, and the input variable are combined in a linear fashion). Completing the comparison between the specific input–output model derived in this example and the general form for a linear stationary system given by Eq. (3.3), one can see that $n = 1, m = 0, a_0 = b, a_1 = m$, and $b_0 = 1$.

The next three examples, 3.2, 3.3, and 3.4, deal with three variations of a mechanical system that is a little more complex than the system considered in Example 3.1, although it is still relatively simple. These examples have been chosen to demonstrate how involved the derivation of input–output models becomes for all except the simplest low-order, single-input, single-output systems.

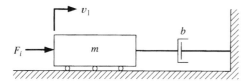

Figure 3.2. Mechanical system considered in Example 3.1.

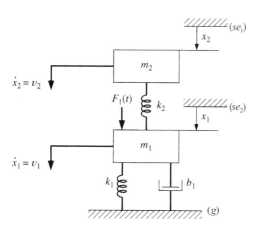

Figure 3.3. Mechanical system considered in Example 3.2.

EXAMPLE 3.2

Derive input–output equations for the mechanical system shown in Fig. 3.3 by using force $F_1(t)$ as the input variable and displacements $x_1(t)$ and $x_2(t)$ as the output variables.

The symbols se_1 and se_2 represent static-equilibrium positions of masses m_1 and m_2.

SOLUTION

The equation of motion for mass m_1 is

$$m_1\ddot{x}_1 + b_1\dot{x}_1 + (k_1 + k_2)x_1 - k_2x_2 = F_1(t).$$

The equation of motion for mass m_2 is

$$m_2\ddot{x}_2 + k_2x_2 - k_2x_1 = 0.$$

It should be noted that, as was demonstrated in Example 2.5, when displacements are measured from static-equilibrium positions, i.e., from positions taken by the two masses when the external force F_1 acting on the system is zero and the system is linear, the gravity terms do not appear in the equations of motion.

Combining the equations for the two masses and eliminating x_1 yields the input–output equation for the system:

$$\frac{d^4x_2}{dt^4} + \left(\frac{b_1}{m_1}\right)\frac{d^3x_2}{dt^3} + \left(\frac{k_2}{m_2} + \frac{k_1}{m_1} + \frac{k_2}{m_1}\right)\frac{d^2x_2}{dt^2}$$
$$+ \left(\frac{b_1k_2}{m_1m_2}\right)\frac{dx_2}{dt} + \left(\frac{k_1k_2}{m_1m_2}\right)x_2 = \left(\frac{k_2}{m_1m_2}\right)F_1(t).$$

The preceding fourth-order differential equation can be solved provided the initial conditions, $x_2(0)$, $(dx_2/dt)\big|_{t=0}$, $(d^2x_2/dt^2)\big|_{t=0}$, $(d^3x_2/dt^3)\big|_{t=0}$, and the input variable, $F_1(t)$ for $t \geq 0$, are known.

Similarly, the input–output equation relating x_1 to $F_1(t)$ is

$$\frac{d^4x_1}{dt^4} + \left(\frac{b_1}{m_1}\right)\frac{d^3x_1}{dt^3} + \left(\frac{k_2}{m_2} + \frac{k_1}{m_1} + \frac{k_2}{m_1}\right)\frac{d^2x_1}{dt^2} + \left(\frac{b_1k_2}{m_1m_2}\right)\frac{dx_1}{dt}$$
$$+ \left(\frac{k_1k_2}{m_1m_2}\right)x_1 = \left(\frac{1}{m_1}\right)\frac{d^2F_1}{dt^2} + \left(\frac{k_2}{m_1m_2}\right)F_1(t).$$

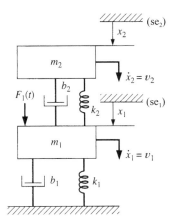

Figure 3.4. Mechanical system considered in Example 3.3.

The process of deriving the input–output equation in Example 3.2 will become considerably more complicated if an additional damper b_2 is included between the two masses, as shown in Fig. 3.4.

EXAMPLE 3.3

Derive an input–output equation for the mechanical system shown in Fig. 3.4, using x_2 as the output variable.

SOLUTION

The equations of motion for masses m_1 and m_2 now take the form

$$m_1 \frac{d^2 x_1}{dt^2} + (b_1 + b_2)\frac{dx_1}{dt} + (k_1 + k_2)x_1 - b_2 \frac{dx_2}{dt} - k_2 x_2 = F_1(t),$$

$$m_2 \frac{d^2 x_2}{dt^2} + b_2 \frac{dx_2}{dt} + k_2 x_2 - b_2 \frac{dx_1}{dt} - k_2 x_1 = 0.$$

The unwanted variable x_1 cannot be eliminated from these equations by use of simple substitutions as in Example 3.2 because the derivatives of both x_1 and x_2 are present in each equation. In such a case, an operator D can be introduced, defined as

$$D^k x(t) = \frac{d^k x(t)}{dt^k}.$$

The D operator transforms differential equations into algebraic equations, which are usually easier to manipulate than the original differential equations.

With the D operator, the differential equations of motion can be rearranged into the following form:

$$m_1 D^2 x_1 + (b_1 + b_2)Dx_1 + (k_1 + k_2)x_1 - b_2 Dx_2 - k_2 x_2 = F_1(t),$$
$$m_2 D^2 x_2 + b_2 Dx_2 + k_2 x_2 - b_2 Dx_1 - k_2 x_1 = 0.$$

From the last equation, x_1 can be expressed as

$$x_1 = \left(\frac{m_2 D^2 + b_2 D + k_2}{b_2 D + k_2} \right) x_2.$$

Substituting into the operator equation for mass m_1 yields

$$m_1 m_2 D^4 x_2 + (m_2 b_1 + m_2 b_2 + m_1 b_2) D^3 x_2 + (m_1 k_2 + m_2 k_1 + m_2 k_2 + b_1 b_2) D^2 x_2$$
$$+ (b_1 k_2 + b_2 k_1) D x_2 + k_1 k_2 x_2 = b_2 D F_1 + k_2 F_1.$$

Using the inverse of the definition of the D operator to transform this equation back to the time domain gives the input–output equation for the system:

$$(m_1 m_2) \frac{d^4 x_2}{dt^4} + (m_2 b_1 + m_2 b_2 + m_1 b_2) \frac{d^3 x_2}{dt^3}$$

$$+ (m_1 k_2 + m_2 k_1 + m_2 k_2 + b_1 b_2) \frac{d^2 x_2}{dt^2}$$

$$+ (b_1 k_2 + b_2 k_1) \frac{dx_2}{dt} + k_1 k_2 x_2 = b_2 \frac{d F_1}{dt} + k_2 F_1.$$

Example 3.3 demonstrates that inserting a damper between the two masses makes the process of deriving the input–output equation considerably more complicated. In the next example, a two-input, two-output system is considered. Two separate input-output equations, one for each output variable, will have to be derived.

EXAMPLE 3.4

Consider again the mechanical system shown in Fig. 3.3. The system is now subjected to two external forces, $F_1(t)$ and $F_2(t)$. The displacements of both masses are of interest, and therefore $x_1(t)$ and $x_2(t)$ will be the two output variables of this system. The system is shown in Fig. 3.5. The differential equations of motion for the two masses are

$$m_1 \frac{d^2 x_1}{dt^2} + b_1 \frac{dx_1}{dt} + (k_1 + k_2) x_1 - k_2 x_2 = F_1,$$

$$m_2 \frac{d^2 x_2}{dt^2} + k_2 x_2 - k_2 x_1 = F_2.$$

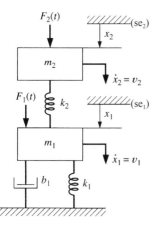

Figure 3.5. Two-input, two-output system considered in Example 3.4.

Figure 3.6. Multi-input, multi-output system.

When these two equations are combined, the separate input–output equations for $x_1(t)$ and $x_2(t)$ are obtained:

$$m_1 m_2 \frac{d^4 x_1}{dt^4} + m_2 b_1 \frac{d^3 x_1}{dt^3} + (m_1 k_2 + m_2 k_1 + m_2 k_2) \frac{d^2 x_1}{dt^2}$$
$$+ b_1 k_2 \frac{dx_1}{dt} + k_1 k_2 x_1 = m_2 \frac{d^2 F_1}{dt^2} + k_2 F_1 + k_2 F_2,$$
$$m_1 m_2 \frac{d^4 x_2}{dt^4} + m_2 b_1 \frac{d^3 x_2}{dt^3} + (m_1 k_2 + m_2 k_1 + m_2 k_2) \frac{d^2 x_2}{dt^2}$$
$$+ b_1 k_2 \frac{dx_2}{dt} + k_1 k_2 x_2 = k_2 F_1 + m_1 \frac{d^2 F_2}{dt^2} + b_1 \frac{d F_2}{dt} + (k_1 + k_2) F_2.$$

Note that the two input–output equations are independent of each other and can be solved separately. *On the other hand, the coefficients of each of the terms on the left-hand sides are the same, regardless of which system variable is chosen as the output.*

In general, a linear system with l inputs and p outputs, shown schematically in Fig. 3.6, is described by p independent input–output equations:

$$f_1\big(y_1^{(n)}, \ldots, \dot{y}_1, y_1, u_1^{(m)}, \ldots, \dot{u}_1, u_1, u_2^{(m)}, \ldots, \dot{u}_2, u_2, \ldots, u_l^{(m)}, \ldots, \dot{u}_l, u_l, t\big) = 0,$$
$$f_2\big(y_2^{(n)}, \ldots, \dot{y}_2, y_2, u_1^{(m)}, \ldots, \dot{u}_1, u_1, u_2^{(m)}, \ldots, \dot{u}_2, u_2, \ldots, u_l^{(m)}, \ldots, \dot{u}_l, u_l, t\big) = 0$$
$$\vdots$$
$$f_p\big(y_p^{(n)}, \ldots, \dot{y}_p, y_p, u_1^{(m)}, \ldots, \dot{u}_1, u_1, u_2^{(m)}, \ldots, \dot{u}_2, u_2, \ldots, u_l^{(m)}, \ldots, \dot{u}_l, u_l, t\big) = 0,$$
$$(3.4)$$

where $m \leq n$ and a superscript enclosed in parentheses denotes the order of a derivative. If a system is assumed to be stationary, time t does not appear explicitly in Eqs. (3.4).

If a system is assumed to be linear, the functions $f_1, f_2, \ldots, f_p$ are linear combinations of terms involving the system inputs, outputs, and their derivatives. The input–output model for a linear, stationary, multi-input, multi-output system can be presented in a more compact form:

$$\sum_{i=0}^{n} a_{1i} y_1^{(i)} = \sum_{j=1}^{l} \sum_{k=0}^{m} b_{1jk} u_j^{(m)},$$

$$\sum_{i=0}^{n} a_{2i} y_2^{(i)} = \sum_{j=1}^{l} \sum_{k=0}^{m} b_{2jk} u_j^{(m)},$$

$$\vdots$$

$$\sum_{i=0}^{n} a_{pi} y_p^{(i)} = \sum_{j=1}^{l} \sum_{k=0}^{m} b_{pjk} u_j^{(m)}. \qquad (3.5)$$

Note that some of the a and b coefficients may be equal to zero; also note that $a_{1i} = a_{2i} = \cdots = a_{pi}, i = 1, 2, \ldots, n$.

The input–output equations, even for relatively simple multi-input, multi-output models, become extremely complicated. Moreover, as will be shown in Chap. 4, analytical methods for solving input–output equations are practical only for low-order, single-input, single-output models. In fact, most numerical methods for solving high-order differential equations, such as input–output equations, require that those equations be replaced with an equivalent set of first-order equations, which is the standard form of state equations. Also, many quite powerful concepts and methodologies based on input–output models, such as the transfer function, are applicable to linear, stationary models only. The conceptual simplicity of using the input–output representation of a dynamic system is lost in the complexity of the mathematical forms with models that are nonlinear, have many inputs, outputs, or both, or simply are of an order higher than third. Moreover, it is not even possible to obtain input–output differential equations for most nonlinear systems because the presence of nonlinearities inhibits the combination of model equations to eliminate unwanted variables.

3.3   STATE MODELS

The concept of the state of a dynamic system was promoted in the 1950s. Its significance has grown since then, and today the state approach to modeling is the most powerful and dominant technique used in analysis of engineering systems.

The term *state* is very similar to the thermodynamic state of a substance. In thermodynamics, there exist a minimum number of variables such as temperature and pressure that, when known, allow one to deduce all important properties of a substance. Similarly, in a given dynamic system, there exists a minimum set of physical variables, called the state variables that, when known, tell everything one needs to know about that system at that instant in time. Moreover, if the state of the system at time t_0 and the inputs for time greater than t_0 are known, one can completely determine the behavior of the system for all time t greater than t_0.

The physical variables, or quantities, that when known completely describe the state of the system are called *state variables*. These variables are usually well-known and measurable quantities, such as displacement and velocities, in mechanical systems.

The set of state variables can be assembled in an ordered grouping, a vector. This is called, not surprisingly, the *state vector*. The state vector is an important concept because it opens the door to a geometric interpretation of the state of a dynamic system. If the state vectors can be interpreted as geometric vectors, then the state variables can be used to describe a geometric space, called the *state space*. It is well known to engineers that any ordered set of two or three numbers has a geometric interpretation of a point in physical space. An ordered set of two numbers determines a point on a plane (the Cartesian plane, for example), whereas a set of three numbers uniquely determines the location of a point in a physical space. However, unlike in planar or spatial geometry, the space described by state variables can be of a dimension higher than three. Although this may make visualization difficult, the vast array of mathematical tools available for geometric analysis can be applied to the analysis of state models of dynamic systems. This property lies at the heart of the power of state models, and methods that take advantage of this property are often referred to as state-space methods.

Because the values of the state variables at any instant in time determine the state of the system at that instant, the state of a system at any instant in time can be viewed as a point in state space. As time progresses and the dynamic system changes, the state variables change, thus moving the system to a new location in state space. A system can therefore be envisioned as a point moving smoothly in time through many locations, following a continuous curve in state space. This curve, which describes the system behavior in time, is called the *state trajectory*. State trajectories are very powerful tools in visualizing and understanding behavior, particularly of complex and nonlinear systems. Recent advances in the field of nonlinear dynamics and what is now known as "chaos theory" have relied heavily on the interpretation of state trajectories.[2]

The basic terms associated with the state models, are now defined:

- *State* of a dynamic system is defined by the smallest set of variables such that the knowledge of these variables at time $t = t_0$, together with the knowledge of the input for $t > t_0$, completely determines the behavior of the system for time $t \geq t_0$.
- *State variables*, $q_1, q_2, \ldots, q_n$, are the elements of the smallest set of variables required for completely describing the state of the system. One important implication of this definition is that state variables are independent of each other. If it were possible to express any of the state variables in terms of others, those variables would not be necessary to uniquely describe the system dynamics, and such a set of variables would not constitute the smallest set of variables as required by the definition of the state of a dynamic system.
- *State vector* of a dynamic system is the column vector **q** whose components are the state variables, $q_1, q_2, \ldots, q_n$.
- *State space* is an n-dimensional space containing the n-system state variables. The state of a dynamic system at any instant of time t is represented by a single point in the state space.
- *State trajectory* is the path over time of the point representing the state of the system in a state space.

The concepts of state and state space are extremely powerful, yet somewhat abstract and difficult for a newcomer to the field to fully appreciate. The following example of a simple mechanical system should help to clarify these concepts.

EXAMPLE 3.5

Imagine you are standing in the center of a large field holding a small rock. You hurl the rock directly upward, stand out of the way, of course, and watch it return to earth. The behavior of the rock is very simple to describe if losses of energy that are due to wind resistance are negligible. This is a well-known ballistics problem of a projectile of a given initial velocity and moving under constant acceleration (gravity). Consider this problem from the state-space standpoint.

First it is necessary to identify the variables that are required for defining the system (the rock) at any instant in time. Clearly velocity is one of these variables, as the kinetic

---

[2] S. H. Strogratz, *Nonlinear Dynamics and Chaos* (Addison-Wesley, Reading, MA, 1994).

energy of the rock is completely determined by velocity. Also, as the rock moves vertically upward in the earth's gravitational field, it gains potential energy because of its height in that field. As the rock moves downward, it loses potential energy and regains its kinetic energy. It is therefore clear that knowing the position (height) of the rock is also necessary to determine its state. It can thus be concluded that the state of the system can be determined by two variables, velocity v and position x. Because two variables are chosen, the system is said to be of a second order and its state space is two dimensional.

The elemental equations that can be used to calculate the two state variables are

$$v(t) = \int_0^t a\, dt + v(0),$$

$$x(t) = \int_0^t v(t)dt + x(0),$$

where $v(0)$ is the initial velocity of the rock and $x(0)$ is its initial position. Given that the acceleration a is constant at one standard gravity g, and assuming that the initial position of the rock is zero, the velocity and height at time t can be found from the integrals

$$v(t) = at + v(0),$$

$$x(t) = \frac{1}{2}at^2 + v(0)t.$$

Consider now a graph of the state of the rock in a coordinate system in which the rock's height is on the horizontal axis and its velocity is on the vertical axis, as shown in Fig. 3.7. At time $t = t_0$, the state is defined as a point at $(0, v_0)$ on the vertical axis. As time progresses, the rock gains height and its velocity decreases. This is illustrated by the trajectory's moving downward and to the right until it intersects the horizontal axis at $(x_f 0)$. This point represents the maximum height x_f achieved by the rock and velocity equal to zero. After reaching this point, the rock begins to fall back to earth, the velocity becomes negative, and the trajectory ends at the point $(0, -v_0)$, representing the point at which the rock strikes the earth with a velocity of the same magnitude but opposite direction of the initial velocity (which is why you had to step out of the way after hurling the rock!).

Mathematically, state models take the form of sets of first-order differential equations, as follows:

$$\dot{q}_1 = f_1(q_1, q_2, \ldots, q_n, u_1, u_2, \ldots, u_l, t),$$
$$\dot{q}_2 = f_2(q_1, q_2, \ldots, q_n, u_1, u_2, \ldots, u_l, t)$$
$$\vdots$$
$$\dot{q}_n = f_n(q_1, q_2, \ldots, q_n, u_1, u_2, \ldots, u_l, t), \tag{3.6}$$

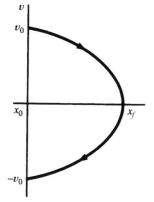

Figure 3.7. State trajectory of a rock thrown vertically from x_0.

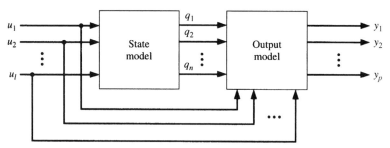

Figure 3.8. Block diagram of a state model.

where $q_1, q_2, \ldots, q_n$ are the system state variables and $u_1, u_2, \ldots, u_l$, are the input variables. If the model is nonlinear, at least one of the functions $f_i, i = 1, 2, \ldots, n$, is nonlinear.

Although the state vector completely represents the system dynamics, the selected state variables are not necessarily the same as the system outputs. However, each output variable, or, in general, any system variable that is of interest, can be expressed mathematically in terms of the system state variables and the input variables. A block diagram of a state representation of a multi-input, multi-output dynamic system is shown in Fig. 3.8.

In general, the system output equations can be written in the following form:

$$
\begin{aligned}
y_1 &= g_1(q_1, q_2, \ldots, q_n, u_1, u_2, \ldots, u_l, t), \\
y_2 &= g_2(q_1, q_2, \ldots, q_n, u_1, u_2, \ldots, u_l, t)
\end{aligned}
$$
$$
\vdots
$$
$$
y_p = g_p(q_1, q_2, \ldots, q_n, u_1, u_2, \ldots, u_l, t). \tag{3.7}
$$

It should be noted that, although the state model given by Eqs. (3.6) provides a complete description of the system dynamics, the output model given by Eqs. (3.7) represents static relations that exist between the selected output variables and the system input and state variables. Consequently, the state equations [Eqs. (3.6)] are first-order differential equations, whereas the output equations [Eqs. (3.7)] are algebraic equations with no derivatives of any of the system variables.

If a system model is linear, all functions on the right-hand sides of state equations (3.6), f_i for $i = 1, 2, \ldots, n$, and all functions on the right-hand sides of output equations (3.7), g_j for $j = 1, 2, \ldots, p$, are linear. Also, in stationary model equations, system parameters do not vary with time, and thus linear, stationary-state model equations take the form

$$
\begin{aligned}
\dot{q}_1 &= a_{11}q_1 + a_{12}q_2 + \cdots + a_{1n}q_n + b_{11}u_1 + b_{12}u_1 + \cdots + b_{1l}u_l, \\
\dot{q}_2 &= a_{21}q_1 + a_{22}q_2 + \cdots + a_{2n}q_n + b_{21}u_1 + b_{22}u_2 + \cdots + b_{2l}u_l
\end{aligned}
$$
$$
\vdots \tag{3.8}
$$
$$
\dot{q}_n = a_{n1}q_1 + a_{n2}q_2 + \cdots + a_{nn}q_n + b_{n1}u_1 + b_{n2}u_2 + \cdots + b_{nl}u_l.
$$

The linear output equations are

$$
\begin{aligned}
y_1 &= c_{11}q_1 + c_{12}q_2 + \cdots + c_{1n}q_n + d_{11}u_1 + d_{12}u_2 + \cdots + d_{1l}u_l, \\
y_2 &= c_{21}q_2 + c_{22}q_2 + \cdots + c_{2n}q_n + d_{21}u_1 + d_{22}u_2 + \cdots + d_{2l}u_l
\end{aligned}
$$
$$
\vdots \tag{3.9}
$$
$$
y_p = c_{p1}q_1 + c_{p2}q_2 + \cdots + c_{pn}q_n + d_{p1}u_1 + d_{p2}u_2 + \cdots + d_{pl}u_l.
$$

Equations (3.8) and (3.9) can be written in matrix–vector notation:

$$
\begin{bmatrix} \dot{q}_1 \\ \dot{q}_2 \\ \vdots \\ \dot{q}_n \end{bmatrix} = \begin{bmatrix} a_{11} & a_{12} & \cdots & a_{1n} \\ a_{21} & a_{22} & \cdots & a_{2n} \\ \vdots & & & \\ a_{nl} & a_{n2} & \cdots & a_{nn} \end{bmatrix} \begin{bmatrix} q_1 \\ q_2 \\ \vdots \\ q_n \end{bmatrix} + \begin{bmatrix} b_{11} & b_{12} & \cdots & b_{1l} \\ b_{21} & b_{22} & \cdots & b_{2l} \\ \vdots & & & \\ b_{n1} & b_{n2} & \cdots & b_{nl} \end{bmatrix} \begin{bmatrix} u_1 \\ u_2 \\ \vdots \\ u_l \end{bmatrix}, \tag{3.10}
$$

$$
\begin{bmatrix} y_1 \\ y_2 \\ \vdots \\ y_p \end{bmatrix} = \begin{bmatrix} c_{11} & c_{12} & \cdots & c_{1n} \\ c_{21} & c_{22} & \cdots & c_{2n} \\ \vdots & & & \\ c_{p1} & c_{p2} & \cdots & c_{pn} \end{bmatrix} \begin{bmatrix} q_1 \\ q_2 \\ \vdots \\ q_n \end{bmatrix} + \begin{bmatrix} d_{11} & d_{12} & \cdots & d_{1l} \\ d_{21} & d_{22} & \cdots & d_{2l} \\ \vdots & & & \\ d_{p1} & d_{p2} & \cdots & d_{pl} \end{bmatrix} \begin{bmatrix} u_1 \\ u_2 \\ \vdots \\ u_l \end{bmatrix}. \tag{3.11}
$$

These equations can be rewritten in more compact forms:

$$
\mathbf{q} = \mathbb{A}\mathbf{q} + \mathbb{B}\mathbf{u}, \tag{3.10a}
$$

$$
\mathbf{y} = \mathbb{C}\mathbf{q} + \mathbb{D}\mathbf{u}, \tag{3.11a}
$$

where $\mathbb{A}$ is an $n \times n$ state matrix, $\mathbb{B}$ is an $n \times l$ input matrix, $\mathbb{C}$ is a $p \times n$ output matrix, $\mathbb{D}$ is a $p \times l$ direct-transmission matrix, $\mathbf{q}$ is a state vector, $\mathbf{u}$ is an input vector, and $\mathbf{y}$ is an output vector.

As mentioned before, the selection of state variables constitutes a nontrivial problem, because for each system there are usually many different sets of variables that uniquely represent the system dynamics. The following sets of variables are most commonly used as state variables:

(1) sets of T-type and A-type variables associated with T and A energy-storing elements of the system;
(2) sets including one variable and its successive derivatives;
(3) sets including two or more variables and their derivatives;
(4) sets including an auxiliary variable and its successive derivatives.

In addition, there are still other, relatively less common types of state variables, such as the variables associated with the roots of the system characteristic equation obtained by means of manipulation of the system matrix or sets of variables obtained as nonredundant algebraic combinations of other state variables.[3]

In general, the state of a dynamic system evolves from the process of storing energy in those system components that are capable of storing it. In fact, the number of state variables is always equal to the number of independent energy-storing elements in the system, regardless of the type of the state variables employed.[4]

In Example 3.6, the first three most common types of the state models just listed are derived. The use of an auxiliary variable and its derivatives as the state variables is demonstrated in Section 3.4.

---

[3] Y. Takahashi. M. J. Rabins, and D. M. Auslander, *Control and Dynamic Systems* (Addison-Wesley, Reading, MA, 1972).

[4] Strictly speaking, the number of state variables is equal to the number of independent energy-storing processes in the system. If an element is involved in two energy-storing processes, two state variables are needed to represent this element in the state model. An example of such an element is the roller in Problem 3.11. The roller stores kinetic energy from its translational motion and, independently, kinetic energy from the rotational motion, thus requiring both translational and rotational velocities to be used as the state variables.

EXAMPLE 3.6

Derive state models of types (1), (2), and (3) for the mechanical system shown in Fig. 3.3.

SOLUTION

Type (1): T-type and A-type variables There are four independent energy-storing elements in the system shown in Fig. 3.3: masses m_1 and m_2, and springs k_1 and k_2. Masses in mechanical systems are A-type elements, which can store kinetic energy, whereas springs are T-type elements, capable of storing potential energy. The respective A-type and T-type variables are the velocities of the two masses, v_1 and v_2 and the forces exerted by the springs, F_{k1} and F_{k2}. Hence the four variables selected to represent the state of the system are $q_1 = F_{k1}, q_2 = v_1, q_3 = F_{k2}$, and $q_4 = v_2$. To derive the state equations, first consider the forces exerted by the springs, F_{k1}, and F_{k2}. The equations defining these forces are

$$F_{k1} = k_1 x_1,$$

$$F_{k2} = k_2(x_2 - x_1).$$

Differentiating both sides of these equations with respect to time gives the first two state-variable equations:

$$\dot{F}_{k1} = (k_1)v_1,$$

$$\dot{F}_{k2} = (-k_2)v_1 + (k_2)v_2.$$

The equations of motion for masses m_1, and m_2 derived in Example 3.2 were

$$m_1 \underbrace{\frac{d^2 x_1}{dt^2}}_{dv_1/dt} + b_1 \underbrace{\frac{dx_1}{dt}}_{v_1} + \underbrace{k_1 x_1}_{F_{k1}} - \underbrace{k_2(x_2 - x_1)}_{F_{k2}} = F_1(t),$$

$$m_2 \underbrace{\frac{d^2 x_2}{dt^2}}_{dv_2/dt} + \underbrace{k_2(x_2 - x_1)}_{F_{k2}} = 0,$$

which give the other two state-variable equations

$$\dot{v}_1 = -\frac{1}{m_1} F_{k1} - \frac{b_1}{m_1} v_1 + \frac{1}{m_1} F_{k2} + \frac{1}{m_1} F_1(t),$$

$$\dot{v}_2 = -\frac{1}{m_2} F_{k2}.$$

Rewriting the state-variable equations in vector–matrix form yields

$$\begin{bmatrix} \dot{F}_{k1} \\ \dot{v}_1 \\ \dot{F}_{k2} \\ \dot{v}_2 \end{bmatrix} = \begin{bmatrix} 0 & k_1 & 0 & 0 \\ -1/m_1 & -b_1/m_1 & 1/m_1 & 0 \\ 0 & -k_2 & 0 & k_2 \\ 0 & 0 & -1/m_2 & 0 \end{bmatrix} \begin{bmatrix} F_{k1} \\ v_1 \\ F_{k2} \\ v_2 \end{bmatrix} + \begin{bmatrix} 0 \\ 1/m_1 \\ 0 \\ 0 \end{bmatrix} F_1(t).$$

The output equations for the displacements x_1 and x_2 are

$$x_1 = \frac{1}{k_1} F_{k1},$$

$$x_2 = \frac{1}{k_1} F_{k1} + \frac{1}{k_2} F_{k2}.$$

In vector–matrix form, the output model equations are written as

$$\begin{bmatrix} x_1 \\ x_2 \end{bmatrix} = \begin{bmatrix} 1/k_1 & 0 & 0 & 0 \\ 1/k_1 & 0 & 1/k_2 & 0 \end{bmatrix} \begin{bmatrix} F_{k1} \\ v_1 \\ F_{k2} \\ v_2 \end{bmatrix} + \begin{bmatrix} 0 \\ 0 \end{bmatrix} F_1(t).$$

Type (2): One variable and its successive derivatives State variables of this type are particularly convenient when an input–output equation is available. Very often in such cases the output variable and successive derivatives are selected as the state variables. In Example 3.2, the input–output equations for the system shown in Fig. 3.3 were derived relating displacements x_1 and x_2 to the input force $F_1(t)$. Both equations are of fourth order, and so four state variables are needed to uniquely represent the system dynamics. Let x_2 and its first three derivatives be selected as the state variables – i.e., $q_1 = x_2$, $q_2 = dx_2/dt$, $q_3 = d^2x_2/dt^2$, and $q_4 = d^3x_2/dt^3$.

The input–output equation relating x_2 to the input force $F_1(t)$ was

$$\frac{d^4x_2}{dt^4} + \left(\frac{b_1}{m_1} \right) \frac{d^3x_2}{dt^3} + \left(\frac{k_2}{m_2} + \frac{k_1}{m_1} + \frac{k_2}{m_1} \right) \frac{d^2x_2}{dt^2}$$

$$+ \left(\frac{b_1 k_2}{m_1 m_2} \right) \frac{dx_2}{dt} + \left(\frac{k_1 k_2}{m_1 m_2} \right) x_2 = \left(\frac{k_2}{m_1 m_2} \right) F_1(t).$$

By use of the preceding equation and the definitions of the selected state variables, the following state equations are formed:

$$\dot{q}_1 = q_2,$$
$$\dot{q}_2 = q_3,$$
$$\dot{q}_3 = q_4,$$
$$\dot{q}_4 = - \left(\frac{k_1 k_2}{m_1 m_2} \right) q_1 - \left(\frac{b_1 k_2}{m_1 m_2} \right) q_2$$

$$- \left(\frac{k_1}{m_1} + \frac{k_2}{m_2} + \frac{k_2}{m_1} \right) q_3 - \left(\frac{b_1}{m_1} \right) q_4 + \left(\frac{k_2}{m_1 m_2} \right) F_1(t).$$

It can be seen that $x_1 = x_2 - (x_2 - x_1) = x_2 - F_{k2}/k_2 = x_2 + (m_2/k_2)(d^2x_2/dt^2)$; then, if x_1 and x_2 are selected as the output variables y_1 and y_2, respectively, the output equations become

$$y_1 = q_1 + \left(\frac{m_2}{k_2} \right) q_3 = x_1,$$

$$y_2 = q_1 = x_2.$$

Type (3): Two or more variables and their derivatives State variables of this type are used most often in modeling mechanical systems. For the system considered in this example, x_1 and x_2 and their derivatives $\dot{x}_1$ and $\dot{x}_2$ are selected as the four state variables. Noting that $\dot{x}_1 = v_1$ and $\dot{x}_2 = v_2$ and using the equations of motion for masses m_1 and m_2 derived in Example 3.2, one obtains the following state equations:

$$\dot{q}_1 = q_3,$$
$$\dot{q}_2 = q_4,$$
$$\dot{q}_3 = -\left(\frac{k_1}{m_1} + \frac{k_2}{m_1}\right)q_1 + \left(\frac{k_2}{m_1}\right)q_2 - \left(\frac{b_1}{m_1}\right)q_3 + \left(\frac{1}{m_1}\right)F_1(t),$$
$$\dot{q}_4 = \left(\frac{k_2}{m_2}\right)q_1 - \left(\frac{k_2}{m_2}\right)q_2,$$

where $q_1 = x_1$, $q_2 = x_2$, $q_3 = v_1$, and $q_4 = v_2$.

The output equations are simply

$$y_1 = q_1,$$
$$y_2 = q_2.$$

3.4 TRANSITION BETWEEN INPUT–OUTPUT AND STATE MODELS

At the beginning of this chapter, both input–output and state models were said to be equivalent in the sense that each form completely represents the dynamics of the same system. It is therefore natural to expect that there is a corresponding state model involving successive derivatives of one state variable for each input–output model, and vice versa. The transition between the two forms of models is indeed possible, although it is not always straightforward.

Consider first a simple case of a single-input, single-output system model with no derivatives of the input variable present in the input–output equation:

$$a_n \frac{d_n y}{dt_n} + \cdots + a_1 \frac{dy}{dt} + a_0 y = b_0 u. \tag{3.12}$$

If the following state variables are selected [as in type (2) in Section 3.3],

$$q_1 = y, q_2 = \frac{dy}{dt}, \ldots, q_n = \frac{d^{n-1}y}{dt^{n-1}}, \tag{3.13}$$

the equivalent set of state-model equations is

$$\dot{q}_1 = q_2,$$
$$\dot{q}_2 = q_3$$
$$\vdots$$
$$\dot{q}_{n-1} = q_n,$$
$$\dot{q}_n = -\left(\frac{a_0}{a_n}\right)q_1 - \left(\frac{a_1}{a_n}\right)q_2 - \cdots - \left(\frac{a_{n-1}}{a_n}\right)q_n + \left(\frac{b_0}{a_n}\right)u. \tag{3.14}$$

The procedure for the transition between an input–output model and an equivalent state model becomes more complicated if derivatives of input variables are

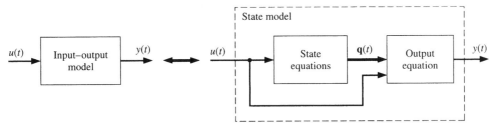

Figure 3.9. Equivalent input–output and state models.

present on the right-hand side of the input–output equation. The input–output equation of a single-input, single-output model in this case is

$$a_n \frac{d^n y}{dt^n} + \cdots + a_1 \frac{dy}{dt} + a_0 y = b_m \frac{d^m u}{dt^m} + \cdots + b_1 \frac{du}{dt} + b_0 u, \qquad (3.15)$$

where $m \le n$ for physically realizable systems. An equivalent state model in this case consists of a set of state-variable equations based on a state variable and $(n - 1)$ of its derivatives together with the output equations, as illustrated in Fig. 3.9.

To obtain an equivalent state model, first the higher derivatives of the input variable in Eq. (3.15) are ignored to yield an auxiliary input–output differential equation:

$$a_n \frac{d^n x}{dt^n} + \cdots + a_1 \frac{dx}{dt} + a_0 x = u, \qquad (3.16)$$

where x is an auxiliary state variable. The set of equivalent state equations for the auxiliary input–output equation [Eq. (3.16)] can be obtained as before, by use of the auxiliary variable and its successive derivatives as the state variables [the type (4) state variables discussed in Section 3.3]:

$$q_1 = x, q_2 = \dot{x}, \ldots, q_n = \frac{d^{n-1} x}{dt^{n-1}}. \qquad (3.17)$$

The state equations become

$$\dot{q}_1 = q_2,$$
$$\dot{q}_2 = q_3$$
$$\vdots$$
$$\dot{q}_{n-1} = q_n,$$
$$\dot{q}_n = -\frac{a_0}{a_n} q_1 - \frac{a_1}{a_n} q_2 - \cdots - \frac{a_{n-1}}{a_n} q_n + \frac{1}{a_n} u. \qquad (3.18)$$

Applying the differentiation operator introduced in Section 3.2 to Eqs. (3.15) and (3.16) gives

$$y(a_n D^n + \cdots + a_1 D + a_0) = u(b_m D^m + \cdots + b_1 D + b_0), \qquad (3.19)$$

$$x(a_n D^n + \cdots + a_1 D + a_0) = u. \qquad (3.20)$$

Substitution of the expression on the left-hand side of Eq. (3.20) for u in Eq. (3.19) yields

$$y = x(b_m D^m + \cdots + b_1 D + b_0). \qquad (3.21)$$

Now use the definition of the D operator to transform Eq. (3.21) for back to the time domain:

$$y = b_m \frac{d^m x}{dt^m} + \cdots + b_1 \frac{dx}{dt} + b_0 x. \qquad (3.22)$$

The auxiliary variable x and its derivatives can be replaced with the state variables defined by Eqs. (3.17) to produce the following output equation:

$$y = b_0 q_1 + b_1 q_2 + \cdots + b_m q_{m+1} \text{ for } m < n. \qquad (3.23)$$

Equation (3.23) holds for $m < n$. If both sides of the input–output equation are of the same order, $m = n$, then q_{m+1} in Eq. (3.23) is replaced with the expression for $\dot{q}_n$ given by the last of Eqs. (3.18) and the output equation becomes

$$y = \left(b_0 - \frac{b_m a_0}{a_n} \right) q_1 + \left(b_1 - \frac{b_m a_1}{a_n} \right) q_2 + \cdots$$

$$+ \left(b_{m-1} - \frac{b_m a_{n-1}}{a_n} \right) q_n + \left(\frac{b_m}{a_n} \right) u \quad \text{for} \quad m = n. \qquad (3.24)$$

The input–output equation involving derivatives of the input variable [Eq. (3.15)] is therefore equivalent to the state model, consisting of the auxiliary state-variable equations [Eqs. (3.18)] and output equations [Eqs. (3.23) and (3.24)] for $m < n$ and $m = n$, respectively.

The procedure for transforming input–output equations into an equivalent state model is illustrated in Example 3.7.

EXAMPLE 3.7

Consider again the mechanical system shown in Fig. 3.4. The input–output equation for this system, derived in Example 3.3, was

$$(m_1 m_2) \frac{d^4 x_2}{dt^4} + (m_2 b_1 + m_2 b_2 + m_1 b_2) \frac{d^3 x_2}{dt^3}$$

$$+ (m_1 k_2 + m_2 k_1 + m_2 k_2 + b_1 b_2) \frac{d^2 x_2}{dt^2}$$

$$+ (b_1 k_2 + b_2 k_1) \frac{dx^2}{dt^2}$$

$$+ k_1 k_2 x_2 = b_2 \frac{d F_1}{dt} + k_2 F_1.$$

Derive an equivalent state model for this system.

SOLUTION

First, the derivative of F_1 is ignored and a simplified input–output equation is obtained by use of the auxiliary output variable x:

$$(m_1 m_2)\frac{d^4 x}{dt^4} + (m_2 b_1 + m_2 b_2 + m_1 b_2)\frac{d^3 x}{dt^3}$$
$$+ (m_1 k_2 + m_2 k_1 + m_2 k_2 + b_1 b_2)\frac{d^2 x}{dt^2}$$
$$+ (b_1 k_2 + b_2 k_1)\frac{dx}{dt} + k_1 k_2 x = F_1.$$

The state variables are

$$q_1 = x, q_2 = \frac{dx}{dt}, q_3 = \frac{d^2 x}{dt^2}, q_4 = \frac{d^3 x}{dt^3},$$

and the state-variable equations are

$$\dot{q}_1 = q_2,$$
$$\dot{q}_2 = q_3,$$
$$\dot{q}_3 = q_4,$$
$$\dot{q}_4 = -\left(\frac{k_1 k_2}{m_1 m_2}\right) q_1 - \left[\frac{(b_1 k_2 + b_2 k_1)}{m_1 m_2}\right] q_2$$
$$- \left[\frac{(m_1 k_2 + m_2 k_1 + m_2 k_2 + b_1 b_2)}{m_1 m_2}\right] q_3$$
$$- \left[\frac{(m_2 b_1 + m_2 b_2 + m_1 b_2)}{m_1 m_2}\right] q_4 + \left(\frac{1}{m_1 m_2}\right) F_1.$$

The output variable in this system is displacement x_2. The form of the output equation is given by Eq. (3.23), which in this case becomes

$$x_2 = k_2 q_1 + b_2 q_2,$$

which completes the system mathematical model in a state form.

<!-- section marker --> **3.5** **NONLINEARITIES IN INPUT–OUTPUT AND STATE MODELS**

Very often in modeling dynamic systems, nonlinear characteristics of some of the system elements cannot be linearized either because the linearization error is not acceptable or because a particular nonlinearity may be essential for the system performance and must not be replaced with a linear approximation. The superiority of state models over input–output models in such cases is particularly pronounced. The derivation of input–output differential equations for systems in which nonlinearities are to be modeled without linearization is usually very cumbersome or even impossible, to say nothing about solving those equations. The derivation of state models, on the other hand, is barely affected by the presence of nonlinearities in the system. Furthermore, most computer programs for solving sets of state-variable equations are capable of handling both linear and nonlinear models.

The effect of nonlinearities on the process of derivation of the two forms of mathematical models is illustrated in Example 3.8.

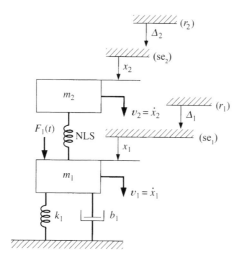

Figure 3.10. Mechanical system with NLS.

EXAMPLE 3.8

The mechanical system shown in Fig. 3.10 is similar to the system considered earlier in Example 3.2, except that the linear spring k_2 is here replaced with a NLS.

The force generated by the NLS, F_{NLS}, is approximated by

$$F_{NLS} = f_{NL}(z_2 - z_1) = c\,|z_2 - z_1|\,(z_2 - z_1),$$

where c is a constant and z_1 and z_2 are displacements of the two masses measured from their respective spring-relaxed positions, r_1 and r_2:

$$z_1 = \Delta_1 + x_1,$$
$$z_2 = \Delta_2 + x_2,$$

where Δ_1 and Δ_2 are displacements that are due to gravity and x_1 and x_2 are displacements that are due to force $F_1(t)$ measured from the static-equilibrium positions of the masses, se_1 and se_2, respectively. Recall that a similar notation was used for vertical displacements in Example 2.5.

The square-law expression embodied in the function f_{NL} is a fairly common type of nonlinearity, although the specific form of the nonlinear function is of no significance here.

Derive an input–output equation by using x_2 as the output variable and a set of state equations for this system.

SOLUTION

The equations of motion for masses m_1 and m_2 are

$$m_1\ddot{z}_1 + b_1\dot{z}_1 + k_1 z_1 - c\,|z_1 - z_2|\,(z_1 - z_2) = F_1(t) + m_1 g,$$
$$m_2\ddot{z}_2 + c\,|z_1 - z_2|\,(z_1 - z_2) = m_2 g.$$

One would normally obtain the input–output equation relating output x_2 to input $F_1(t)$ by combining the preceding equations of motion and eliminating z_1 (and subtracting Δ_2 to get x_2). However, in this case, the substitution is impossible because of the nonlinear term present in both equations!

To obtain a state model, select displacements z_1 and z_2 and their derivatives v_1 and v_2 as the state variables. Based on the definitions of the state variables and the equations of motion, the state-variable equations are

$$\dot{z}_1 = v_1,$$
$$\dot{v}_1 = -\frac{k_1}{m_1}z_1 - \frac{b_1}{m_1}v_1 + \frac{c}{m_1}|z_2 - z_1|(z_2 - z_1) + \frac{1}{m_1}F_1(t) + g,$$
$$\dot{z}_2 = v_2,$$
$$\dot{v}_2 = -\frac{c}{m_2}|z_2 - z_1|(z_2 - z_1) + g,$$

or, with general symbols used for the state variables,

$$\dot{q}_1 = q_2,$$
$$\dot{q}_2 = -\frac{k_1}{m_1}q_1 - \frac{b_1}{m_1}q_2 + \frac{c}{m_2}|q_3 - q_1|(q_3 - q_1) + \frac{1}{m_1}F_1(t) + g,$$
$$\dot{q}_3 = q_4,$$
$$\dot{q}_4 = -\frac{c}{m_2}|q_3 - q_1|(q_3 - q_1) + g,$$

where $q_1 = z_1$, $q_2 = v_1$, $q_3 = z_2$, and $q_4 = v_2$. Note that two inputs are used in this model, $F_1(t)$ and g.

The preceding state-variable equations can be solved numerically by use of one of the methods described in Chap. 5.

EXAMPLE 3.9

Linearize the state-variable equations derived in Example 3.8 for the system shown in Fig. 3.10 in a small vicinity of a normal operating point determined by the input force, $F_1(t) = \bar{F}_1 = \text{const}$.

SOLUTION

You solve this problem by following the five-step procedure introduced in Section 2.4.

Step 1. Derive the nonlinear model. The following state-variable equations were derived in Example 3.8:

$$\dot{z}_1 = v_1,$$
$$\dot{v}_1 = -\frac{k_1}{m_1}z_1 - \frac{b_1}{m_1}v_1 + \frac{c}{m_1}|z_2 - z_1|(z_2 - z_1) + \frac{1}{m_1}F_1(t) + g,$$
$$\dot{z}_2 = v_2,$$
$$\dot{v}_2 = -\frac{c}{m_2}|z_2 - z_1|(z_2 - z_1) + g.$$

Step 2. Determine the normal operating point. The normal operating point is determined by the constant input force, $F_1(t) = \bar{F}_1 = \text{const}$, and the corresponding unknown values of the four state variables, $z_1 = \bar{z}_1 = const$, $v_1 = \bar{v}_1 = const$, $z_2 = \bar{z}_2 = const$, and $v_2 = \bar{v}_2 = const$. The unknown normal operating–point values of the state variables can be found by their substitution into the state-variable equations, which yields

$$0 = \bar{v}_1,$$
$$0 = -k_1 \bar{z}_1 + c\,|\bar{z}_2 - \bar{z}_1|\,(\bar{z}_2 - \bar{z}_1) + \bar{F}_1 + m_1 g,$$
$$0 = \bar{v}_2,$$
$$0 = -c\,|\bar{z}_2 - \bar{z}_1|\,(\bar{z}_2 - \bar{z}_1) + m_2 g.$$

The preceding equations describe the system at the normal operating point established in the system by the constant input force, $\bar{F}_1$. It can be seen that at the normal operating point the system is at rest (the velocities of both masses are zero), whereas one can find the displacements of the two masses by solving the preceding equations and assuming that $c > 0$ to obtain

$$\bar{z}_1 = \frac{1}{k_1}(\bar{F}_1 + m_1 g + m_2 g),$$
$$\bar{z}_2 = \frac{1}{k_1}(\bar{F}_1 + m_1 g + m_2 g) + \sqrt{\frac{m_2 g}{c}}.$$

Step 3. Introduce incremental variables. Substitute $z_1 = \bar{z}_1 + \hat{z}_1$, $z_2 = \bar{z}_2 + \hat{z}_2$, $v_1 = \bar{v}_1 + \hat{v}_1$, $v_2 = \bar{v}_2 + \hat{v}_2$, and $F_1(t) = \bar{F}_1 + \hat{F}_1$ into the original state-variable equations:

$$\dot{\bar{z}}_1 + \dot{\hat{z}}_1 = \bar{v}_1 + \hat{v}_1,$$
$$\dot{\bar{v}}_1 + \dot{\hat{v}}_1 = -\frac{k_1}{m_1}(\bar{z}_1 + \hat{z}_1) - \frac{b_1}{m_1} - (\bar{v}_1 + \hat{v}_1) + \frac{1}{m_1} f_{NL}(\bar{z}_{21} + \hat{z}_{21}) + \frac{1}{m_1}(\bar{F}_1 + \hat{F}_1) + g,$$
$$\dot{\bar{z}}_2 + \dot{\hat{z}}_2 = \bar{v}_2 + \hat{v}_2,$$
$$\dot{\bar{v}}_2 + \dot{\hat{v}}_2 = -\frac{1}{m_2} f_{NLS}(\bar{z}_{21} + \hat{z}_{21}) + g,$$

where $z_{21} = z_2 - z_1$ and $f_{NLS}(z_{21})$ is the NLS characteristic. Substituting zeros for terms involving $\bar{v}_1$, $\bar{v}_2$ and derivatives of all constant components yields

$$\dot{\hat{z}}_1 = \hat{v}_2,$$
$$\dot{\hat{v}}_1 = -\frac{k_1}{m_1}(\bar{z}_1 + \hat{z}_1) - \frac{b_1}{m_1}\hat{v}_1 + \frac{1}{m_1} f_{NL}(\bar{z}_{21} + \hat{z}_{21}) + \frac{1}{m_1}(\bar{F}_1 + \hat{F}_1) + g,$$
$$\dot{\hat{z}}_2 = \hat{v}_2,$$
$$\dot{\hat{v}}_2 = -\frac{1}{m_2} f_{NLS}(\bar{z}_{21} + \hat{z}_{21}) + g.$$

Step 4. Linearize the nonlinear terms. Expand $f_{NLS}(\bar{z}_{21} + \hat{z}_{21})$ into Taylor series, neglecting second- and higher-order terms:

$$f_{NLS}(\bar{z}_{21} + \hat{z}_{21}) \approx f_{NLS}(\bar{z}_{21}) + \hat{z}_{21}\left(\frac{d f_{NLS}}{d z_{21}}\right)_{\bar{z}_{21}}$$
$$= c\,|\bar{z}_2 - \bar{z}_1|\,(\bar{z}_2 - \bar{z}_1) + (\hat{z}_2 - \hat{z}_1)2c\,|\bar{z}_2 - \bar{z}_1|.$$

Step 5. Arrange the linearized equations into a final form. Substitute the linearized expression obtained in Step 4 into the equations derived in Step 3:

$$\dot{\hat{z}}_1 = \hat{v}_1,$$
$$\dot{\hat{v}}_1 = -\frac{k_1}{m_1}(\bar{z}_1 + \hat{z}_1) - \frac{b_1}{m_1}\hat{v}_1 + \frac{c}{m_1}\left[|\bar{z}_2 - \bar{z}_1|(\bar{z}_2 - \bar{z}_1)\right]$$
$$+ \frac{2c}{m_1}|\bar{z}_2 - \bar{z}_1|(\hat{z}_2 - \hat{z}_1) + \frac{1}{m_1}(\bar{F}_1 + \hat{F}_1) + g,$$
$$\dot{\hat{z}}_2 = \hat{v}_2,$$
$$\dot{\hat{v}}_2 = -\frac{c}{m_2}|\bar{z}_2 - \bar{z}_1|(\bar{z}_2 - \bar{z}_1) - \frac{2c}{m_2}|\bar{z}_2 - \bar{z}_1|(\hat{z}_2 - \hat{z}_1) + g.$$

Referring to the equations for the normal operating point derived in Step 2, it can be seen that all constant terms in the precedings equations cancel out to give the linearized state-variable equations:

$$\dot{\hat{z}}_1 = \hat{v}_1,$$
$$\dot{\hat{v}}_1 = -\frac{k_1}{m_1}\hat{z}_1 - \frac{b_1}{m_1}\hat{v}_1 + \frac{2c}{m_1}|\bar{z}_2 - \bar{z}_1|(\hat{z}_2 - \hat{z}_1) + \frac{1}{m_1}\hat{F}_1,$$
$$\dot{\hat{z}}_2 = \hat{v}_2,$$
$$\dot{\hat{v}}_2 = -\frac{2c}{m_2}|\bar{z}_2 - \bar{z}_1|(\hat{z}_2 - \hat{z}_1).$$

An incremental spring stiffness that approximates the stiffness of the NLS in a small vicinity of the normal operating point is defined as

$$k_{\text{inc}} = 2c\,|\bar{z}_2 - \bar{z}_1|\,.$$

The linearized state-variable equations can now be written in a more compact form:

$$\dot{\hat{z}}_1 = \hat{v}_1,$$
$$\dot{\hat{v}}_1 = -\frac{(k_1 + k_{\text{inc}})}{m_1}\hat{z}_1 - \frac{b_1}{m_1}\hat{v}_1 + \frac{k_{\text{inc}}}{m_1}\hat{z}_2 + \frac{1}{m_1}\hat{F}_1,$$
$$\dot{\hat{z}}_2 = \hat{v}_2,$$
$$\dot{\hat{v}}_2 = \frac{k_{\text{inc}}}{m_2}\hat{z}_1 - \frac{k_{\text{inc}}}{m_2}\hat{z}_2.$$

The linearized equations can also be arranged into a matrix form:

$$
\begin{bmatrix} \dot{\hat{z}}_1 \\ \dot{\hat{v}}_1 \\ \dot{\hat{z}}_2 \\ \dot{\hat{v}}_1 \end{bmatrix}
=
\begin{bmatrix}
0 & 1 & 0 & 0 \\
-(k_1 + k_{\text{inc}})/m_1 & -b_1/m_1 & k_{\text{inc}}/m_1 & 0 \\
0 & 0 & 0 & 1 \\
k_{\text{inc}}/m_2 & 0 & -k_{\text{inc}}/m_2 & 0
\end{bmatrix}
\begin{bmatrix} \hat{z}_1 \\ \hat{v}_1 \\ \hat{z}_2 \\ \hat{v}_2 \end{bmatrix}
+
\begin{bmatrix} 0 \\ 1/m_1 \\ 0 \\ 0 \end{bmatrix}
\hat{F}_1.
$$

The examples presented in this chapter illustrate the effectiveness of the state approach to system modeling. The state equations are much simpler, and the entire process of derivation is less vulnerable to so-called "stupid mistakes," which may often be made when the input–output model is developed. Although, as the next chapter will show, there are certain advantages associated with using input–output models in analysis of low-order linear systems, the state-variable approach is in general superior and is used as much as possible in modeling systems throughout the rest of this book.

3.6   SYNOPSIS

Two types of mathematical models of dynamic systems were presented, input–output models and state models. Although both models are essentially equivalent with regard to the information about the system behavior incorporated in the model equations, the techniques used in their derivation and solution are principally different. In most cases the state models, consisting of sets of first-order differential equations, are much easier to derive and to solve by computer simulation than the input–output equations. This is especially true when the mathematical models involve nonlinearities or when the systems modeled have many inputs and outputs and are to be simulated on the computer.

The state models are based on the concept of state variables. The choice of state variables for a given system is not unique. Four different types of state variables were used in example problems presented in this chapter. It was shown that the number of state variables necessary to describe the system dynamics is always equal to the number of independent energy-storing elements, regardless of the type of state variables used. Also, the number of state variables is the same as the order of the input–output equation.

PROBLEMS

3.1 Derive a complete set of state-model equations for the mechanical rotational system shown in Fig. P3.1. Select the following state variables:

(a) T-type and A-type variables,

(b) one variable and its derivative.

Use torque $T_i(t)$ as the input variable and angular displacement $\theta_l(t)$ as the input variable in each model.

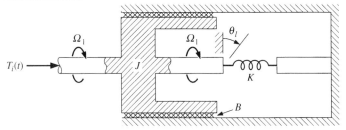

Figure P3.1. Mechanical system considered in Problems 3.1 and 3.2.

3.2 Derive an input–output model equation for the system shown in Fig. P3.1, using torque $T_i(t)$ as the input variable and angular displacement $\theta_l(t)$ as the output variable.

3.3 Derive complete sets of state-model equations for the system shown in Fig. P3.3 by using three different sets of state variables. The input variable is torque $T_m(t)$ and the output variables are displacements $x_1(t)$ and $x_2(t)$. Present the state models in matrix form.

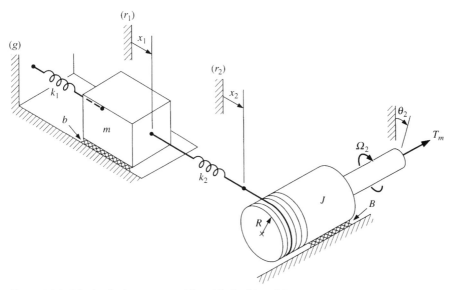

Figure P3.3. Mechanical system considered in Problem 3.3.

3.4 A nonlinear dynamic system has been modeled by the state-variable equations

$$\dot{q}_1 = -4q_1 + 10q_2 + 4u_1,$$

$$\dot{q}_2 = 0.5q_1 - 2\,|q_2|\,q_2,$$

where q_2 is also the output variable and u_1 is the input variable.

(a) Linearize the state-variable equations.

(b) Find the input–output equation for the linearized model.

3.5 Linearize the state-variable equations derived in Example 2.7 for the power transmission system [Eqs. (2.55)–(2.57)]. Combine the linearized equations to obtain an input–output equation for the system using torque T_K as the output variable.

3.6 The state-model matrices of a single-input, single-output linear dynamic system are

$$\mathbb{A} = \begin{bmatrix} -3 & -19 \\ 1 & -2 \end{bmatrix}, \mathbb{B} = \begin{bmatrix} 0 \\ -1 \end{bmatrix},$$

$$\mathbb{C} = [0, 2], \mathbb{D} = 0$$

The column vector $\mathbf{q}$ of the system state variables contains q_1 and q_2. Find the input–output model for this system.

3.7 A linear dynamic system is described by the differential equations

$$\ddot{y} + 4\dot{y} + 4y = 2\dot{x} + 2x,$$

$$2\dot{x} + x - y = u(t).$$

(a) Derive the input–output model for this system using $y(t)$ as the output variable and $u(t)$ as the input variable.

(b) Derive a state model and present it in matrix form.

3.8 Derive a set of state-variable equations for the mechanical system shown in Fig. P3.8.

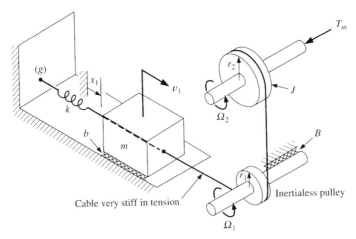

Figure P3.8. Mechanical system considered in Problems 3.8. and 3.9.

3.9 Derive the input–output model equation for the system shown in Fig. P3.8. using torque $T_m(t)$ as the input variable and angular velocity Ω_2 as the output variable.

3.10 A lumped model of a machine-tool drive system is shown in Fig. P3.10. The system parameters are $J_m = 0.2$ N m s^2/rad, $B_m = 30$ N m s/rad, $m = 16$ kg, $k = 20$ N/m, $R = 0.5$ m. The stiffness of the shaft is represented by a nonlinear torque, $T_{\text{NLS}} = 2|\theta_1 - \theta_2|(\theta_1 - \theta_2)$. The force of the nonlinear friction device NLD is $F_{\text{NLD}} = f_{\text{NL}}(v_1) = 2v_1^3 + 4v_1$. The system is driven by a torque consisting of a constant and an incremental component, $T_i(t) = 0.8 + 0.02 \sin(0.1t)$ N m.

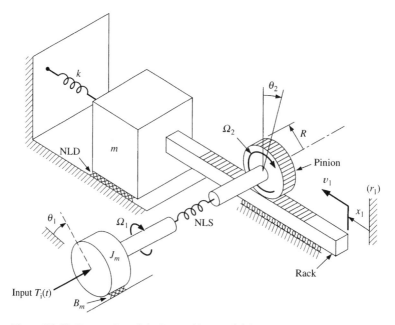

Figure P3.10. Lumped model of a machine-tool drive system.

(a) Select state variables and derive nonlinear state-variable equations.

(b) Find the normal operating-point values for all state variables.

(c) Linearize the state-model equations in the vicinity of the normal operating point. Present the linearized state model in a matrix form.

(d) Derive the input–output model equation for the linearized system using the incremental torque $\hat{T}_1(t)$ as the input variable and the incremental displacement $\hat{x}_1(t)$ as the output variable.

3.11 Derive state-variable equations for the system shown in Fig. P2.10. Use velocities v_1 and v_2 and the spring force F_{kl} as the state variables. Note that the two springs in this system are not independent as long as the lever is massless. Combine the state-variable equations to obtain an input–output equation relating velocity v_2 to force $F_s(t)$.

3.12 The mechanical system shown in Fig. P3.12 is driven by two inputs: force F_i and torque T_i. The roller, whose mass is m and whose moment of inertia is J, rolls and/or slides on the surface of the carriage having mass m_c. The contact between the roller and the carriage is lubricated, and the friction at the contact point is viscous in nature with a coefficient b_c. Select state variables and derive the state-variable equations for the system.

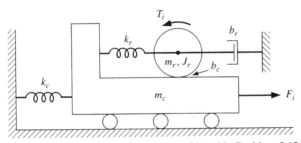

Figure P3.12. Mechanical system considered in Problem 3.12.

3.13 This problem refers to the spring–mass–damper problem, 2.18.

(a) Derive a state-equation model of the system and compare it with the input–output model derived for Problem 2.18.

(b) Comment on the process of deriving the model using the two methods.

(c) Convert the state model to an input–output representation and verify that it is equivalent to the model derived for Problem 2.18.

3.14 Figure P3.14 shows a schematic representation of a mechanical drive system. Derive two state models for this system:

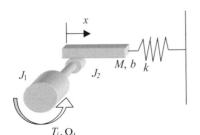

Figure P3.14. Schematic of a mechanical drive system.

(a) Consider the shaft between J_1 and J_2 to be rigid.

(b) Consider the shaft between J_1 and J_2 to be flexible with a torsional spring stiffness of K.

(c) Compare and contrast the two models and comment on their utility for various applications.

4

Analytical Solutions of System Input–Output Equations

LEARNING OBJECTIVES FOR THIS CHAPTER

4–1 To use analytical solution methods for ODEs to predict the response of first- and second-order systems to nonzero initial conditions and typical input signals.

4–2 To estimate key parameters (i.e., time constant, natural frequency, and damping ratio) in system responses.

4–3 To use solution methods for ODEs to derive the relationship between the complex roots of underdamped second-order systems and the natural frequency and damping ratio.

4–4 To use the concept of "dominant poles" to estimate the response of higher-order systems when one or two poles dominate the system's dynamic behavior.

4.1 INTRODUCTION

In Chap. 3, the state representation of system dynamics was introduced and the derivation of state equations was shown to be a relatively simple and straightforward process. Moreover, the state models take the form of sets of first-order differential equations that can be readily solved by use of one of many available computer programs. Having these unquestionable advantages of state-variable models in mind, one might wonder whether devoting an entire chapter to the methods for solving the old-fashioned input–output model equations is justified. Despite all its limitations, the classical input–output approach still plays an important role in analysis of dynamic systems because many of the systems to be analyzed are neither very complex nor nonlinear. Such systems can be adequately described by low-order linear differential equations. Also, even in those cases in which a low-order linear model is too crude to produce an accurate solution and a computer-based method is necessary, an analytical solution of an approximate linearized input–output equation can be used to verify the computer solution.

Section 4.2 gives a brief review of methods for solving linear differential equations. The next three sections deal with application of analytical methods to systems represented by first-, second-, and higher-order linear differential equations.

4.2   ANALYTICAL SOLUTIONS OF LINEAR DIFFERENTIAL EQUATIONS

An input–output equation for a linear, stationary, single-input, single-output model has the following general form:

$$a_n \frac{d^n y}{dt^n} + a_{n-1} \frac{d^{n-1} y}{dt^{n-1}} + \cdots + a_1 \frac{dy}{dt} + a_0 y = f(t), \tag{4.1}$$

where the model parameters, $a_0, a_1, \ldots, a_n$, are constant. Function $f(t)$, which is sometimes referred to as a forcing function, represents a linear combination of terms involving input signal $u(t)$ and its derivatives that appear on the right-hand side of Eq. (3.3). Equation (4.1) can be solved for a given $f(t)$ for $t \geq 0$ if the initial conditions for y and its $n - 1$ derivatives just after the input starts, i.e., at $t = 0^+$, are known:

$$y(0^+) = y_0, \quad \frac{dy}{dt}\bigg|_{t=0^+} = \dot{y}_0, \ldots, \quad \frac{d^{n-1} y}{dt^{n-1}}\bigg|_{t=0^+} = y_0^{(n-1)}. \tag{4.2}$$

Usually the initial conditions are known only as they existed just before the input started to change, i.e., at $t = 0^-$. The required initial conditions at $t = 0^+$ can usually be readily determined by use of the initial conditions at $t = 0^-$ and information about the system model and its inputs, through use of the elemental equations, the interconnecting equations, and energy-storage limitations when they are all available. To simplify notation, all initial conditions throughout the remainder of this book are shown at $t = 0$, even though they will always be meant to be taken at $t = 0^+$.

A complete solution, $y(t)$, of Eq. (4.1) consists of two parts:

$$y(t) = y_h(t) + y_p(t) \tag{4.3}$$

where $y_h(t)$ represents a homogeneous (characteristic) solution and $y_p(t)$ is a particular (forced) solution. The homogeneous solution gives the model response for the input signal equal to zero, $f(t) = 0$, and thus satisfies

$$a_n \frac{d^n y_h}{dt^n} + a_{n-1} \frac{d^{n-1} y_h}{dt^{n-1}} + \cdots + a_1 \frac{dy_h}{dt} + a_0 y_h = 0. \tag{4.4}$$

The particular solution $y_p(t)$ satisfies the nonhomogeneous equation

$$a_n \frac{d^n y_p}{dt^n} + a_{n-1} \frac{d^{n-1} y_p}{dt^{n-1}} + \cdots + a_1 \frac{dy_p}{dt} + a_0 y_p = f(t). \tag{4.5}$$

Note that when Eqs. (4.4) and (4.5) are added, the original model equation, Eq. (4.1), is obtained:

$$a_n \frac{d^n (y_h + y_p)}{dt^n} + a_{n-1} \frac{d^{n-1} (y_h + y_p)}{dt^{n-1}} + \cdots$$

$$+ a_1 \frac{dy(y_h + y_p)}{dt} + a_0 (y_h + y_p) = f(t). \tag{4.6}$$

Both the homogeneous and particular solution components have to be found. First, consider the homogeneous solution. One obtains a characteristic equation from Eq. (4.4) by setting $d^k y_h / dt^k = p^k$ for $k = 0, 1, \ldots, n$, which yields

$$a_n p^n + a_{n-1} p^{n-1} + \cdots + a_1 p + a_0 = 0. \tag{4.7}$$

This nth-order algebraic equation has n roots, some of which may be identical. Assume that there are m distinct roots $(0 \le m \le n)$ and therefore the number of multiple roots is $n - m$. To simplify further derivations, all multiple roots of the characteristic equation are left to the end and are denoted by p_n, where p_n is the value of the identical roots:

$$p_{m+1} = p_{m+2} = \cdots = p_n. \tag{4.8}$$

The general form of the homogeneous solution is

$$y_h(t) = C_1 e^{p_1 t} + C_2 e^{p_2 t} + \cdots + C_m e^{p_m t} + C_{m+1} e^{p_n t}$$
$$+ C_{m+2} t e^{p_n t} + \cdots + C_n t^{n-m-1} e^{p_n t}, \tag{4.9}$$

or, in more compact form,

$$y_h(t) = \sum_{i=1}^{m} C_i e^{p_i t} + \sum_{i=m+1}^{n} C_i t^{i-m-1} e^{p_i t}. \tag{4.10}$$

The integration constants $C_1, C_2, \ldots, C_n$ will be determined after the particular solution is found.

Solving for the particular solution is more difficult to generalize because it always depends on the form of the forcing function $f(t)$. Generally speaking, the form of $y_p(t)$ has to be guessed, based on the form of input, its derivatives, or both.[1] A method of undetermined coefficients provides a more systematic approach to solving for the particular solution. In many cases in which the forcing function $f(t)$ reaches a steady-state value, say f_{ss}, for time approaching infinity, that is, if

$$\lim_{t \to \infty} f(t) = f_{ss} \tag{4.11}$$

and if $a_0 \ne 0$, the particular solution can be calculated simply as

$$y_p(t) = f_{ss}/a_0. \tag{4.12}$$

Once the general forms of both parts of the solution are found, the constants that appear in the two expressions must be determined.

If there are unknown constants in the particular solution – that is, if Eq. (4.12) cannot be applied – the general expression for the particular solution is substituted into Eq. (4.5) and the constants are found by the solution of equations created by equating corresponding terms on both sides of the equation.

To find the integration constants $C_1, C_2, \ldots, C_n$ in the homogeneous solution, the set of n initial conditions, given by Eqs. (4.2), is used to form the following n equations:

$$y_h(0) = y_p(0) = y(0),$$

$$\left(\frac{dy_h}{dt} \right)\Big|_{t=0} + \left(\frac{dy_p}{dt} \right)\Big|_{t=0} = \left(\frac{dy}{dt} \right)\Big|_{t=0},$$

$$\vdots$$

$$\left(\frac{d^{n-1} y_h}{dt^{n-1}} \right)\Big|_{t=0} + \left(\frac{d^{n-1} y_p}{dt^{n-1}} \right)\Big|_{t=0} = \left(\frac{d^{n-1} y}{dt^{n-1}} \right)\Big|_{t=0}. \tag{4.13}$$

where the terms on the right-hand sides represent the initial conditions.

---

[1] When the input is of the form e^{pt}, where p is one of the roots of the system characteristic equations, the particular solution may contain an exponential term of the form Cte^{pt}, where one determines C by substituting Cte^{pt} in the system differential equation and solving for C.

The method for solving differential equations of the general form given by Eq. (4.1) can be summarized by the following step-by-step procedure.

Step 1. Obtain the characteristic equation [Eq. (4.7)].
Step 2. Find roots of the characteristic equation, $p_1, p_2, \ldots, p_n$.
Step 3. Write the general expression for the homogeneous solution, Eq. (4.10).
Step 4. Determine the general expression for the particular solution.
Step 5. Determine constants in the particular solution by equating corresponding terms on both sides of Eq. (4.5). This step can be skipped if the particular solution was found with Eq. (4.12).
Step 6. Find integration constants $C_1, C_2, \ldots, C_n$ by solving the set of equations involving the initial conditions [Eqs. (4.13)].

Several examples involving analytical solution of linear input–output equations are given in Sections 4.3, 4.4, and 4.5.

4.3   FIRST-ORDER MODELS

A linear, stationary, single-input, single-output system model is described by a first-order differential equation,

$$a_1 \dot{y} + a_0 y = f(t), \tag{4.14}$$

with the initial condition $y(0) = y_0$. The characteristic equation is

$$a_1 p + a_0 = 0. \tag{4.15}$$

A general form for the model homogeneous solution is

$$y_h(t) = C_1 e^{p_1 t}, \tag{4.16}$$

where a single root of the characteristic equation, p_1, is

$$p_1 = \frac{-a_0}{a_1}. \tag{4.17}$$

A particular solution of Eq. (4.14) depends on the type of the input signal $f(t)$. Three types of input signals will be considered: zero, step function, and impulse function. The corresponding system outputs will be free response,[2] step response, and impulse response, respectively.

First, an input will be assumed to be equal to zero, $f(t) = 0$ for $t \geq 0$. The homogeneous solution constitutes, in this case, a complete solution given by the expression

$$y(t) = C_1 e^{-(a_0/a_1)t}. \tag{4.18}$$

---

[2] The term "free response," adapted from vibration theory, is somewhat of a misnomer. The response is free only in the sense that the system input is zero for the time interval of interest – in this case, for $t > 0$. In all cases, a variation of the output occurs only in response to some past input forcing function or system change, such as switch or circuit breaker activation, shaft failure, pipe rupture, etc. In this case, the system is responding to a nonzero initial output condition, which must be the remainder of a response to a previous input.

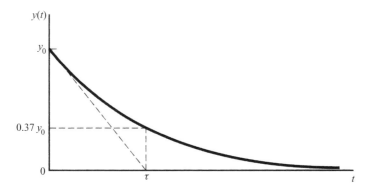

Figure 4.1. Free response of first-order model.

One can find the integration constant C_1 by setting $t = 0$ in Eq. (4.18) and using the initial condition to obtain

$$C_1 = y_0, \qquad (4.19)$$

where $y_0 = y(0)$. Hence the free response of the first-order model is

$$y(t) = y_0 e^{-(a_0/a_1)t}. \qquad (4.20)$$

The free response curve is shown in Fig. 4.1. The curve starts at time $t = 0$ from the initial value y_0 and decays exponentially to zero as time approaches infinity. The rate of the exponential decay is determined by the model time constant τ, defined as

$$\tau = a_1/a_0. \qquad (4.21)$$

The time constant is equal to the time during which the first-order model free response decreases by 63.1% of its initial value. A normalized free response of the first-order model, $y(t)/y_0$, versus normalized time, t/τ, is plotted in Fig. 4.2, and its numerical values are given in Table 4.1.

The second type of input signal considered here is a step perturbation of $f(t)$ expressed by the unit step function $U_s(t)$:

$$f(t) = AU_s(t), \qquad (4.22)$$

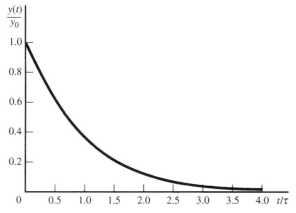

Figure 4.2. Normalized free response of first-order model.

Table 4.1. Normalized free response of first-order model

t/τ	0	0.5	1.0	1.5	2.0	2.5	3.0	4.0
$y(t)/y_0$	1	0.6065	0.3679	0.2231	0.1353	0.0821	0.0498	0.0183

where A is the magnitude of the step function. The unit step function $U_s(t)$ is defined by

$$U_s(t) = \begin{cases} 0 & \text{for} \quad t < 0 \\ 1 & \text{for} \quad t > 0 \end{cases}. \tag{4.23}$$

The homogeneous solution in this case is, of course, of the same form as before [Eq. (4.18)]. Here, the particular solution must satisfy

$$a_1 \dot{y}_p + a_0 y_p = A \quad \text{for} \quad t > 0. \tag{4.24}$$

By use of Eq. (4.12), the particular solution for a step input of magnitude A is

$$y_p(t) = \frac{A}{a_0}. \tag{4.25}$$

Hence a complete solution for $y(t)$ is

$$y(t) = y_h(t) + y_p(t) = C_1 e^{-\frac{t}{\tau}} + \frac{A}{a_0}. \tag{4.26}$$

When $y(t) = y_0$ is substituted for $t = 0$ in Eq. (4.26), the constant C_1 is found to be

$$C_1 = y_0 - \frac{A}{a_0}, \tag{4.27}$$

where C_1 now includes a term that is due to $u(t)$. The complete response of the first-order model is finally obtained as

$$y(t) = \underbrace{y_0 e^{-\frac{t}{\tau}}}_{\text{free response part}} + \underbrace{\frac{A}{a_0}\left(1 - e^{-\frac{t}{\tau}}\right)}_{\text{step response part}}. \tag{4.28}$$

From Eq. (4.26), the steady-state value of the step response, y_{ss}, is

$$y_{ss} = \lim_{t \to \infty} y(t) = \frac{A}{a_0}. \tag{4.29}$$

The step response can now be expressed in terms of its initial and steady-state values,

$$y(t) = y_{ss} - (y_{ss} - y_0)e^{-\frac{t}{\tau}}, \tag{4.30}$$

or in terms of its initial value and the system's steady-state gain,

$$y(t) = KA - (KA - y_0)e^{\frac{t}{\tau}}, \tag{4.31}$$

where the steady-state gain K is defined as the ratio of the magnitude of output over the magnitude of input at steady state, which in this case is

$$K = \frac{y_{ss}}{f_{ss}} = \frac{y_{ss}}{A}, \tag{4.32}$$

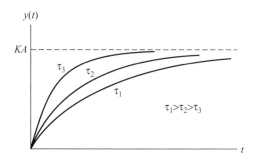

Figure 4.3. Step response curves of first-order model with zero initial condition $y_0 = 0$.

or, by use of Eq. (4.29),

$$K = \frac{1}{a_0}. \tag{4.33}$$

When the initial condition is zero, $y_0 = 0$, the step response of the first-order model is

$$y(t) = y_{ss}\left(1 - e^{-\frac{t}{\tau}}\right) = KA\left(1 - e^{-\frac{t}{\tau}}\right). \tag{4.34}$$

Several step response curves for different values of time constant and zero initial condition, $y_0 = 0$, are plotted in Fig. 4.3.

The third type of input to be considered here is an impulse function $f(t)$ expressed by

$$f(t) = PU_i(t), \tag{4.35}$$

where P is the strength of the impulse function (in other words, its area). The unit impulse function $U_i(t)$ occurring at time $t = 0$ is defined by

$$U_i(t) \begin{cases} \infty & \text{for} \quad t = 0 \\ 0 & \text{for} \quad t \neq 0 \end{cases}, \tag{4.36}$$

$$\int_{-\infty}^{+\infty} U_i(t)dt = \int_{0^-}^{0^+} U_i(t)dt = 1.0. \tag{4.37}$$

In general, a unit impulse, also called Dirac's delta function, which occurs at time $t = t_1$, is denoted by $U_i(t - t_1)$ and is defined by

$$U_i(t - t_1) = \begin{cases} \infty & \text{for} \quad t = t_1 \\ 0 & \text{for} \quad t \neq t_1 \end{cases}, \tag{4.38}$$

$$\int_{-\infty}^{+\infty} U_i(t - t_1)dt = \int_{t_1^-}^{t_1^+} U_i(t - t_1)dt = 1.0. \tag{4.39}$$

An ideal unit impulse function as just defined cannot be physically generated. It can be thought of as a limit of a unit pulse function, $U_p(t)$, of amplitude $1/T$ with the pulse duration T approaching zero, as illustrated in Fig. 4.4.

The unit impulse function is widely used in theoretical system analysis because of its useful mathematical properties, which make it a very desirable type of input.

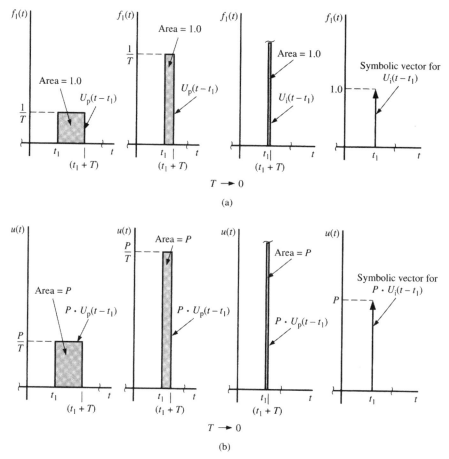

Figure 4.4. Transition from a pulse to an impulse: (a) unit pulse to a unit impulse and (b) pulse of strength P to an impulse of strength P, together with symbolic vector representations of impulses.

One such property, called a filtering property, is mathematically expressed by the sifting integral,

$$\int_{-\infty}^{+\infty} U_i(t - t_1) f(t) dt = f(t_1). \tag{4.40}$$

Equation (4.40) implies that

$$\int_{-\infty}^{+\infty} U_i(t) f(t) dt = f(0). \tag{4.41}$$

Although a rigorous mathematical proof is difficult, a unit impulse function is often considered as the derivative of a unit step function:

$$U_i(t) = \frac{dU_s(t)}{dt} \tag{4.42}$$

Equation (4.42) implies that the dimension of the amplitude of an impulse function is (1/time) if the amplitude of a step function is dimensionless. However, because the

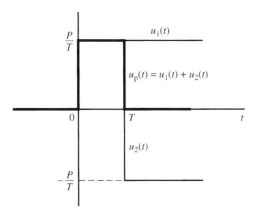

Figure 4.5. Pulse $u_p(t)$ presented as a sum of two step functions, $u_1(t)$ and $u_2(t)$.

amplitude of $U_i(t)$ is infinite at $t = 0$, an impulse is measured instead by its strength, P, equal to its integral from $t = 0^-$ to $t = 0^+$ [in other words, the area under the "spikelike" curve representing $PU_i(t)$].

To find a response of a first-order model to an impulse input, it is convenient to consider an impulse function as a limit of a pulse function with the pulse duration approaching zero, as illustrated in Fig. 4.4. The system impulse response can then be determined as a limit of a pulse response with the pulse duration approaching zero. Consider the response of a first-order model with zero initial condition, $y(0) = 0$, to the pulse input shown in Fig. 4.5. The pulse input, $u_p(t)$, is shown in Fig. 4.5 as a sum of two step functions, one having amplitude P/T and beginning at $t = 0$ and the other having amplitude $-P/T$ and beginning at $t = T$:

$$u_p(t) = u_1(t) + u_2(t), \tag{4.43}$$

where

$$u_1(t) = \frac{P}{T} U_s(t), \tag{4.44}$$

$$u_2(t) = -\frac{P}{T} U_s(t - T). \tag{4.45}$$

Because the system is linear and the principle of superposition applies, the system pulse response will also consist of two components:

$$y_p(t) = y_1(t) + y_2(t), \tag{4.46}$$

where $y_1(t)$ is the response to step $u_1(t)$ and $y_2(t)$ is the response to step $u_2(t)$. By use of Eq. (4.28), the two components of the pulse response are found to be

$$y_1(t) = \frac{P}{a_0 T} \left(1 - e^{-\frac{t}{\tau}}\right), \tag{4.47}$$

$$y_2(t) = -\frac{P}{a_0 T} \left(1 - e^{-\frac{t-T}{\tau}}\right). \tag{4.48}$$

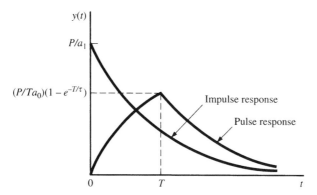

Figure 4.6. Impulse response of a first-order model together with response to a pulse of strength P and duration T.

The complete pulse response is

$$y_p(t) = \frac{P}{a_0 T}\left(1 - e^{-\frac{t}{\tau}}\right) - \frac{P}{a_0 T}\left(1 - e^{-\frac{t-T}{\tau}}\right) = \frac{P}{a_0 T}e^{-\frac{t}{\tau}}\left(e^{\frac{T}{\tau}} - 1\right). \tag{4.49}$$

Now the impulse response can be obtained as a limit of the pulse response for T approaching zero. Taking the limit of the expression obtained in Eq. (4.49) and using l'Hôpital's rule yields

$$y_i(t) = \lim_{T \to 0} y_p(t) = \frac{P}{a_1}e^{-\frac{t}{\tau}}. \tag{4.50}$$

If the initial condition is not zero, $y(0) = y_0$, the complete response of the first-order model is

$$y_i(t) = \underbrace{y_0 e^{-\frac{t}{\tau}}}_{\text{free response part}} + \underbrace{\frac{P}{a_1}e^{-\frac{t}{\tau}}}_{\text{impulse response part}}, \tag{4.51}$$

or, equally, in terms of the steady-state gain, by use of Eqs. (4.21) and (4.33),

$$y_i(t) = y_0 e^{-\frac{t}{\tau}} + \frac{KP}{\tau}e^{-\frac{t}{\tau}} \tag{4.52}$$

Figure 4.6 shows an impulse response curve together with the response to a pulse of width T and strength P for a first-order model.

By use of the principle of superposition, which applies to all linear systems, the response to several inputs and/or nonzero initial conditions can be obtained as the sum of the responses to the individual inputs and/or nonzero initial conditions found separately.

EXAMPLE 4.1

In the mechanical system shown in Fig. 4.7, mass m is initially subjected to a constant force $\overline{F}$ and is moving with initial velocity v_0. At time t_0 the force changes suddenly from $\overline{F}$ to $\overline{F} + \Delta F$. Find the velocity and acceleration of the mass for time $t > t_0$.

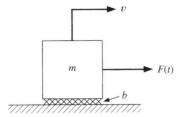

Figure 4.7. Mechanical system considered in Example 4.1.

SOLUTION

Using force $F(t)$ applied to the mass as the input and using the velocity of the mass $v(t)$ as the output variable, we can write the input–output equation as

$$m\dot{v} + bv = F(t).$$

Because the velocity of the mass cannot change suddenly, the initial condition is

$$v(t_0) = v_0.$$

A mathematical form describing the input force occurring at $t = t_0$ is

$$F(t) = \overline{F} + \Delta F U_s(t - t_0).$$

The input–output equation can be rewritten in the form

$$m\dot{v} + bv = \overline{F} + \Delta F U_s(t - t_0).$$

The general expression for the step response of a first-order model given by Eq. (4.30) can be used here but it has to be modified to reflect the initial time as t_0, rather than as 0. The modified form of Eq. (4.30) is

$$y(t) = y_{ss} - (y_{ss} - y_0)e^{-\frac{(t-t_0)}{\tau}},$$

or, when velocity $v(t)$ is substituted as the output variable,

$$v(t) = v_{ss} - (v_{ss} - v_0)e^{-\frac{(t-t_0)}{\tau}},$$

with the general form of the solution known, the problem is reduced to finding three unknowns in the preceding equation: initial velocity v_0, steady-state velocity v_{ss}, and time constant τ. The initial velocity resulting from a constant initial force $\overline{F}$ can be found from the input–output equation when both velocity and force are made constant and equal to their respective initial values:

$$m\dot{v}_0 + bv_0 = \overline{F}.$$

The first term on the left-hand side is zero and thus the initial velocity is

$$v_0 = \frac{\overline{F}}{b}.$$

One can find the steady-state velocity of the mass by taking limits of both sides of the input–output equation for time approaching infinity:

$$\lim_{t\to\infty}(m\dot{v} + bv) = \lim_{t\to\infty}\left[\overline{F} + \Delta F U_s(t - t_0)\right],$$

and hence

$$m \lim_{t \to \infty} \dot{v} + b \lim_{t \to \infty} v = \overline{F} + \Delta F.$$

Because the velocity reaches a constant value of v_{ss} for time approaching infinity, the first term on the left-hand side is zero and thus the steady-state velocity is

$$v_{ss} = \frac{(\overline{F} + \Delta F)}{b}.$$

The time constant, defined by Eq. (4.21), is

$$\tau = \frac{m}{b}.$$

Substituting the expressions found for the three unknowns into the general form of the solution gives

$$v(t) = \frac{(\overline{F} + \Delta F)}{b} - \left[\frac{(\overline{F} + \Delta F)}{b} - \frac{\overline{F}}{b} \right] e^{-(b/m)(t-t_0)},$$

and hence

$$v(t) = \frac{\overline{F}}{b} + \left(\frac{\Delta F}{b} \right) \left[1 - e^{-(b/m)(t-t_0)} \right].$$

It should also be noted that the steady-state gain for this system, defined by Eq. (4.33), is

$$K = \frac{1}{b}.$$

With the steady-state gain, the velocity of the mass for $t > t_0$ can be expressed in yet another form:

$$v(t) = K \left[\overline{F} + \Delta F \left(1 - e^{-\frac{t-t_0}{\tau}} \right) \right].$$

The response of acceleration for $t > t_0$ can be obtained by differentiation of the velocity step response :

$$a(t) = \left(\frac{\Delta F}{m} \right) e^{-(b/m)(t-t_0)} = \left(\frac{K}{\tau} \right) \Delta F e^{-\frac{(t-t_0)}{\tau}}.$$

Both the velocity and acceleration step response curves are plotted in Fig. 4.8.

A conclusion from these considerations is that a step response of a first-order model is always an exponential function of time involving three parameters: initial value, steady-state value, and time constant. If the initial value is specified in the problem statement, the other two parameters can be found simply by inspection of the model input–output equation, by use of the relations defined in Eqs. (4.21) and (4.29). A step response of a first-order model can thus be found in a very simple manner, without solving the differential equation.

4.4   SECOND-ORDER MODELS

An input–output equation for a stationary, linear second-order model is

$$a_2 \ddot{y} + a_1 \dot{y} + a_0 y = f(t). \tag{4.53}$$

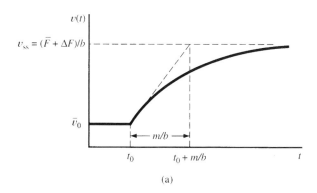

(a)

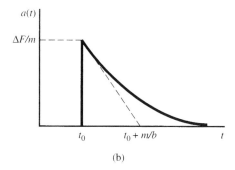

(b)

Figure 4.8. (a) Velocity and (b) acceleration step response curves of the system considered in Example 4.1.

The initial conditions are

$$y(0) = y_0, \quad \dot{y}(0) = \dot{y}_0. \tag{4.54}$$

The characteristic equation is

$$a_2 p^2 + a_1 p + a_0 = 0. \tag{4.55}$$

4.4.1 Free Response

The form of the homogeneous solution representing the system free response depends on whether the two roots of the characteristic equation [Eq. (4.55)], p_1 and p_2, are distinct or identical. If the roots are distinct, $p_1 \neq p_2$, the homogeneous solution is of the form

$$y_h(t) = C_1 e^{p_1 t} + C_2 e^{p_2 t}. \tag{4.56}$$

If the roots are identical, $p_1 = p_2$, the free response is

$$y_h(t) = C_1 e^{p_1 t} + C_2 t e^{p_1 t}. \tag{4.57}$$

If the roots are complex, they occur as pairs of complex-conjugate numbers, i.e.,

$$p_1 = a + jb, \qquad p_2 = a - jb. \tag{4.58}$$

Substitution of these expressions for the complex roots into Eq. (4.56) yields the homogeneous solution for this case:

$$y_h(t) = C_1 e^{(a+jb)t} + C_2 e^{(a-jb)t}, \tag{4.59}$$

or use of the trigonometric forms of the complex numbers gives

$$
\begin{aligned}
y_h(t) &= C_1 e^{at}(\cos bt + j \sin bt) + C_2 e^{at}(\cos bt - j \sin bt) \\
&= e^{at}(C_3 \cos bt + C_4 \sin bt),
\end{aligned}
\tag{4.60}
$$

where the constants C_3 and C_4 are a different, but corresponding, set of integration constants:

$$C_3 = C_1 + C_2, \tag{4.61}$$

$$C_4 = j(C_1 - C_2). \tag{4.62}$$

Note that, if the roots of the characteristic equation are complex, C_1 and C_2 are also complex-conjugate numbers, but C_3 and C_4 are real constants.

The model free response has been shown to depend on the type of roots of the model characteristic equation, and for a second-order model the free response takes the form of Eq. (4.56), Eq. (4.57), Eq. (4.59) or (4.60) if the roots are real and distinct, real and identical, or complex conjugate, respectively. Moreover, the character of the free response depends on whether real roots or real parts of complex roots are positive or negative. Table 4.2 illustrates the effect of location of the roots in a complex plane on the model impulse response.

Several general observations can be made on the basis of Table 4.2. First, it can be seen that the impulse response of a second-order model is oscillatory when the roots of the model characteristic equation are complex and nonoscillatory when the roots are real. Furthermore, the impulse response approaches zero as time approaches infinity only if the roots are either real and negative or complex and have negative real parts. The systems that have all the roots of their characteristic equations located in the left-hand side of a complex plane are referred to as stable systems. However, if at least one root of the model characteristic equation lies in the right-hand side of a complex plane, the model impulse response grows without bound with time; such a system is considered unstable. Marginal stability occurs when there are no roots in the right-hand side of a complex plane and at least one root is located on the imaginary axis. System stability constitutes one of the most important problems in analysis and design of feedback systems and will be treated more thoroughly in Chap. 13.

Two important parameters widely used in characterizing the responses of second-order systems are the damping ratio ζ and the natural frequency ω_n. These two parameters appear in the modified input–output equation

$$\ddot{y} + 2\zeta \omega_n \dot{y} + \omega_n^2 y = \frac{1}{a_2} f(t). \tag{4.63}$$

The damping ratio ζ represents the amount of damping in a system, whereas the natural frequency ω_n is a frequency of oscillations in an idealized system with zero

Table 4.2. Locations of roots of characteristic equations and the corresponding impulse response curves

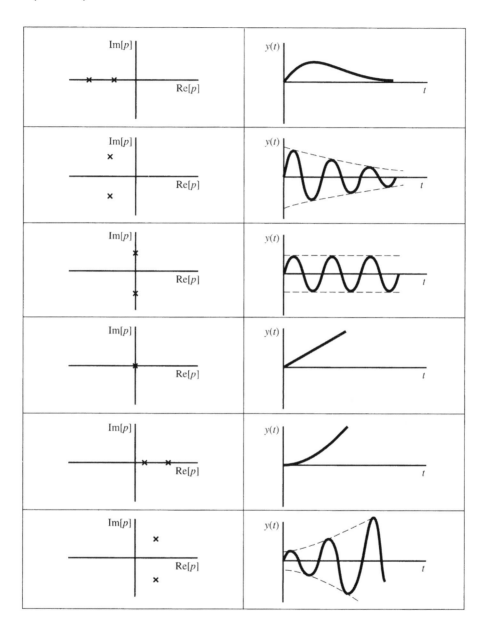

damping. In other words, because the amount of damping is related to the rate of dissipation of energy in the system, ω_n represents the frequency of oscillations in an idealized system that does not dissipate energy.

The input–output equation [Eq. (4.53)] can be rewritten in the following form:

$$\ddot{y} + (a_1/a_2)\dot{y} + (a_0/a_2)y = (1/a_2)f(t). \tag{4.64}$$

By comparing Eqs. (4.63) and (4.64), one can express the natural frequency and the damping ratio in terms of the coefficients of the input–output equation:

$$\omega_n = \sqrt{\frac{a_0}{a_2}}, \tag{4.65}$$

$$\zeta = \frac{a_1}{2\sqrt{a_0 a_2}}. \tag{4.66}$$

The system characteristic equation is

$$p^2 + 2\zeta \omega_n p + \omega_n^2 = 0. \tag{4.67}$$

The roots of the characteristic equation can be expressed in terms of the coefficients a_0, a_1, and a_2,

$$p_1, p_2 = \frac{\left[-a_1 \pm \sqrt{\left(a_1^2 - 4a_0 a_2\right)}\right]}{2a_2}, \tag{4.68}$$

or, equivalently, in terms of ζ and ω_n for the underdamped case,

$$p_1, p_2 = -\zeta \omega_n \pm j\omega_n\sqrt{1 - \zeta^2}, \tag{4.69}$$

or, similarly, in terms of τ_1 and τ_2 for the overdamped case,

$$p_1, p_2 = -1/\tau_1, -1/\tau_2, \tag{4.70}$$

where

$$\tau_1, \tau_2 = \frac{2a_2}{\left[a_1 \pm \sqrt{\left(a_1^2 - 4a_0 a_2\right)}\right]}. \tag{4.71}$$

It was pointed out earlier that a system is stable if real parts of the complex roots of the characteristic equation are negative, that is, if

$$\mathrm{Re}\,[p_1] = \mathrm{Re}\,[p_2] = -\zeta \omega_n < 0. \tag{4.72}$$

Because the natural frequency is not negative, for stability of a second-order system the damping ratio must be positive so that $\zeta > 0$. From Eqs. (4.66) and (4.69) it can be deduced that the roots are complex; thus the system response is oscillatory if

$$0 < \zeta < 1. \tag{4.73}$$

A system is said to be underdamped when the damping ratio satisfies condition (4.73). The frequency of oscillation of an underdamped system is called a damped natural frequency and is equal to

$$\omega_d = \omega_n\sqrt{1 - \zeta^2}. \tag{4.74}$$

If the damping ratio is equal to or greater than 1, the expressions on the right-hand side of Eq. (4.69) become real, and thus the system free response is nonoscillatory. A system for which the damping ratio is greater than 1 is referred to as an overdamped system. The damping is said to be critical if the damping ratio is equal to 1.

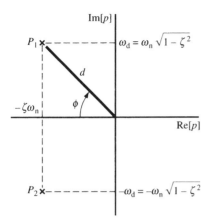

Figure 4.9. Pair of complex roots and corresponding values of ζ, ω_n, and ω_d.

Table 4.2 shows how the nature of the system impulse response depends on the locations of roots of the characteristic equation. In addition, the dynamics of a second-order model can be uniquely described in terms of the natural frequency and the damping ratio. Therefore the conclusion can be drawn that there must exist a unique relationship between pairs of roots of the characteristic equation and pairs of ω_n and ζ values. To determine this relationship, consider a second-order model having two complex roots, p_1 and p_2, located as shown in Fig. 4.9.

The real parts of both roots are

$$\text{Re}\,[p_1] = \text{Re}\,[p_2] = -\zeta\,\omega_n \tag{4.75}$$

and the imaginary parts are

$$\text{Im}\,[p_1] = \omega_n\sqrt{1 - \zeta^2} = \omega_d, \tag{4.76}$$

$$\text{Im}\,[p_2] = \omega_n\sqrt{1 - \zeta^2} = -\omega_d. \tag{4.77}$$

The damped natural frequency is therefore equal to the ordinates of the points p_1 and p_2 with a plus or minus sign, respectively. To identify a corresponding natural frequency ω_n, consider the distance d between the points p_1 and p_2 and the origin. In terms of the real and imaginary parts of the two roots, d can be calculated as

$$d = \sqrt{(\text{Re}\,[p_1])^2 + (\text{Im}\,[p_1])^2}$$

$$= \sqrt{(\text{Re}\,[p_2])^2 + (\text{Im}\,[p_2])^2}. \tag{4.78}$$

Substituting the expressions for the real and imaginary parts of p_1 and p_2 from Eqs. (4.75), (4.76), and (4.77) yields

$$d = \omega_n. \tag{4.79}$$

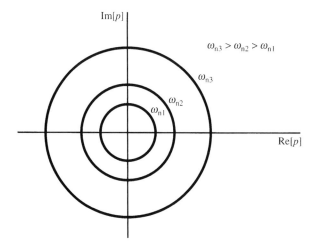

Figure 4.10. Loci of constant natural frequency.

The natural frequency of a second-order underdamped system is thus equal to the distance between the locations of the system characteristic roots and the origin of the coordinate system in the complex plane.

Finally, by a comparison of Eqs. (4.75) and (4.79), the damping ratio can be expressed as

$$\zeta = -\mathrm{Re}\,[p_1]\,/d, \tag{4.80}$$

which, after inspection of Fig. 4.9, can be rewritten as

$$\zeta = \cos \varphi, \tag{4.81}$$

where φ is the acute angle measured from the negative real axis.

The results of the preceding considerations are presented graphically in Figs. 4.10, 4.11, and 4.12. Figure 4.10 shows loci of constant natural frequency. The loci are concentric circles with radii proportional to ω_n. The farther from the origin of the coordinate system the roots of the characteristic equation are, the higher the value of the corresponding natural frequency.

In Fig. 4.11 the horizontal lines represent loci of constant damped natural frequency. The greater the distance between the roots of the characteristic equation and the real axis, the higher the value of w_d.

The loci of constant damping ratio for an underdamped system, shown in Fig. 4.12, take the form of straight lines described by Eq. (4.81). Practically, the use of the damping ratio is limited to stable systems having both roots of the characteristic equation in the left-hand side of a complex plane, and the loci in Fig. 4.12 represent only stable underdamped systems. When the system is overdamped, the roots lie along the real axis, and there is no oscillation in the response.

4.4.2 Step Response

A step response equation for a second-order model is now derived for the three cases of an underdamped, a critically damped, and an overdamped system.

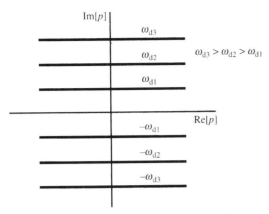

Figure 4.11. Loci of constant damped natural frequency.

The input–output equation for the system subjected to a step input of magnitude A is

$$a_2\ddot{y} + a_1\dot{y} + a_0 y = AU_s(t). \tag{4.82}$$

Both initial conditions are zero, $y(0) = 0$ and $\dot{y}(0) = 0$.

Underdamped case, $0 < \zeta < 1$. The input–output equation in a parametric form [shown in Eq. (4.63)] is

$$\ddot{y} + 2\zeta\omega_n\dot{y} + \omega_n^2 y = (A/a_2)U_s(t). \tag{4.83}$$

The characteristic equation is

$$p^2 + 2\zeta\omega_n p + \omega_n^2 = 0. \tag{4.84}$$

For $0 < \zeta < 1$, Eq. (4.84) has two complex-conjugate roots,

$$p_1 = -\zeta\omega_n - j\omega_d, \tag{4.85}$$

$$p_2 = -\zeta\omega_n + j\omega_d, \tag{4.86}$$

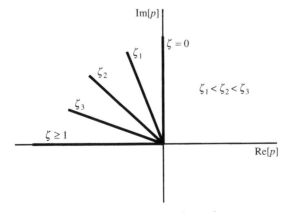

Figure 4.12. Loci of constant damping ratio.

where ω_d is defined by Eq. (4.74). The general form of the homogeneous solution is given by Eq. (4.60). Substituting real and imaginary parts of p_1 and p_2 from Eqs. (4.85) and (4.86) yields

$$y_h(t) = e^{-\zeta \omega_n t}(C_3 \cos \omega_d t + C_4 \sin \omega_d t), \qquad (4.87)$$

where C_3 and C_4 are unknown real constants. The particular solution for a unit step input, in accordance with Eq. (4.12), is

$$y_p(t) = \frac{A}{a_2 \omega_n^2} = \frac{A}{a_0}, \qquad (4.88)$$

or, by use of the definition of the steady-state gain given by Eq. (4.33),

$$y_p(t) = KA. \qquad (4.89)$$

The complete solution is a sum of the right-hand sides of Eqs. (4.87) and (4.89):

$$y(t) = KA + e^{-\zeta \omega_n t}(C_3 \cos \omega_d t + C_4 \sin \omega_d t). \qquad (4.90)$$

By use of the initial conditions, the two constants C_3 and C_4 are found:

$$C_3 = -KA, \qquad (4.91)$$

$$C_4 = -\frac{KA\zeta}{\sqrt{1 - \zeta^2}}. \qquad (4.92)$$

Hence the step response of an underdamped second-order model is

$$y(t) = KA\left[1 - e^{-\zeta \omega_n t}\left(\cos \omega_d t + \frac{\zeta}{\sqrt{1 - \zeta^2}} \sin \omega_d t\right)\right]. \qquad (4.93)$$

Critically damped case, $\zeta = 1$. One can find the step response in this case simply by taking a limit of the right-hand side of Eq. (4.93) for ζ approaching unity, which yields

$$y(t) = KA[1 - e^{-\omega_n t}(1 + \omega_n t)]. \qquad (4.94)$$

One can also obtain the same result by following the general procedure for solving linear differential equations presented in Section 4.2. The characteristic equation in this case has a double root:

$$p_1 = p_2 = -\omega_n. \qquad (4.95)$$

The general form of the homogeneous solution, as given in Eq. (4.10), is

$$y_h(t) = C_1 e^{-\omega_n t} + C_2 t e^{-\omega_n t}. \qquad (4.96)$$

The particular solution for a step input is, as before,

$$y_p(t) = KA, \qquad (4.97)$$

and hence the complete solution is

$$y(t) = KA + C_1 e^{-\omega_n t} + C_2 t e^{-\omega_n t}. \qquad (4.98)$$

By use of the initial conditions, $y(0) = 0$ and $\dot{y}(0) = 0$, the integration constants are found to be

$$C_1 = -KA, \tag{4.99}$$

$$C_2 = -\omega_n KA. \tag{4.100}$$

Substitution of the expressions for the constants C_1 and C_2 into Eq. (4.98) yields the complete solution obtained earlier, Eq. (4.94).

Overdamped case, $\zeta > 1$. When $\zeta > 1$, the characteristic equation [Eq. (4.84)] has two distinct real roots given by Eq. (4.68). The homogeneous solution takes the form

$$y_h(t) = C_1 e^{-t/\tau_1} + C_2 e^{-t/\tau_2}. \tag{4.101}$$

The particular solution for a step input of magnitude A is

$$y_p(t) = KA. \tag{4.102}$$

Combining Eqs. (4.101) and (4.102) gives the complete solution:

$$y(t) = KA + C_1 e^{-t/\tau_1} + C_2 e^{-t/\tau_2}. \tag{4.103}$$

The integration constants are found by use of the zero initial conditions:

$$C_1 = KA \frac{\tau_1}{(\tau_1 - \tau_2)}, \tag{4.104}$$

$$C_2 = KA \frac{\tau_2}{(\tau_1 - \tau_2)}. \tag{4.105}$$

The complete solution for a step response of an overdamped second-order model is

$$y(t) = KA \left[1 - \frac{1}{(\tau_1 - \tau_2)} (\tau_1 e^{-t/\tau_1} - \tau_2 e^{-t/\tau_2}) \right]. \tag{4.106}$$

A family of step response curves for a second-order model with zero initial conditions normalized with respect to the steady-state gain is plotted in Fig. 4.13 for different values of the damping ratio and the same natural frequency.

A dynamic behavior of a second-order model described by Eq. (4.82) can also be described in terms of selected specifications of the model step response. Figure 4.14 shows the most common specifications of a step response of an underdamped second-order model with zero initial conditions, although the the use of these step response specifications is not limited to a single type of model.

The period of oscillations T_d is related to the damped natural frequency by

$$T_d = \frac{2\pi}{\omega_d}. \tag{4.107}$$

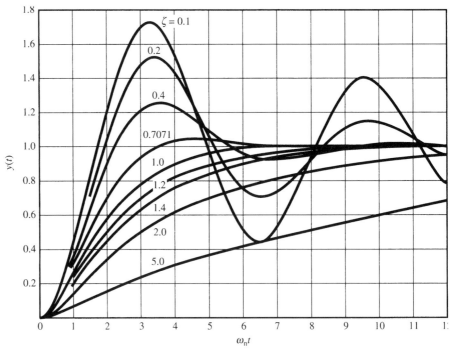

Figure 4.13. Unit step response curves of a second-order model.

The peak time t_p is the time between the start of the step response and its first maximum. It is equal to half of the period of oscillations:

$$t_p = \frac{\pi}{\omega_d}. \tag{4.108}$$

The peak time is a measure of the speed of response of an underdamped system. For a critically damped or overdamped system, the speed of response is usually represented by a delay time or a rise time.

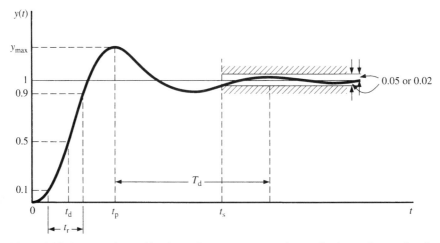

Figure 4.14. Parametric specifications of a step response of an underdamped second-order model.

The delay time t_d is the time necessary for the step response to reach a point halfway between the initial value and the steady-state value, which can be expressed mathematically as

$$y(t)\big|_{t=t_d} = 0.5\,[y_{ss} - y(0)] + y(0). \tag{4.109}$$

The rise time t_r is the time necessary for the step response to rise from 10 percent to 90 percent of the difference between the initial value and the steady-state value.

Settling time t_s is defined as the time required for the step response to settle within a specified percentage of the steady-state value. A 2 percent settling time is the time for which the following occurs:

$$\big|y(t) - y_{ss}\big| \leq 0.02[y_{ss} - y(0)], \quad \text{for} \quad t \geq t_s. \tag{4.110}$$

An oscillatory character of a system step response is represented by a maximum overshoot M_P, defined as

$$M_P = \frac{y_{max} - y_{ss}}{y_{ss} - y(0)}. \tag{4.111}$$

A percent maximum overshoot, $M_P^{\%}$ is used more often.

$$M_P^{\%} = \left[\frac{y_{max} - y_{ss}}{y_{ss} - y(0)}\right] 100\%. \tag{4.112}$$

For a second-order model, the percentage of maximum overshoot can be expressed as a function of the damping ratio:

$$M_P^{\%} = 100e^{-\frac{\pi\zeta}{\sqrt{1-\zeta^2}}}. \tag{4.113}$$

This relationship is presented in graphical form in Fig. 4.15.

Another useful specification of the transient response of an underdamped second-order system is a decay ratio (DR), defined as the ratio of successive

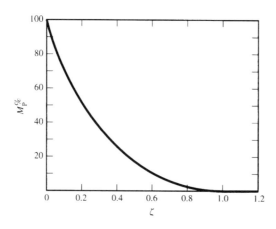

Figure 4.15. Maximum percentage of overshoot versus damping ratio.

amplitudes of the system step response. Referring to Fig. 4.14, we find that the DR can be expressed as

$$DR = \frac{\left[y(t_p + T_d) - y_{ss}\right]}{\left[y(t_p) - y_{ss}\right]}. \tag{4.114}$$

For a system described by Eq. (4.83), the DR can be related to the damping ratio by the formula

$$DR = e^{-\frac{2\pi\zeta}{\sqrt{1-\zeta^2}}}. \tag{4.115}$$

A logarithmic DR (LDR) is sometimes used instead of the DR, where

$$LDR = \ln(DR) = \frac{-2\pi\zeta}{\sqrt{1-\zeta^2}}. \tag{4.116}$$

The DR is useful in determining the system damping ratio from the system oscillatory step response.

EXAMPLE 4.2

In the system shown in Fig. 4.16(a), mass $m = 9$ kg is subjected to force $F(t)$ acting vertically and undergoing a step change from 0 to 1.0 N at time $t = 0$. The mass, suspended on a spring of constant $k = 4.0$ N/m, is moving inside an enclosure with a coefficient of friction between the surfaces of $b = 4.0$ N s/m. Using force $F(t)$ as the input variable and the position of mass $x(t)$ as the output variable, sketch an approximate step response of the system. If this response is oscillatory, determine the necessary modification to make the system critically damped.

SOLUTION

The input–output equation of the system is

$$m\ddot{x} + b\dot{x} + kx = F(t).$$

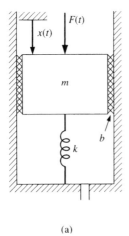
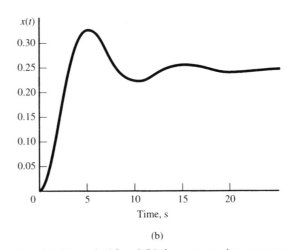

(a)   (b)

Figure 4.16. (a) Original system considered in Example 4.2 and (b) the system unit step response.

Substitution of the numerical values for the system parameters yields

$$9\ddot{x} + 4\dot{x} + 4x = F(t).$$

From Eq. (4.66), the damping ratio is

$$\zeta = \frac{4}{2\sqrt{(9)(4)}} = 0.3333.$$

Because $\zeta < 1$, the system step response will be oscillatory. Other step response specifications useful in sketching the step response curve can be determined from Eqs. (4.65), (4.74), (4.107), and (4.113) as follows:

$$\omega_n = \sqrt{4/9} = 0.6667 \text{ rad/s},$$
$$\omega_d = 0.6667\sqrt{1 - 0.3333^2} = 0.6286 \text{ rad/s},$$
$$T_d = 10.0 \text{s},$$
$$M_p^\% = 100e^{-\frac{\pi 0.3333}{\sqrt{1-0.3333^2}}} = 32.94\%.$$

Given these parameters, the step response can be sketched as shown in Fig. 4.16(b).
    To make the system critically damped, another damper, b_{ad}, is added, as shown in Fig. 4.17(a). The input–output equation of the modified system is

$$9\ddot{x} + (4 + b_{ad})\dot{x} + 4x = F(t).$$

The damping ratio is now given by

$$\zeta = \frac{(4 + b_{ad})}{2\sqrt{4 - 9}}.$$

For critical damping, the term on the right-hand side of the preceding equation must be equal to 1.0, which yields the value of additional damping, $b_{ad} = 8$ N s/m. The step response curve of the system with additional damping is shown in Fig. 4.17(b).

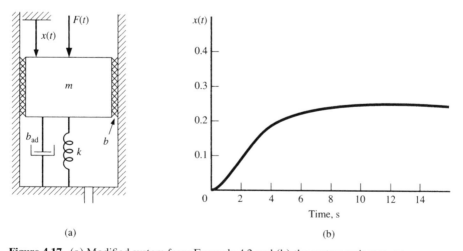

(a)   (b)

Figure 4.17. (a) Modified system from Example 4.2 and (b) the system unit step response.

4.5   THIRD- AND HIGHER-ORDER MODELS

Theoretically, the analysis of linear models of third and higher orders should be nothing more than a simple extension of the methods developed for first- and second-order models, presented in earlier sections. Practically, however, methods that are fast and easy to use with lower-order models become excessively involved and cumbersome in applications involving third- and higher-order models. One can obtain an analytical solution of an nth-order input–output equation for a linear, stationary model by simply following the same general procedure described in Section 4.2; however, the algebra involved in the solution is considerably more complex. First- and second-order linear differential equations are not only easy to solve analytically but in addition, the parameters used in these equations have straightforward physical meanings, which allow evaluation of general system characteristics even without solving the model equation. The physical meanings of the parameters in third- and higher-order model equations are considerably less clear and more difficult to interpret in terms of the system dynamic properties.

Another apparently insignificant problem occurs in obtaining the homogeneous solution, which requires determination of the roots of a third- or higher-order characteristic equation. There are many computer programs for solving higher-order algebraic equations; however, in such situations the following question arises: If it becomes necessary to use a computer to solve part of a problem (find roots of a characteristic equation), wouldn't it be worthwhile to use the computer to solve the entire problem (find the solution of an input–output equation)?

Computer programs for solving linear differential equations, or rather for solving equivalent sets of first-order equations, are almost as readily available and easy to use as the programs for finding roots of a characteristic equation. Some of these programs are described in Chap. 6. Generally speaking, the higher the order of the input–output equation, the more justified and more efficient the use of a computer. As has been and still will be stressed throughout this text, however, it is always necessary to verify the computer-generated results with a simple approximate analytical solution. In this section, the concept of the so-called dominant roots of a characteristic equation is presented. The dominant roots are very useful in obtaining approximate solutions of third- or higher-order differential equations.

The general form of a homogeneous solution of an nth-order differential equation having m distinct and $n - m$ multiple roots, presented in Section 4.2, is

$$y_h(t) = \sum_{i=1}^{m} C_i e^{p_i t} + \sum_{i=m+1}^{n} C_i t^{i-m-1} e^{p_i t}. \tag{4.117}$$

If a system is stable, all roots $p_1, p_2, \ldots, p_n$ are real and negative or complex and have negative real parts. The rate at which the exponential components on the right-hand side of Eq. (4.117) decay depends on the magnitudes of the real parts of $p_1, p_2, \ldots, p_n$. The larger the magnitude of a negative real root and/or the larger the magnitude of a negative real part of a complex root, the faster the decay of the corresponding exponential terms in the model free response. In other words, the roots of the characteristic equation located farther from the imaginary axis in the left half of the complex plane affect the model free response relatively less than do the roots closer the imaginary axis. The roots of the characteristic equation nearest the imaginary axis in the complex plane are called the dominant roots.

The concept of dominant roots is now illustrated in an example of a third-order model.

The model differential input–output equation is

$$a_3 \frac{d^3 y}{dt^3} + a_2 \frac{d^2 y}{dt^2} + a_1 \frac{dy}{dt} + a_0 y = f(t). \tag{4.118}$$

The model is assumed to have distinct, real, and negative roots, p_1, p_2, p_3, with the last root, p_3, being much farther away from the imaginary axis of the complex plane than p_1 and p_2, so that

$$|p_3| \gg |p_1|, |p_3| \gg |p_2|. \tag{4.119}$$

The model equation can be rewritten as follows:

$$\frac{d^3 y}{dt^3} - (p_1 + p_2 + p_3) \frac{d^2 y}{dt^2} + (p_1 p_2 + p_1 p_3 + p_2 p_3) \frac{dy}{dt}$$
$$- p_1 p_2 p_3 y = \frac{1}{a_3} f(t). \tag{4.120}$$

Hence the characteristic equation is

$$p^3 - (p_1 + p_2 + p_3)p^2 + (p_1 p_2 + p_1 p_3 + p_2 p_3)p - p_1 p_2 p_3 = 0. \tag{4.121}$$

Dividing both sides of the characteristic equation by p_3 yields

$$\left(\frac{1}{p_3} \right) p^3 - \left(\frac{p_1}{p_3} + \frac{p_1}{p_3} + 1 \right) p^2 + \left(\frac{p_1 p_2}{p_3} + p_1 + p_2 \right) p - p_1 p_2 = 0. \tag{4.122}$$

The terms having p_3 in the denominator can be neglected, based on assumptions (4.119), to yield the following approximation of the characteristic equation:

$$p^2 - (p_1 + p_2)p + p_1 p_2 = 0. \tag{4.123}$$

The corresponding differential equation, with the right-hand side term as in Eq. (4.120), is

$$\frac{d^2 y}{dt^2} - (p_1 + p_2) \frac{dy}{dt} + p_1 p_2 y = \frac{1}{a_3} f(t). \tag{4.124}$$

Although the left-hand side of this equation was derived to approximate the dynamics of the system described by Eq. (4.118), it can be seen that the steady-state behavior of the approximating model does not match the steady-state performance of the original system. The steady-state response of the original system for the forcing function reaching a constant value f_{ss} is

$$y_{ss} = -\frac{1}{p_1 p_2 p_3 a_3} f_{ss}, \tag{4.125}$$

and the steady-state response of the approximating second-order system described by Eq. (4.124) is

$$y_{ss} = \frac{1}{p_1 p_2 a_3} f_{ss}. \tag{4.126}$$

To eliminate this discrepancy, we must adjust (calibrate) the approximating model to match the steady-state performance of the original system by dividing the right-hand side of Eq. (4.124) by $(-p_3)$ to obtain

$$\frac{d^2 y}{dt^2} - (p_1 + p_2)\frac{dy}{dt} + p_1 p_2 y = -\frac{1}{p_3 a_3} f(t). \tag{4.127}$$

The division of the right-hand side by a constant has no effect on the characteristic equation of the system, so the dynamic performance of the system remains the same, but the steady-state performance of the approximating model given by Eq. (4.127) matches exactly the response of the original system described by Eq. (4.118).

It should be noted that, although in deriving approximating equation (4.127) it was assumed that all three roots of the characteristic equation were real negative numbers, it can be shown that the approximation is just as good if the dominant roots, p_1 and p_2, are complex-conjugate numbers. In general, the approximation improves as the distance between the dominant roots, p_1 and p_2, and the other root, p_3, measured in a horizontal direction in the complex plane, increases. This is illustrated in Fig. 4.18, which shows the step response curves of three third-order systems having the same pair of dominant roots, p_1 and p_2, $(-1 \pm 4j)$ but a different value of the third root, p_3. The figure also shows the step response of a second-order system with the dominant roots only. The input–output equations for the four systems along with the values of the roots of their characteristic equations are shown in Table 4.3.

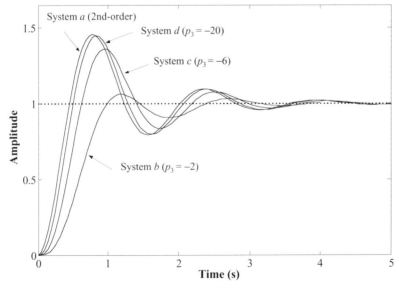

Figure 4.18. Step responses of three third-order system (*b, c, d*) with the same dominant roots and different third root and a second-order system with the dominant roots only (system *a*).

Table 4.3. Equations and roots of three third-order systems and an approximating second-order system based on dominant roots

System	Input–output equation	p_1	p_2	p_3
a	$\ddot{y} + 2\dot{y} + 17y = u$	$-1 + 4j$	$-1 - 4j$	–
b	$\dddot{y} + 4\ddot{y} + 21\dot{y} + 34y = 2u$	$-1 + 4j$	$-1 - 4j$	-2
c	$\dddot{y} + 8\ddot{y} + 29\dot{y} + 102y = 6u$	$-1 + 4j$	$-1 - 4j$	-6
d	$\dddot{y} + 22\ddot{y} + 57\dot{y} + 340y = 20u$	$-1 + 4j$	$-1 - 4j$	-20

It can be seen that the step response of the second-order system based on dominant roots gets closer to the step response of a third-order system as the third "nondominant" root moves further away from the dominant roots. That nondominant root, which is a negative real number in this case, corresponds to a time constant equal to a negative inverse of the root:

$$\tau = -\frac{1}{p_3}.$$

For system b, $p_3 = -2$, which is very close to the dominant roots (which are not really dominant here!), and the corresponding time constant, $\tau = 0.5$ s, has a strong damping effect on the step response. For systems c and d, the time constants corresponding to root p_3 are 0.1667 and 0.05 s, respectively. These smaller time constants have considerably weaker impact on the system step response.

In summary, the second-order system based on dominant roots (system a) is not a good approximation of system b but appears to be a reasonably good approximation of system c and a very good approximation of system d.

The simplification procedure just described should be used with caution because it leads to a system model that may be deficient in some applications. For instance, the small time constant associated with the rejected root p_3, $\tau_3 = -1/p_3$, may still be significant in some closed-loop systems that use high-gain feedback.

In addition to simplifying the analysis, the rejection of a nondominant root (or roots) makes it easier to run a check solution on the computer and verify it before proceeding with the more complete model when it is required.

4.6 SYNOPSIS

This chapter reviewed the classical methods frequently used to solve for the dynamic responses of linear systems modeled by ODEs. Greater emphasis was placed on finding the responses of first- and second-order systems to step inputs; however, responses to nonequilibrium initial conditions (so-called free response) and impulse response were also covered.

In each case the solution was shown to be of the form

$$y(t) = y_p(t) + y_h(t),$$

where $y_p(t)$ is the particular or forced part of the response, usually having the same form as the input, its derivatives, or both, and $y_h(t)$ is the homogeneous or natural part of the response consisting of exponential terms that uses the roots of the system characteristic equation.

Thus the solution for a first-order system is

$$y(t) = y_p(t) + C_1 e^{pt} = y_p(t) + C_1 e^{-t/\tau}.$$

where $p = -1/\tau$ is the single root of the first-order characteristic equation. The constant C_1 is then found by use of the initial condition equation with the values of the particular part, the homogeneous part, and the output at $t = 0$,

$$y(0) = y_p(0) + C_1 e^0$$

or

$$C_1 = y(0) - y_p(0).$$

For a second-order characteristic equation of the form

$$a_2 p^2 + a_1 p + a_0 = 0,$$

the roots may be real or conjugate complex. For the underdamped case, the complex-conjugate roots, involving the undamped natural frequency and damping ratio, are expressed by Eq. (4.68). For the overdamped and critically damped cases, the real roots, involving two time constants τ_1 and τ_2, are expressed by Eq. (4.71).

The solution for a second-order system is of the form

$$y(t) = y_p(t) + y_h(t) = y_p(t) + \begin{cases} e^{-\zeta \omega_n t}(C_3 \cos \omega_d t + C_4 \sin \omega_d t), & 0 < \zeta < 1 \\ C_1 e^{p_1 t} + C_2 t e^{p_1 t}, & \zeta = 1 \\ C_1 e^{p_1 t} + C_2 e^{p_2 t}, & \zeta > 1 \end{cases},$$

where ω_n, ζ, and $\omega_d = \omega_n\sqrt{1-\zeta^2}$ are the undamped natural frequency, the damping ratio, and the damped natural frequency, respectively, of an underdamped $(0 < \zeta < 1)$ system having complex-conjugate roots, p_1 repeated are the identical roots of a critically damped $(\zeta = 1)$ system, and $p_1 = -1/\tau_1$ and $p_2 = -1/\tau_2$ are the real roots of an overdamped $(\zeta > 1)$ system.

To assess each new situation as it as arises, solving first for $\zeta = a_1/(2\sqrt{a_0 a_2})$ determines immediately which case is to be dealt with. Then, when $y_p(t)$ has been determined from the form of the forcing function, the constants C_1 and C_2 or C_3 and C_4 are readily determined from the two initial condition equations [Eqs. (4.54)].

The greatest emphasis was placed on first- and second-order systems in order to simplify the illustration of the classical method of solving differential equations. In addition, first-and second-order models exhibit the major features encountered in the responses of higher-order systems: exponential decay with time constant/s and/or oscillatory response with decaying amplitude of oscillation, which are also present in the responses of third- and higher-order systems. Having a good understanding of first-and second-order system response behavior facilitates the verification of computer simulations and the "debugging" of computer programs when they are being developed for computer simulation.

Use of a simplified model that uses only the dominant roots of a higher-order system makes it possible to uncover the most significant dynamic response characteristics of the higher-order model through use of a lower-order model. Also, running a reduced-order check solution as part of a simulation study helps to find program errors and "glitches" that seem to plague even the most experienced programmers. Once this simplified model is running properly on the computer, it is usually a simple matter to reinsert the less-dominant roots and then to produce a "full-blown" solution that includes all of the higher-order effects inherent in the higher-order system differential equation model of the system.

PROBLEMS

4.1 A first-order model of a dynamic system is

$$2\dot{y} + 5y = 5f(t).$$

(a) Find and sketch the response of this system to the unit step input signal $f(t) = U_s(t)$, for $y(0) = 2$.

(b) Repeat part (a) for zero initial condition $y(0) = 0$.

(c) Repeat part (a) for a unit impulse input $f(t) = U_i(t)$.

(d) Repeat part (b) for a unit impulse input $f(t) = U_i(t)$.

4.2 The roots of a second-order model are $p_1 = -1 + j$ and $p_2 = -1 - j$.

(a) Find and sketch the system unit step response assuming zero initial conditions, $\dot{y}(0) = 0$ and $y(0) = 0$.

(b) Repeat part (a) for the roots of the characteristic equation, $p_1 = 1 + j$ and $p_2 = 1 - j$. Explain the major difference between the step responses found in parts (a) and (b).

4.3 A rotational mechanical system has been modeled by the equation

$$J\dot{\Omega} + B\Omega = T(t).$$

Determine the values of J and B for which the following conditions are met:
○ Steady-state rotational velocity for a constant torque, $T(t) = 10$ N m, is 50 rpm (revolutions per minute).
○ The speed drops below 5% of its steady-state value within 160 ms after the input torque is removed, $T(t) = 0$.

4.4 Output voltage signals $y_1(t)$ and $y_2(t)$ of two linear first-order electrical circuits were measured, as shown in Figs. P4.4(a) and P4.4(b). Write analytical expressions describing the two signals.

4.5 A mass $m = 1.5$ lb s^2/ft sliding on a fixed guideway is subjected to a suddenly applied constant force $F(t) = 100$ lb at time $t = 0$. The coefficient of linear friction between the mass and the guideway is $b = 300$ lb s/ft. Find the system time constant. Write the system model equation and solve it for the response of mass velocity v as a function of time, assuming $v(0) = 0$. Sketch and label the system response versus time.

4.6 The electric generator in the steam turbine drive system shown in Fig. P4.6 has been running steadily at speed $(\Omega_1)_0$ with a constant input steam torque, T_{steam}. At time $t = t_0$, the shaft, which has a developing fatigue crack, breaks suddenly. This problem is

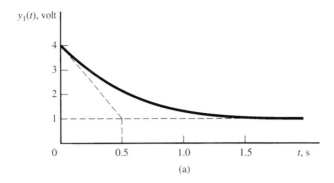

(a)

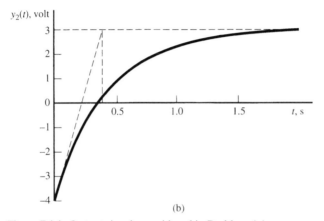

(b)

Figure P4.4. Output signals considered in Problem 4.4.

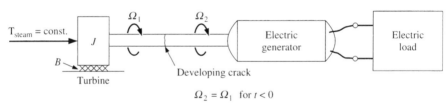

$$\Omega_2 = \Omega_1 \quad \text{for } t < 0$$

Figure P4.6. Steam turbine drive system considered in Problem 4.6.

concerned with how the speed Ω_1 varies with time after the shaft breaks (what happens to the generator is of no concern here).

(a) Find the steady torque $(T_g)_0$ in the shaft before the shaft breaks.

(b) Find the time constant of the remaining part of the system to the left of the crack for $t > t_0$.

(c) Determine the initial condition $\Omega_1(t_0)$.

(d) Solve for the response $\Omega_1(t)$ for $t > t_0$ and sketch the response.

(e) Determine how long it takes (in terms of the system time constant) for $\Omega_1(t)$ to reach a 5% overspeed condition.

4.7 A first-order system is modeled by the equation

$$\dot{y}(t) = ay(t) + bf(t).$$

With the system initially at rest, a unit step input $f(t) = U_s(t)$ is applied. Two measurements of the output signal are taken:

$$y(0.5) = 1.2, \qquad \lim_{t \to \infty} y(t) = 2.0.$$

Find a and b.

4.8 A mechanical system is described by the following set of state-variable equations:

$$\dot{q}_1 = -6q_1 + 2q_2,$$
$$\dot{q}_2 = -6q_2 + 5f.$$

The state variable q_1 is also the output variable. Sketch a unit step response of this system.

4.9 The mechanical device shown schematically in Fig. P4.9 is used to measure the coefficient of friction between a rubber shoe and a pavement surface. The value of mass m is 4 kg, but the value of the spring constant is not exactly known. In a measuring procedure, mass m is subjected to force $F(t)$, which changes suddenly from 0 to 10 N at time $t = 0$. The position of the mass, $x(t)$, is recorded for $t \geq 0$. From the system step response curve recorded as shown in Fig. P4.9, find the unknown system parameters k and b.

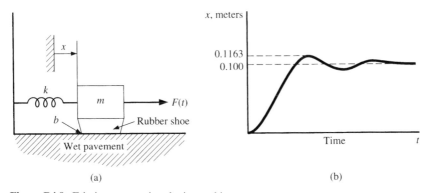

Figure P4.9. Friction-measuring device and its step response curve.

4.10 Consider again the mechanical device shown in Fig. P4.9 but with mass m of 5 kg. The response of this system to a step change in force $F(t)$ was found to be very oscillatory (Fig. P4.10). The only measurements obtained were two successive amplitudes, A_1 and A_2, equal to 55 and 16.5 cm, respectively, and the period of oscillation T_d equal to 1 s. Determine the values of the spring constant k and the coefficient of friction b in this case.

4.11 For each of the three mechanical systems shown in Figs. P4.11(a), P4.11(b), and P4.11(c), do the following:

(a) Derive an input–output equation.

(b) Write an expression for the unit step response.

(c) Sketch and carefully label the unit step response curve.

4.12 An input–output model of a third-order system was found to be

$$\dddot{y} + 12\ddot{y} + 25\dot{y} + 50y = f(t).$$

The system step response for $f(t) = 50U_s(t)$ is plotted in Fig. P4.12. Compute the roots for this third-order system and then find an approximating second-order model for this system by using the dominant roots of the third-order model. Sketch the response of

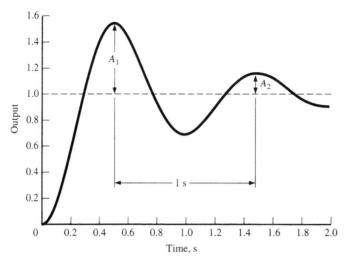

Figure P4.10. Step response curve of the device considered in Problem 4.10.

the approximating second-order model to input $f(t) = 50\ U_s(t)$ and compare it with the curve shown in Fig. P4.12.

4.13 A schematic of the mechanical part of a drive system designed for use in a drilling machine is shown in Fig. P4.13. The driving torque T_m supplied by an electric motor is applied through a gear reduction unit having ratio R_1/R_2 to drive a drilling spindle represented here by mass m_s. The gears' moments of inertia are J_1 and J_2 and their equivalent coefficient of rotational friction is B_{eq}. The spindle is suspended on air bearings of negligible friction, and it is pulled by a steel cable, which is assumed to be massless and has a spring constant k_s.

To examine the basic dynamic characteristics of the mechanical part of the system, the motor was shut off, $T_m = 0$, and an impact force F_{imp} was applied to the spindle. The

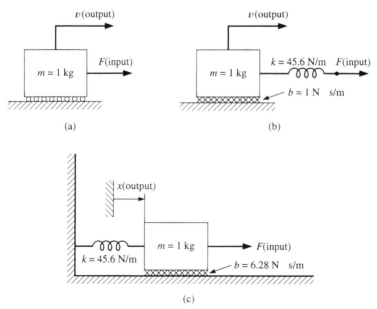

Figure P4.11. Mechanical systems considered in Problem 4.11.

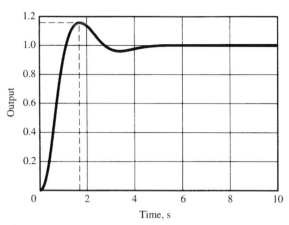

Figure P4.12. Step response of the third-order system considered in Problem 4.12.

response of the system, measured as the position $x(t)$ of the spindle, was found to be excessively oscillatory. To identify the source of this oscillation, determine the locations of the roots of the system characteristic equation and suggest how the system parameters should be changed, relative to their values during the test, to provide more damping.

Hint: Find approximate analytical expressions for the real roots and/or ζ and ω_n associated with complex-conjugate roots. Assume that the one real root of the equation,

$$p^3 + a_2 p^2 + a_1 p + a_0 = 0,$$

is approximated by $p_1 = -a_0/a_1$.

4.14 Figure P4.14 shows a schematic of a mechanical drive system in which a servomotor (inertia J_1) drives a pinion gear (inertia J_2, radius R) that is meshed with a rack (mass m). The rack moves laterally against a spring element (k). The linear bearing that guides the rack is well lubricated and produces a viscous friction effect (b). Assume that the shaft between the motor (J_1) and the gear (J_2) is rigid.

(a) Use the values in the table for the parameters and solve for the poles of the system.

(b) Sketch the response of the system for a unit step input.

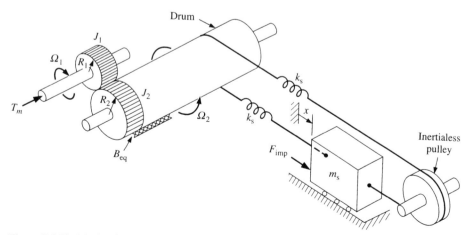

Figure P4.13. Mechanical part of the drive system considered in Problem 4.13.

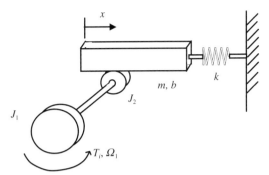

Figure P4.14. Mechanical drive system.

(c) As a design engineer, how would you characterize the major effect of the radius on the system performance?

(d) If the viscous friction (damping) factor could be adjusted, what value would lead to a state of critical damping?

Parameter	Value	Units
J_1	0.1	in. lb$_f$ s^2
J_2	0.05	in. lb$_f$ s^2
R	3.0	in.
b	10.0	in. lb$_f$ s
m	0.05	lb$_f$ s^2/in
k	1000.0	lb$_f$/in

4.15 Sketch the step responses for systems with the following pole locations shown in Fig. P4.15. Plot the four responses on the same time scale.

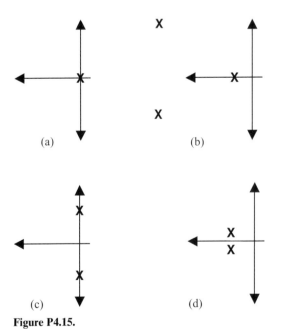

Figure P4.15.

4.16 Consider the mechanical system sketched in Fig P4.16. It represents a motor driving a heavy inertial load through a belt drive. The belt is flexible and is modeled as a torsional spring. Also, the friction inherent in the motor is modeled as a linear damping factor between the motor inertia and the ground. Develop a state model of this system and, for the following values of the parameters, find the system poles and the steady-state gain of this system

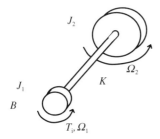

Figure P4.16. Schematic of a mechanical drive system with two inertias and compliant drive system between them.

Parameter	Value	Units
J_1	2.0	kg m^2
J_2	5.0	kg m^2
B	10.0	N s/m
K	20,000	N m/rad

4.17 Refer to Problem 2.10 and Fig. P2.10.

(a) Derive a state-space model of the system shown in Fig. P2.10.

(b) For the parameter values shown in Table P4.16, what are the values of the system poles?

Parameter	Value	Units
m_1	1.0	kg
m_2	1.5	kg
k_1	2500	N/m
k_2	3500	N/m
b_1	50	N s/m
b_2	25	N s/m
a	0.1	m
b	0.2	m

(c) Using MATLAB, graphically represent how the poles of the system change as the ratio of a/b changes through a range of values from 0.25 to 4.0

4.18 For the four systems shown in Fig. P4.18, find the frequencies at which the indicated (with a ?) variables oscillate when the system is disturbed.

4.19 A simple mechanical system, shown in Fig. P4.19(a), is used to measure torque. The equivalent inertia of the rotation parts of the system is 0.3 N m s^2/rad. The spring and viscous friction are both assumed to be linear, although the exact values of K and B are unknown. An unknown constant torque T_1 is suddenly applied at $t = 0$ after the system has been at rest. The resulting angular displacement θ is measured and recorded as shown in Fig. P4.19(b).

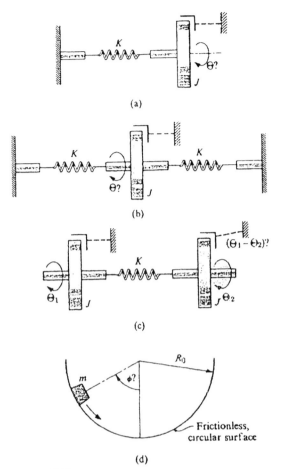

(a)

(b)

(c)

(d)

Figure P4.18.

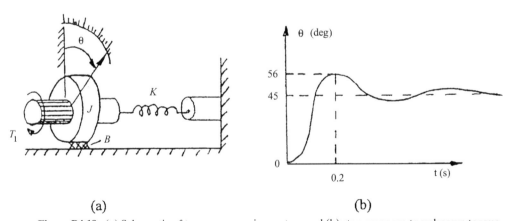

(a)   (b)

Figure P4.19. (a) Schematic of torque measuring system and (b) step response to unknown torque.

(a) Find the values of K and B.

(b) Using the values of K and B from part (a), determine the input torque T_1 that produced the system response shown.

(c) How would you change the system parameters to reduce the maximum overshoot in the system response without changing the steady-state relationship between torque T_1 and displacement θ?

4.20 The roots of a system characteristic equation were found to be

$$
\begin{aligned}
p_1 &= -6.2 + j12.5, \\
p_2 &= -6.2 - j12.5, \\
p_3 &= -75.1.
\end{aligned}
$$

(a) How many independent energy-storing elements are in the system?

(b) What are the approximate values of the maximum overshoot and period of oscillation for this system?

(c) Derive the characteristic equation for this system.

5

Numerical Solutions of Ordinary Differential Equations

LEARNING OBJECTIVES FOR THIS CHAPTER

5-1 To explain the fundamental principle of numerical integration as a finite sum of approximate areas.

5-2 To implement simple numerical integration methods by use of MATLAB or a similar computing platform.

5-3 To articulate the advantages of higher-order approximation methods and adaptive-step-size algorithms as more accurate and more efficient methods for integration.

5-4 To recognize numerically stiff systems and use methods that alleviate the difficulties that they present in computer solutions.

5.1 INTRODUCTION

For centuries, engineers and scientists have sought help from calculating machines of all kinds in solving mathematical equations that model dynamic systems. A digital computer, the most recent version of the calculating machine, has come a long way from the more than 5000-year-old Babylonian abacus. Most engineers would probably prefer a computer to an abacus because of its superior computational power, but both devices are capable of performing only those tasks that engineers already know how to perform but either choose not to do, for some reason, or cannot do because of lack of sufficient speed or memory or both. Despite today's fascination with computers, it is important to remember that they can do only what they are programmed to do within their vast yet finite performance limits.

Generally speaking, correctness of computer results depends on two conditions: the correctness of the formulation of the problem and the computational capability of the computer to solve it. It does not seem widely recognized how often at least one of these conditions is not met, leading to worthless computer results. Great care must therefore always be taken in considering computer output. These authors' advice is never to accept computer results unless they can be fully understood and verified against the basic laws of physics.

In previous chapters, it was shown how mathematical models of systems are expressed in terms of ODEs. In particular, state-space representations take the form of sets of first-order ODEs. It should come as no surprise that numerical solutions of these equations are implemented by numerical integration of the first derivatives of the state variables. Therefore numerical solutions of the responses of dynamic systems are numerical integrations. In this chapter, the fundamental principles of numerical integration are introduced.

In the next two sections, selected computer methods for solving ODEs are presented. In Section 5.2, the classical Euler's method is described. The method of Euler has more historical than practical significance today because of its large computational error. More accurate methods, including an improved Euler method and the fourth-order Runge–Kutta method, are introduced in Section 5.3. In Section 5.4, the issue of integration step size and its implication are discussed.

Section 5.5 extends the discussion of numerical methods to systems of differential equations and discusses certain problems that may arise. In Section 5.6, the problems associated with stiff systems, differential equations dealing with widely varying time scales, are presented.

Numerical integration is the core technology of computer simulations. Chapter 6 describes the common features of computer simulation packages and introduces Simulink, an extension of MATLAB that allows engineers to quickly produce complex computer models with a graphical interface.

5.2 EULER'S METHOD

In 1768, Leonhard Euler, the most prolific mathematician of the 18th century, and perhaps of all time, published the first numerical method for solving first-order differential equations of the general form

$$\frac{dx}{dt} = f(x, t), \tag{5.1}$$

with an initial condition $x(t_0) = x_0$. Note that system state-variable equations [Eq. (3.6)] as well as first-order input–output equations [Eq. (4.14)] can be presented in the form of Eq. (5.1). In the case of the input–output model, the dependent variable is y and Eq. (5.1) becomes

$$\frac{dy}{dt} = f(y, t) = -\left(\frac{a_0}{a_1}\right) y + \left(\frac{1}{a_1}\right) f(t). \tag{5.2}$$

The method proposed by Euler was based on a finite-difference approximation of a continuous first derivative dx/dt that uses a formula derived by another great mathematician, contemporary with Euler, Brook Taylor. Taylor's approximation of a first-order continuous derivative defined as

$$\frac{dx}{dt} = \lim_{\Delta t \to 0} \frac{[x(t_0 + \Delta t) - x(t_0)]}{\Delta t} \tag{5.3}$$

is

$$\frac{dx}{dt} \approx \frac{[x(t_0 + \Delta t) - x(t_0)]}{\Delta t}. \tag{5.4}$$

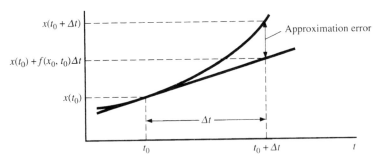

Figure 5.1. Geometric interpretation of Euler's method.

Hence the estimate of the value of function x at time $t_0 + \Delta t$ is

$$x(t_0 + \Delta t) \approx x(t_0) + \left.\frac{dx}{dt}\right|_{t_0} \Delta t. \tag{5.5}$$

Substituting for dx/dt from Eq. (5.1) gives the solution approximation

$$x(t_0 + \Delta t) \approx x(t_0) + f(x_0, t_0)\Delta t. \tag{5.6}$$

Figure 5.1 shows a geometric interpretation of Euler's method.

Figure 5.1 offers insight into the nature of the approximation that is made during numerical simulation. The function being approximated, $x(t)$, is extrapolated from time $t = t_0$ to $t = t_0 + \Delta t$ with the assumption that the function is a straight line during that time interval. Recall that the very form of Eq. (5.1) indicates that the function $f(x, t)$ is the slope of $x(t)$. Hence Euler's method for numerical integration assumes that the slope of $x(t)$ will remain constant at its value at $t = t_0$ for the entire integration interval. This is equivalent to using the first term of the Taylor's series approximation of the function and is therefore classified as a first-order method.

Approximation (5.6) provides a recursive algorithm for computation of $x(t_0 + k\Delta t)$, $k = 1, 2, \ldots, N$. The Euler's solution procedure is marching from the initial time t_0 to the final time $t_f = t_0 + N\Delta t$ with a constant time step Δt. The following equations are solved successively in the computational process:

$$
\begin{aligned}
x(t_0 + \Delta t) &= x_0 + f(x, t_0)\Delta t, \\
x(t_0 + 2\Delta t) &= x(t_0 + \Delta t) + f\left[x, (t_0 + \Delta t)\right]\Delta t \\
&\;\;\vdots \\
x(t_0 + N\Delta t) &= x\left[t_0 + (N - 1)\Delta t\right] + f\left[x, \left[t_0 + (N - 1)\Delta t\right]\right]\Delta t.
\end{aligned} \tag{5.7}
$$

In general, in the case of an nth-order system represented by a set of n state-variable equations [Eqs. (3.6)], a corresponding set of difference equations, one for each state variable, similar to approximation (5.6), has to be provided. The state-variable equations can be rearranged into the form of Eq. (5.1), as follows:

$$\frac{dq_i}{dt} = f_i(\mathbf{q}, t), \quad i = 1, 2, \ldots, n, \tag{5.8}$$

where the state vector $\mathbf{q}$ is a column vector containing $q_1, q_2, \ldots, q_n$ and $f_i(\mathbf{q}, t)$ includes the inputs $u_j(t)$, ($j = 1, 2, \ldots, l$), as well as the state and input matrix coefficients. At each step of the numerical solution process, a set of equations of the following form will be solved:

$$q_i(t_0 + \Delta t) = q_i(t_0) + f_i(\mathbf{q}, t_0)\Delta t, \quad i = 1, 2, \ldots, n. \quad (5.9)$$

Euler's method is illustrated by Example 5.1.

EXAMPLE 5.1

Use Euler's method to obtain a numerical solution of the differential equation

$$4\frac{dx}{dt} + x = 4$$

over a period of time from 0 to 12 s. The initial condition is $x(0) = 10$, and the system time constant is 4 s.

SOLUTION

First rewrite the differential equation in the same form as that of Eq. (5.1):

$$\frac{dx}{dt} = -0.25x + 1.$$

By use of Euler's formula (5.6), the successive values of x are calculated as follows:

$$x(1) = x(0) + [-0.25x(0) + 1]\,\Delta t,$$
$$x(2) = x(1) + [-0.25x(1) + 1]\,\Delta t$$
$$\vdots$$

The first step in implementing the numerical solution is to choose a value of the integration interval Δt. Although rules of thumb and implications associated with this choice are discussed in greater detail in Section 5.4, the reader can gain some appreciation for the problem by referring to Fig. 5.1. The Euler method approximates the function as a series of straight lines. Because it is well known that the solution to this equation is an exponential function (see Section 4.3), it would seem reasonable to choose a time step that is somehow related to the exponential time constant τ. For this system, the time constant is 4 s, and to illustrate this relationship, the Euler method is implemented for Δt's of 0.5, 1.0, 2.0, and 4.0 s.

A simple script was written for MATLAB to solve this equation and is shown in Table 5.1. Note that MATLAB script is fairly easy to understand and it would be rather straightforward to go from this simple script to another programming environment such as BASIC or C .

As already stated, this program was run for various values of integration time step (dt in the script file) and the results were compared with the exact solution, which can be found to be

$$x(t) = 6e^{-t/4} + 4.$$

Table 5.2 summarizes the results for various integration intervals.

Note that, in all cases, some error is apparent, even at the very first time step. The program listed in Table 5.1 (and all scripts listed in this text) is available for downloading at the authors' web site (see preface for details). The script can easily be modified to solve other first-order differential equations by modifying the line

```
k1 = -1/4 * x (i - 1) + 1;
```

Table 5.1. MATLAB script for the solution of Example 5.1 use of the Euler method of integration

```
% File EULMETH.M
%
%   MATLAB script to integrate a first order
%   ODE using the Euler method.
%
dt = 0.5;                        % time step
t(1) = 0.0;                      % initial time
tf = 12.0;                       % final time
%
x(1) = 10.0;                     % Set initial condition
%
for i = 2: tf/dt + 1
    k1 = - 1/4*x(i - 1)+1;
    x(i) = x(i - 1) + k1*dt;
    t(i) = t(i - 1)+ dt;         % increment time end
end
```

This line of the script computes the value of the derivative (k1 for this time step). For a new differential equation (linear or nonlinear), one simply changes this line to compute the appropriate value of the derivative for the value of x(i−1). This points to the real power of numerical integration for engineering analysis. Once the integration routines are written and debugged, new systems can be analyzed with a minimum of new programming.

5.3   MORE ACCURATE METHODS

The Euler method presented in the previous section is very simple and easy to use, but it is seldom used in serious computation because of its poor accuracy, as illustrated in Table 5.2. In general, the accuracy of a numerical integration method can be improved in two ways: by use of a more sophisticated algorithm for numerical approximation

Table 5.2. Comparison of Euler method solution for various step sizes

Time (s)	$\Delta t = 0.5$ s	$\Delta t = 1.0$ s	$\Delta t = 2.0$ s	$\Delta t = 4.0$ s	Exact
0.0	10.0000	10.0000	10.0000	10.0000	10.0000
1.0	8.5938	8.5000			8.6728
2.0	7.5171	7.3750	7.0000		7.6392
3.0	6.6928	6.5312			6.8342
4.0	6.0617	5.8984	5.5000	4.0000	6.2073
5.0	5.5785	5.4238			5.7190
6.0	5.2085	5.0679	4.7500		5.3388
7.0	4.9253	4.8009			5.0426
8.0	4.7084	4.6007	4.3750	4.0000	4.8120
9.0	4.5424	4.4505			4.6324
10.0	4.4153	4.3379	4.1875		4.4925
11.0	4.3179	4.2534			4.3836
12.0	4.2434	4.1901	4.0938	4.0000	4.2987

of the derivatives and by reduction of the integration interval. More sophistication usually implies the use of more terms of the Taylor series approximation of the function; hence these are called "higher-order" methods. Reducing the integration step size will, up to a limit, improve the estimation. However, as will be made clear in this section, higher-order methods have a greater impact than simply decreasing the step size. In this section, higher-order methods are introduced. Section 5.4 discusses the issue of step size in more detail.

It might be worthwhile to point out the explicit constraints of numerical integration. The equations that are being solved (first-order differential equations) express a functional relationship between the derivative of some unknown function and the function itself. This implies that, if the value of the function is known at some instant in time t_0, the value of its derivative at that instant can be computed from the differential equation. That relationship is exploited by use of the Euler method illustrated in the previous section. To improve accuracy and efficiency, higher-order methods are used that require estimates of derivatives at time(s) greater than time t_0. Unfortunately, the function, and hence its derivative, is unknown for times greater than t_0 (this is the very problem being solved). It is this apparent piece of circular logic that is the essence of the numerical integration problem: How does one estimate values of derivatives that are dependent on those very functional values that are being computed?

5.3.1 Improved Euler Method

Different variations of this method can be found in the literature (sometimes called the midpoint method, Heun's method, or second-order Runge–Kutta), and they share the common theme of using two evaluations of the derivative function to find a better approximation of the average slope through the interval. This is achieved in two steps. First, Euler's method is used to approximate the integrand at the end of the time step:

$$\hat{x}(t_0 + \Delta t) = x(t_0) + k_1 \Delta t, \tag{5.10}$$

$$k_1 = f(x_0, t_0). \tag{5.11}$$

Then the value of $\hat{x}$ is used to approximate the slope of the function at the end of the interval:

$$k_2 = f[\hat{x}(t_0 + \Delta t), t_0 + \Delta t]. \tag{5.12}$$

The approximation then uses the average of the two slopes to compute the integrand:

$$x(t_0 + \Delta t) \cong x(t_0) + \frac{k_1 + k_2}{2} \Delta t. \tag{5.13}$$

This method is only marginally more complicated than Euler's method (requiring an additional evaluation of the function for each step), but the improved accuracy more than justifies this additional effort. The method is easily implemented in a MATLAB script (left as an exercise for the student in Problem 5.3). Table 5.3 lists the numerical

Table 5.3. Comparison of the improved Euler method for the solution of a first-order differential equation by use of various time steps

Time (s)	$\Delta t = 0.5$ s	$\Delta t = 1.0$ s	$\Delta t = 2.0$ s	$\Delta t = 4.0$ s	Exact
0	10.0000	10.0000	10.0000	10.0000	10.0000
1.0	8.6761	8.6875			8.6728
2.0	7.6444	7.6621	7.7500		7.6392
3.0	6.8403	6.8610			6.8342
4.0	6.2136	6.2352	6.3438	7.0000	6.2073
5.0	5.7252	5.7462			5.7190
6.0	5.3445	5.3642	5.4648		5.3388
7.0	5.0479	5.0658			5.0426
8.0	4.8167	4.8327	4.9155	5.5000	4.8120
9.0.	4.6365	4.6505			4.6324
10.0	4.4960	4.5082	4.5722		4.4925
11.0	4.3866	4.3970			4.3836
12.0	4.3013	4.3102	4.3576	4.7500	4.2987

solution of the differential equation from Example 5.1 in which the improved Euler method is used for the same time steps for Table 5.2.[1]

Tables 5.2 and 5.3 invite comparison between the two methods. In particular, look at the results for the time step of 2.0 s. After 12 s, Euler's method has an accumulated error of approximately 0.2049 when compared with the exact solution. The improved Euler method has an error of 0.0589. Although the improved Euler method required twice as many function evaluations, the error at $t = 12$ s was reduced by a factor of 3.5. One can make a similar comparison by noting that Euler's method for a time step of 1.0 s (which requires the same number of function evaluations as the improved method for a time step of 2.0 s) leads to an error of 0.1086, which is still almost two times greater than the error introduced by the improved Euler method for the same number of function evaluations. This comparison leads to the conclusion that the higher-order methods are more effective in increasing integration accuracy than simply making the time step smaller. The next subsection introduces a method that is widely considered to be the best trade-off between higher-order approximations and ease of implementation.

5.3.2 Runge–Kutta Method

In the classical Euler method, a continuous derivative is approximated by use of the first-order term from the Taylor series. The improved Euler method uses the first and second terms from the Taylor series. The accuracy of the approximation of the first derivative, and eventually of the estimate of $x(t_0 + \Delta t)$, can be improved by the inclusion of higher-order terms from the Taylor series, and this is essentially the idea behind most of the modern numerical integration techniques. The trick in the Runge–Kutta method is that the higher derivatives are approximated by finite-difference expressions and thus do not have to be calculated from the original differential

---

[1] W. S. Dorn and D. D. McCracken, *Numerical Methods with FORTRAN IV Case Studies* (Wiley, New York, 1972), pp. 368–72.

equation. The approximating expressions are calculated by use of data obtained from tentative steps taken from t_0 toward $t_0 + \Delta t$. The number of steps used to estimate $x(t_0 + \Delta t)$ determines the order of the Runge–Kutta method. In the most common version of the method, four tentative steps are made within each time step, and the successive value of the dependent variable is calculated as

$$x(t_0 + \Delta t) = x(t_0) + \left(\frac{\Delta t}{6}\right)(k_1 + 2k_2 + 2k_3 + k_4), \qquad (5.14)$$

where

$$k_1 = f\left[x(t_0), t_0\right], \qquad (5.15)$$

$$k_2 = f\left[\left(x(t_0) + \Delta t \frac{k_1}{2}\right), \left(t_0 + \frac{\Delta t}{2}\right)\right], \qquad (5.16)$$

$$k_3 = f\left[\left(x(t_0) + \Delta t \frac{k_2}{2}\right), \left(t_0 + \frac{\Delta t}{2}\right)\right], \qquad (5.17)$$

$$k_4 = f\left[\left(x(t_0) + \Delta t\, k_3\right), (t_0 + \Delta t)\right]. \qquad (5.18)$$

For higher-order systems, a set of n equations, one for each state variable, of the form of Eq. (5.14) is solved at every time step of the numerical solution.

Table 5.4 presents a MATLAB script that implements the fourth-order Runge–Kutta method for the solution of the differential equation of Example 5.1. Note that this script is substantially the same as the one shown in Table 5.1, with additional

Table 5.4. MATLAB script for implementation of the fourth-order Runge–Kutta numerical integration scheme

```
%
% File RKMETH.M
%
% MATLAB script to integrate a first order
% ODE using the Runge–Kutta method.
%
dt = 1.0;                                    % time step
t(1) = 0.0;                                  % initial time
tf = 12.0;                                   % final time
%
x(1) = 10.0;                                 % Set initial condition
%
for i = 2:tf/dt + 1
   k1 = -1/4 * x(i - 1) + 1;
   xhat1 = x(i - 1) + k1*dt/2;
   k2 = -1/4 * xhat1 + 1;
   xhat2 = x(i-1) + k2*dt/2;
   k3 = -1/4 * xhat2 + 1;
   xhat3 = x(i-1) + k3*dt;
   k4 = -1/4 * xhat3 + 1;
   x(i) = x(i-1) + dt/6* (k1+2*k2+2*k3+k4);
   t(i) = t(i-1) + dt;                       % increment time
end
plot(t,x)                                    % plot response
```

Table 5.5. Comparison of the fourth-order Runge–Kutta method for the solution of a first-order differential equation by use of various time steps

Time (s)	$\Delta t = 0.5$ s	$\Delta t = 1.0$ s	$\Delta t = 2.0$ s	$\Delta t = 4.0$ s	Exact
0	10.0000	10.0000	10.0000	10.0000	10.0000
1.0000	8.6728	8.6729			8.6728
2.0000	7.6392	7.6393	7.6406		7.6392
3.0000	6.8342	6.8343			6.8342
4.0000	6.2073	6.2074	6.2090	6.2500	6.2073
5.0000	5.7190	5.7191			5.7190
6.0000	5.3388	5.3389	5.3404		5.3388
7.0000	5.0426	5.0427			5.0426
8.0000	4.8120	4.8121	4.8133	4.8438	4.8120
9.0000	4.6324	4.6325			4.6324
10.0000	4.4925	4.4926	4.4935		4.4925
11.0000	4.3836	4.3836			4.3836
12.0000	4.2987	4.2988	4.2994	4.3164	4.2987

lines added to compute the new terms. Table 5.5 shows the results of the computation for the same set of time steps used in the previous tables.

The results obtained with the Runge–Kutta method show a very dramatic increase in accuracy for the number of additional function evaluations. The functions that describe the system derivatives must be evaluated four times for each time step. Yet the accuracy achieved for a time step of 2.0 s is almost 140 times better than the improved Euler method after 12 s and about 300 times better when compared with Euler's method. Clearly, the effort invested in deriving and implementing a more sophisticated integration scheme pays off impressively.

These concepts are also illustrated in Fig. 5.2, in which the root-mean-square (rms) errors between the analytical and numerical solutions are plotted for the three

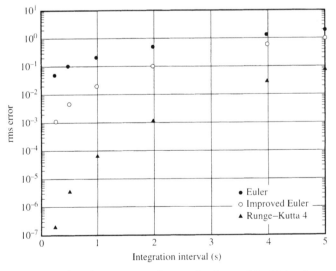

Figure 5.2. Plot of rms error vs. integration interval for Euler, improved Euler, and Runge–Kutta integration methods, using the problem stated in Example 5.1.

methods as a function of integration interval. (The rms error is the square root of the average of the error squared for each step in the integration.) Note that the error is plotted on a logarithmic scale so that the relative sizes of the errors are more clearly seen.

These comparisons are by no means conclusive, but they are indicative of the relative merits of the various methods. It is important to note that the example used in this case is a linear first-order system. The true strength of numerical integration lies in its ability to solve highly complex and nonlinear differential equations. The next section deals with issues of integration step size, particularly in regard to more complicated equations.

5.4 INTEGRATION STEP SIZE

Comparison of the various integration methods discussed in the previous sections shows a steady improvement in the accuracy of the numerical estimate as smaller and smaller step sizes are used. The essential question of just which step size to use is still open, however. Although many approaches to this problem exist, a reasonable approach is the following two-step process.

First, estimate an appropriate step size based on available knowledge of the system dynamics. In nearly all cases, the engineer has some *a priori* information regarding the system. Either a linearized model has already been used, or some actual experimental work on the physical system has been carried out. In either case, some knowledge about the time constants is available. The basic rule of thumb is to choose an integration step size that is one-fourth to one-eighth of the smallest time constant in the system. In Example 5.1, the time constant was 4 s. Note that integration step sizes of 1 and 0.5 s yielded quite good results, especially when the Runge–Kutta method was used.

Second, test the candidate time step by trying a smaller one. Once an approximate step size is chosen, carry out the integration for a representative period of time and typical input. Then cut the integration time in half and carry out the integration again. If the solution is not significantly improved, the original step size is appropriate. On the other hand, if the solution is vastly different, the smaller step size must be considered as the candidate step size and this step must be repeated.

Although this is a time-proven approach to computer simulation, it has certain drawbacks. First, it requires at least two "dry runs" of the simulation. For large-scale models that may be run on supercomputers, this can be very expensive. Most applications, however, are more modest in size and can be quickly and cheaply run on desktop workstations. Second, this approach is satisfactory only in assessing the appropriate time step for the conditions under which it has been tested. In other words, if the input is changed or the system is run through different operating conditions, the time step may be inappropriate. Even for linear systems, if the input function changes in a stepwise fashion, the time step must be very small to accurately predict the behavior of the system in that region in time. Similarly, for nonlinear systems, the response may be very different for different regions of the operating space.

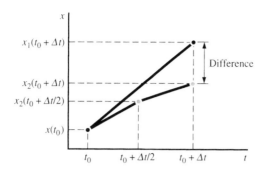

Figure 5.3. Illustration of the procedure used in adaptive-step-size integration algorithms.

There is a more sophisticated approach to numerical integration that addresses these concerns. The integration interval can be adjusted within the algorithm, with an estimate of the integration error used to guide the interval selection. The adaptive-step-size algorithms are very efficient and have become the standard in numerical integration of ODEs.

There are many variations on the theme of adaptive-step-size algorithms, and the interested reader and programmer is encouraged to read further in this area.[2] Although a detailed treatment of the algorithms is beyond the scope of this text, the concepts are rather straightforward and usually involve the following steps:

Step 1. Starting at some time t_0, integrate to $t_0 + \Delta t$, using a default time step, Δt (see Fig. 5.3).

Step 2. Return to the original time, and integrate to $t_0 + \Delta t$ by using two time steps of $\Delta t/2$.

Step 3. Compare the difference between the results of Steps 1 and 2 to a user-defined tolerance.

Step 3. If the tolerance is exceeded, decrease the step size and return to Step 1; if not, nominally increase the step size and move to the next time step.

Although the interval can theoretically become arbitrarily large or small, realistic implementations of the algorithm require that the user specify limits for the largest and smallest time steps, as well as the relative tolerance for the adaptation. Typically, the tolerance is specified as a relative number, to be computed relative to the values of the variables themselves. Therefore, if a tolerance of 0.001 is specified, the adaptation algorithm will switch to a smaller step size if the difference between the two values is greater than one part in one thousand.

The incremental amounts of increase and decrease of the time step are suggested by a theoretical treatment of the integration errors and are described in detail elsewhere.[3]

The following example illustrates the utility of variable integration-step-size approaches.

---

[2] W. H. Press, B. P. Flannery, S. A. Teukolsky, and W. T. Vetterling, *Numerical Recipes in C* (Cambridge University Press, New York, 1988), Chap. 15.
[3] *Ibid.*

EXAMPLE 5.2

Consider the following nonlinear differential equation:

$$4\frac{dx}{dt} + x + f_{NL}(x) = u(t),$$

where

$$f_{NL}(x) = \begin{cases} 0.4 & \text{for } x > 0 \\ -0.4 & \text{for } x \le 0 \end{cases},$$

and $u(t)$ is a time-varying function with the following characteristics:

$$u(t) = \begin{cases} -1 & \text{for} \quad 0 \le t < 15 \\ +1 & \text{for} \quad 15 \le t < 30. \\ -2 & \text{for} \quad 30 \le t < 50 \end{cases}$$

Note that the particular nonlinearity exhibited by f_{NL} can be used to represent the phenomenon known as Coulomb (dry) friction, which is very common in mechanical systems. Both the nonlinear friction load and the input function exhibit discontinuities. The input u is discontinuous at $t = 15$ and $t = 30$ s. The friction load, f_{NL}, is discontinuous when the variable x changes sign.

Figure 5.4 shows the solution of this differential equation when the fourth-order Runge-Kutta method and a constant integration interval of 2.0 s are used. In this plot, the result of each time step is shown as an individual point.

Note in particular the apparent erratic behavior of the response in the few steps following $t = 10$ s and at $t = 30$ s. Although it is difficult to predict the behavior of a nonlinear system, this type of output should be considered suspicious, and an experienced engineer will consider attempting a smaller step size when evaluating the solution.

Figure 5.5 shows the response of the system when a fourth-order Runge–Kutta method with an adaptive-step-size algorithm is used. In this case, the tolerance was set to 10^{-4}.

Note that the response is much smoother, indicating that the overshoot at 30 s predicted in Fig. 5.4 was erroneous. Note also that the adaptive-step-size algorithm used

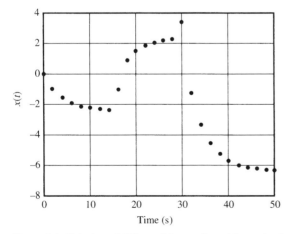

Figure 5.4. Solution of differential equation with varying inputs by use of constant-step-size Runge–Kutta algorithms with a time step of 2.0 s.

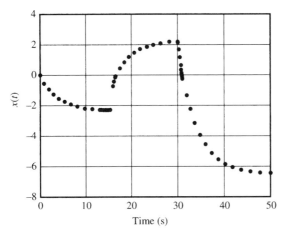

Figure 5.5. Solution of the differential equation considered in Example 5.2, by use of a variable-time-step approach; Runge–Kutta with $\Delta t_{\max} = 5$ s, $\Delta t_{\min} = 10^{-5}$ s, and tolerance $= 10^{-4}$.

much smaller steps as the solution approached the discontinuities in the input (15 and 30 s) and also as the variable x changed sign (at approximately 16 and 31 s). Figure 5.6 shows a plot of the actual step size used versus time for the solution. Note that the step size is plotted on a logarithmic scale so that the total spread of the step sizes can be appreciated.

Approximately 72 steps were taken with the variable-step-size algorithm, ranging from 0.007 to 2.0 s. Roughly half of the time steps were taken at steps greater than or equal to 1.0 s. Note that, to achieve the same relative accuracy using a fixed-time-step algorithm, the entire problem would have to be integrated with the smallest time step of 0.007 s, which corresponds to almost 7200 integration steps. Therefore the solution would have taken 100 times longer to compute in that manner. Clearly, the adaptive-step-size algorithm is an important modification to numerical solutions of problems of this class.

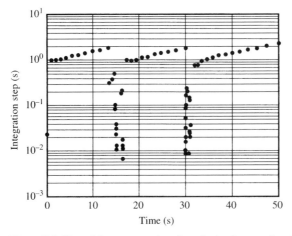

Figure 5.6. Plot of time step vs. time for solution by use of variable-step-size Runge–Kutta algorithm.

5.5 SYSTEMS OF DIFFERENTIAL EQUATIONS

The previous sections presented numerical methods for the solution of ODEs. Examples 5.1 and 5.2 showed the methods for single differential equations, that is, first-order systems. The methods described are easily expanded to a system of differential equations. This section looks at the implementation of numerical solutions for multiple differential equations and some of the problems that may arise for such systems.

First, consider some of the important implications of state-space theory outlined in Chap. 3. The essence of state-space theory is that any lumped-parameter system (variables are not dependent on spatial coordinates, only time) can be represented by a set of first-order differential equations:

$$\dot{q}_i = f_i(\mathbf{q}, \mathbf{u}, t) \ i = 1, 2, \dots, n. \tag{5.19}$$

Although previous discussions of these equations focused on the various forms they take, the implications of these equations for computer solution are now considered. The equations previously shown, together with the theory of state space, imply that if the values of the state variables are known at any given instant in time, the derivatives of the states can be computed, assuming that the functional relationships are known. In the field of computer programming, this means that one can write a subroutine or function that takes as an argument the values of the state variables at some time t_0, and the value of t_0 itself. It would return the values of all the state derivatives at that instant. Note that all of the techniques described in this chapter are structured in this manner. It should be clear therefore that the process of computer simulation reduces to writing the proper computer subroutine to compute the state derivatives given values of the state variables and the independent variable, time.

EXAMPLE 5.3

Recall the spring–mass–damper system of Example 4.2. Application of Newton's second law gave rise to a second-order differential equation relating applied force to mass displacement:

$$m\ddot{x} + b\dot{x} + kx = F(t).$$

This differential equation can be expressed as two first-order differential equations, taking velocity and displacement as the states q_1 and q_2, respectively:

$$\dot{q}_1 = \frac{1}{m}\left[F(t) - bq_1 - kq_2\right],$$
$$\dot{q}_2 = q_2.$$

The MATLAB script for implementing Euler's method in Table 5.1 can be easily modified to solve these equations. The resultant program is shown in Table 5.6. Verification of this program is left as an exercise.

5.6 STIFF SYSTEMS OF DIFFERENTIAL EQUATIONS

A sizable majority of simulations that are encountered by engineers are adequately solved by the well-tested variable-step-size algorithms that are currently available in

Table 5.6. MATLAB script file `MEULER.M` for implementation of the Euler method for the integration of sets of differential equations

```
%
% File: MEuler.m
%
% MATLAB script to numerically integrate
% multiple ODE's for a specified time step
% for model in Example 4.2
%
m = 9.0;
b = 4.0;                                    % define model parameters
k = 4.0;
%
f = 1.0;                                    % unit step input
%
dt = 0.2;                                   % Time step
t(1) = 0.0;                                 % Initial time
tf = 12.0;                                  % Final time
%
x1(1) = 0.0;
x2(1) = 0.0;                                % set initial conditions
%
for i = 2: tf/dt +1
   % Compute the derivatives
   xd1 = 1/m* (f - b*x1(i-1) - k*x2(i-1));
   xd2 = x1(i-1);
   % compute the states
   x1(i) = x1(i-1) + xd1*dt;
   x2(i) = x2(i-1) + xd2*dt;
   % increment time
   t(i) = t(i-1) + dt;
end;
plot (t, x2)
```

commercial packages. Some systems, however, exhibit behaviors that defy quick and accurate solution by these means, and specialized algorithms are required.

So-called "stiff systems" arise in many engineering applications. Stiff systems are characterized by two or more components whose typical time responses vary by several orders of magnitude. For example, if a system has a fast-acting actuator (first-order time constant of 0.2 ms) acting on a large mass that is constrained by a soft spring (natural frequency of 0.2 rad/s), such a system would be characterized as a stiff system. One method of characterizing stiff systems is by examination of the ratio of the slowest to the fastest system pole magnitude. In this case, that ratio would be $5/0.0002 = 25{,}000$. In general, ratios of 1000 or greater lead to systems that are considered stiff.

Even adaptive-step-size algorithms will be constantly adjusting the time step to accommodate the very fast-acting portion of the system, even though the slower portions could be solved with much larger time steps. The problem is more complicated than simply requiring more computer time than might be practical. The methods described in the previous sections sometimes fail altogether for stiff systems. Other approaches are called for.

A detailed presentation of these other approaches is outside the scope of this text, but a brief overview is in order. Approaches to solving stiff systems generally rely on a somewhat indirect technique to solve for the values of the state variables at any given time step. To see why this is an efficient approach, consider the following example.

A linear first-order system with no input (unforced) can be represented by the differential equation

$$\dot{q} = -\frac{1}{\tau}q. \tag{5.20}$$

Numerical solutions of this equation will estimate values of $q(t)$ at specific, discrete instants of time, $q(k\Delta t)$, where k is an integer and Δt is the integration interval. For the sake of clarity, the argument of q is given as $q(k)$, with the Δt implied. The Euler method computes the value of q at the $k + 1$ interval as a function of the value at the k interval:

$$q(k+1) = q(k) + \Delta t \dot{q}(k). \tag{5.21}$$

Substituting Eq. (5.20), for $t = k\Delta t$, into Eq. (5.21) and rearranging yields

$$q(k+1) = \left(1 - \frac{\Delta t}{\tau}\right) q(k). \tag{5.22}$$

Equation (5.22) is an important result. Equations of this form are common in mathematics and engineering, and general statements about their behavior are possible based on simple observations. In particular, it can be shown that the sequence of numbers $q(k)$ given by Eq. (5.22) will converge (be stable) as long as the coefficient $(1 - \Delta t/\tau)$ has an absolute value of less than 1. From this, it can be concluded that the Euler method will not converge and thus will give erroneous results if the time step Δt becomes greater than 2τ. Note that this result applies only to the stability of the method, which assesses whether or not the solution grows without bound. Accuracy demands that Δt be much smaller, of the order of $\tau/4$ to $\tau/8$.

Consider now a different approach to applying Euler's method to this simple problem. Equation (5.23) shows an expression that results if the integration is performed with a value of the derivative evaluated at the end of the integration interval, the $k + 1$ time step:

$$q(k+1) = q(k) + \Delta t \left[\dot{q}(k+1)\right]. \tag{5.23}$$

Substituting Eq. (5.20), for $t = (k + 1) \Delta t$, into Eq. (5.23) yields the difference equation

$$q(k+1) = \frac{1}{1 + \Delta t/\tau} q(k). \tag{5.24}$$

This equation is of the same form as that of Eq. (5.15), but, because of the different arrangement of the terms, the coefficient is always less than 1, regardless of the choice of Δt. This method is then inherently stable and will always yield a convergent sequence of numbers. When Eq. (5.20) is substituted into Eq. (5.24), the result is an implicit equation for $q(k +1)$. Hence methods that take advantage of this property are termed "implicit integration methods," and it can be seen that implicit methods are much more robust to numerical stability issues than are explicit methods such as Euler and Runge–Kutta.

A large number of these implicit methods are available and are especially well suited for stiff systems of differential equations. In general, these techniques require more computation time or more information about the system, or both. However, for systems that fit into this category, they are the only choice for numerical solutions. Of these methods, some of the better known are Gear, Adams, and Rosenbrock methods, named after the originators of the techniques.[4]

A brief discussion about the philosophy of modeling is perhaps in order at this point. The astute student may recognize the fact that the contribution of the very fast system components to the overall response of the system will be negligible, because a fast actuator will appear to be instantaneous relative to the motion of a large mass. An appropriate approach would be to ignore the dynamics of the actuator and model it as an algebraic equation. This approach is appropriate only for linear models. Systems with significant nonlinearities will require that we consider all dynamics.

Typical applications that give rise to stiff systems are kinetics encountered in the chemical processing field and in detailed dynamics of rotating equipment in which the engineer is interested in the dynamics of a high-speed shaft throughout its rotation.

Solving the differential equations that describe stiff systems by use of the methods discussed in this chapter will require far smaller steps, more computer time, and ultimately lead to less accurate results than are possible with algorithms specifically designed for stiff systems. The details of these algorithms, first proposed by Gear[5] and later refined by many others, are outside the scope of this text. MATLAB and Simulink provide a group of algorithms specifically designed for stiff systems and, in the following example, the advantage of these algorithms is illustrated.

EXAMPLE 5.4

The subsequent equations describe a second-order nonlinear system that can be characterized as stiff:

$$\dot{q}_1 = u(t) - q_1, \tag{5.25}$$

$$\dot{q}_2 = 10^4 \left[(q_1 + 1)^3 - q_2 \right]. \tag{5.26}$$

The application requires that the solution be found for 1.5 s, zero initial conditions, and an input function $u(t)$ that is a square wave with a unit amplitude at 1 Hz.

Table 5.7 summarizes the nature of the solution computed by two different algorithms available in MATLAB: RK45, a variable-step-size Runge–Kutta approach, and ODE15S,

---

[4] *Ibid.*

[5] C. W. Gear, "The automatic integration of ordinary differential equations," *Commun. ACM* **14**, 176–9 (1971).

Table 5.7. Details of the numerical solution

Algorithm	Number of steps	Maximum step size	Minimum step size
RK45	4523	0.001	0.00001
ODE15S	155	0.08	0.000001

a low-order algorithm specifically designed for stiff systems, which also uses variable step sizes.

When the RK45 routine is used, the solution requires 4523 time steps, ranging in size from 10^{-3} to 10^{-5} s. By comparison, ODE15S solves the same problem in only 155 time steps, using steps ranging from 0.08 to 10^{-6} s. Because the majority of the computational load of solving differential equations is the evaluation of the system equations, a rough comparison of the computational load (and computer execution time) between the two methods can be made by examination of the ratio of numbers of time steps taken. By that comparison, the stiff-system method is approximately 30 times faster than the traditional variable-step-size approach.

The execution time, however, does not tell the complete story. Figure 5.7 shows a comparison of q_2 computed by the two solutions over a brief period of the response. Note that the two responses are not the same. The traditional approach (RK45) exhibits high-frequency small-amplitude oscillations throughout the response whereas the algorithm for stiff systems does not exhibit these oscillations. Further investigation into this discrepancy would show that the oscillations are artifacts of numerical instabilities in the nonstiff algorithms and that the approach for stiff systems is not only more efficient, but more accurate as well.

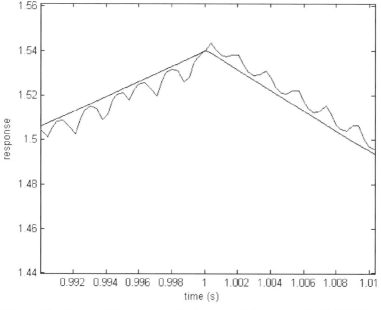

Figure 5.7. Portion of the response for Example 5.4 computed with RK45 (oscillating response) and ODE15S (smooth response).

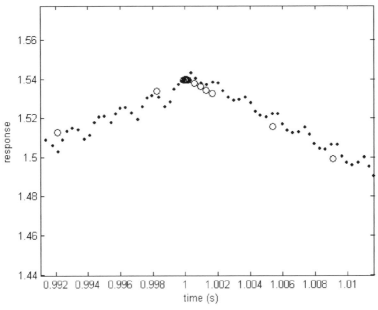

Figure 5.8. Portion of the computed response in Example 5.4 showing the individual time steps for RK45 (points) and ODE15S (circles).

Finally, Fig. 5.8 shows the same response, but in a manner that indicates the individual integration time steps. Note that the adaptive-step-size feature of RK45 appears to be ineffective because the steps are uniformly small. On the other hand, the stiff-system solver is using very large steps when the solution is changing slowly, smaller steps when it is changing rapidly, as one would expect.

5.7 SYNOPSIS

In this chapter, the core technology of computer simulations – numerical integration – was discussed. The fact that any lumped-parameter system can be represented by a set of first-order differential equations (Chap. 3) allows for the construction of truly general numerical algorithms for the solution of these equations. The principles of numerical integration were illustrated with the Euler method. One interpretation of the Euler method is that it approximates the derivative of the function through a fixed integration interval by means of the first term of a Taylor series expansion of the derivative. The Euler method is therefore a first-order approximation method. Higher-order methods were introduced, including a modified Euler (second-order) method and the well-known fourth-order Runge–Kutta method. The examples in this chapter showed how higher-order approximations can dramatically improve the accuracy and efficiency of a computation.

Once the numerical method is chosen, the size of the integration step must also be chosen. Rules of thumb that relate the integration step size to known system parameters such as time constants were discussed. The concept of a variable- or adaptive-step-size algorithm was introduced, and its utility was demonstrated with a simple example.

Most engineering systems encountered in practice are described by more than one differential equation and therefore require the solution of sets of differential equations. The extension of these methods to multiple differential equations is straightforward, but the possibility exists that the system might be numerically ill-conditioned. Systems with a large dynamic range of time constants (i.e., the ratio of the largest to the smallest time constant is 1000 or greater) are known as stiff systems and require the use of implicit integration methods for efficient solution.

Although information provided in this chapter could be used to write new computer subroutines to carry out these numerical methods, there are many commercially available software packages that implement these methods in a highly efficient and error-free manner. In the next chapter, the subject of computer simulations will be discussed, and one such package, MATLAB Simulink, is used as an example.

PROBLEMS

5.1 Starting with the MATLAB script listed in Table 5.1, obtain the numerical solution of Problem 4.1(a). Use the integration time steps $\Delta t = 0.1$ and 0.02 s.

5.2 Starting with the MATLAB script listed in Table 5.1, obtain the numerical solution of Problem 4.1(a) for each of the following values of integration time step: $\tau/10, \tau/5, \tau/2,$ $\tau, 2\tau,$ and 5τ, where τ is the time constant of the dynamic system considered in Problem 4.1. Compare the numerical solutions obtained for different sizes of integration time step with the exact analytical solution.

5.3 Starting with the MATLAB script listed in Table 5.1, modify the program to implement the improved Euler method and obtain a numerical solution to Problem 4.1(a).

5.4 Use the MATLAB script lisited in Table 5.4 to obtain the numerical solution of Problem 4.1(a).

5.5 Use the Runge–Kutta method to verify the analytical solution of Problem 4.3.

5.6 Compare simulations by using the improved Euler method and the Runge–Kutta method to obtain the numerical solution to Problem 4.5.

5.7 Starting with either the Euler method from Table 5.1 or the improved Euler method (solved in Problem 5.3), modify your program to solve a system of two differential equations. Test the program by computing the step response of the original system considered in Example 4.2. Compare performance specifications (percentage of overshoot t_p, DR) obtained in the analytical and numerical solutions.

5.8 Modify the computer program (or MATLAB script) from Problem 5.7 to solve a system of three equations and obtain a numerical solution of Problem 4.12.

5.9 A mechanical system has been modeled by the following nonlinear state-variable equations:

$$\dot{x} = v,$$
$$\dot{v} = -\frac{k}{m}x - \frac{1}{m}F_{\text{NLD}} + \frac{1}{m}F_a(t).$$

The nonlinear friction force F_{NLD} is approximated by the expression

$$F_{\text{NLD}} = f_{\text{NL}}(v) = v^2 + v + 1.$$

The input force $F_a(t)$ has a constant component $\overline{F}_a = 15\,\text{N}$, which has been acting on the system for a very long time, and an incremental component $\hat{F}_a(t)$ equal to $\Delta F_a U_s(t)$. The values of the system parameters are $m = 10\,\text{kg}$ and $k = 5\,\text{N/m}$.

(a) Determine the normal operating-point values for the state variables $\overline{x}$ and $\overline{v}$ corresponding to $\overline{F}_a$.

(b) Linearize the system model in the vicinity of the normal operating point. Find natural frequency ω_n, damping ratio ζ, and period of damped oscillations T_d for the linearized model.

(c) Use the fourth-order Runge–Kutta method shown in Table 5.4 to compute the nonlinear system response to $F_a(t)$ for the step change $\Delta F_a(t) = 1.5\,\text{N}$. A suggested integration time step for the computer program is $\Delta t \leq 0.05\,T_d$. Compare the specifications of the computer-generated step response with those obtained analytically for the linearized model in part (b).

(d) Repeat part (c) for a magnitude of the input step change of $\Delta F_a(t) = 15\,\text{N}$.

(e) Repeat part (c) for the linearized model.

(f) Repeat part (d) for the linearized model.

(g) Compare the agreement between the results obtained for nonlinear and linearized models for small- and large-magnitude inputs.

5.10 (a) Use the fourth-order Runge–Kutta method shown in Table 5.4 to simulate the response of the system having the following input–output equation to a unit step change in $u(t)$ at $t = 0$:

$$0.025\frac{d^3x}{dt^3} + 0.25\frac{d^2x}{dt^2} + 0.1\frac{dx}{dt} + 1 = 0.5u(t).$$

Assume that all initial conditions at $t = 0^-$ are zero.

(b) Modify the subroutine for using the improved Euler method and run the modified subroutine for comparison with part (a) by using the same time step as in part (a).

(c) Compare both simulations with the exact solution obtained by the classical method described in Chap. 4.

5.11 The following differential equations are known as the Lorenz equations, after Ed Lorenz, an atmospheric scientist who was the first person to recognize chaos in nonlinear systems:

$$\dot{x}_1 = 10(x_2 - x_1),$$
$$\dot{x}_2 = 28x_1 - x_2 - x_1x_3,$$
$$\dot{x}_3 = x_1x_2 - 2.67x_3.$$

Write a program to integrate these equations by using the Runge–Kutta method for an arbitrary set of initial conditions for 100 s. Use the plot3 command in MATLAB to examine the state trajectory of this system for various initial conditions.

5.12 Another well-known nonlinear equation is the Van der Poll equation, a second-order nonlinear equation that results in an oscillatory response for some values of the parameters:

$$\frac{d^2y}{dt^2} - \mu(1 - y^2)\frac{dy}{dt} + x = 0.$$

Write a script to solve this equation by using the Runge–Kutta method. Use $\mu = 3$ and experiment with a wide variety of initial conditions for both y and dy/dt.

6

Simulation of Dynamic Systems

LEARNING OBJECTIVES FOR THIS CHAPTER

6–1 To know the common elements of computer simulations.

6–2 To be able to build block diagrams of systems based on the model equations.

6–3 To become familiar with the basic block set of Simulink.

6–4 To understand different approaches to simulation utilizing different features of Simulink.

6–5 To be able to simulate both linear and nonlinear systems.

6–6 To be able to simulate configuration-dependent systems.

6–7 To be able to conduct parametric studies of systems through scripting multiple runs of the simulations.

6.1 INTRODUCTION

System simulation is one of the most widely used tools in modern society. From weather forecasting to economic analysis, from robotics to computer animation, simulation is becoming a commonplace tool for analysts and designers of all types, not just engineers. Yet, as widespread as these applications are, the fundamental basis of system simulations is common. A computer simulation is a numerical solution of a set of differential equations that are intended to model the way in which a particular system evolves in time.

In Chap. 4, techniques for the analytical solution of differential equations were presented. These methods represent a powerful set of tools for the study of linear dynamic systems. It was shown that by simply inspecting the coefficients of a first- or second-order characteristic equation, the entire characteristic of the step response could be predicted. At this point, it would be fair to ask the question: Why are computer-based methods for finding the responses of dynamic systems needed? There are several important answers to this question; three of the more compelling arguments are summarized below.

1. *Nonlinear models:* In the discussion of analytical solutions, only equations that were linear in their variables were solved. However, few real-world systems exhibit consistently linear behavior. In fact, portions of Chap. 2 are devoted to the techniques needed to linearize nonlinear differential equations. In reality, all real-world systems are nonlinear. Although linear analytical methods are powerful

and allow the engineer to approximate system response of many systems, details of system behavior must be revealed through study of the nonlinear model. Analytical solutions exist for a small number of special nonlinear differential equations, but most must be solved numerically through computer simulation.

2. *High-order linear models:* The methods for solution of system responses of linear systems that were described in Chap. 4 apply to linear models of arbitrary order. However, it is very difficult to manipulate such solutions for models of order greater than 3. In such cases, it is often more expedient to use computer simulation to solve for the response of a system.

3. *Arbitrary forcing functions:* Although analytical methods are shown to be very powerful for characterizing the response to initial conditions (homogeneous response) or the response to simple inputs (e.g., the unit step), response to arbitrary inputs is more problematic. The convolution integral provides the necessary capability to solve analytically for the response of a system to arbitrary inputs, but its use still requires that the input be analytically defined, which is not always possible.[1]

Although computer simulations can be used to model a large variety of systems, it can be seen that all computer simulations must embody the following components:

1. *The structure of the mathematical model:* This is the complete set of differential equations that describe the system behavior and reflect the fundamental physical laws governing the behavior of the system.

2. *Model parameter values:* Model parameters refer to numerical constants that usually do not change over the course of the simulation. Typical parameters for mechanical systems are mass, damping coefficient, and spring stiffness. Note that these values may sometimes change over time, but they usually change at a much slower rate than the dynamic variables that are being computed by the simulation.

3. *Initial conditions:* In previous chapters, the importance of initial conditions for the solution of differential equations was discussed. This topic is of equal importance for simulations.

4. *Inputs:* Typically, a system responds to one or more inputs. The simulation must embody the inputs as well.

5. *Outputs:* Although a simulation does not require that the user explicitly define outputs, it is assumed that the goal of a computer simulation of a dynamic system is the time history of specific physical variables in the system under study. The time history of output variables can be stored to the computer hard drive for later analysis or displayed as a graph on the screen.

6. *Simulation solution control parameters:* Simulation solution control parameters define the values and choices made by the engineer that dictate how the numerical methods behind the simulation operate. These include values that determine the step size, output interval, error tolerance, and choice of integration algorithm.

---

[1] D. Inman, *Mechanical Vibrations*, 2nd ed. (Prentice-Hall, Englewood Cliffs, NJ, 2001).

At the root of computer simulation is the process of integration. Chapter 3 described the theory of state space for dynamic systems. The most important implication of that theory is that any lumped-parameter system can be described as a set of first-order differential equations of this form:

$$\dot{q}_i = f_i(\mathbf{q}, \mathbf{u}, t) \quad i = 1, 2, \ldots, n. \tag{6.1}$$

An important implication of this form is that values of $q_i(t)$ can be found by integration of the function $f_i(t)$ over time. This is complicated by the fact that each function f_i is a function of the values of the state variables at the time for which the derivatives are being computed. Chapter 5 discusses integration algorithms that untangle this loop, allowing for numerical solutions of the state equations. The techniques outlined in Chap. 5, and more sophisticated methods based on those techniques, underlie all computer simulation packages.

6.2   SIMULATION BLOCK DIAGRAMS

Although the dynamic interactions between system elements are properly described by mathematical equations of the type developed previously (Newton's second law, state-variable, or input–output differential equations), the use of functional block diagrams is often very helpful in visualizing how these interactions occur. Each block of such a diagram represents a single mathematical operation used in the describing equations. These block diagrams were first used when analog computers were developed to simulate the performance of dynamic systems. Later they became very helpful in both visualizing and making a preliminary analyses of a system without using a computer. As computer-based tools were developed to analyze and simulate dynamic systems, many people used the block diagram as the basis for representing systems and based their model representation on block diagrams. Currently, all popular computer packages designed for system simulation use some form of block diagrams as the primary means of user input of system structure. Appendix 4 is a brief tutorial of MATLAB's version of this approach, Simulink. The simulation case study in Section 6.5 utilizes Simulink in carrying out the study.

In this section, each of the functional blocks is shown and defined for its corresponding mathematical function; then simple connections or combinations are used to represent typical describing equations. Finally, examples have been chosen to illustrate how these diagrams are used to represent complete mechanical systems, showing all the significant interactions involved in the response of the system model to input disturbances. Thus, in addition to specific output variables, all the other system variables are shown as well. Applications to other systems, such as electrical, thermal, fluid, and mixed systems, are left to illustrations, examples, and end-of-chapter problems in later chapters.

6.2.1 Coefficient Blocks

When a system variable is multiplied by a coefficient, this function is represented by a coefficient block, as shown in Fig. 6.1. This figure includes the mathematical

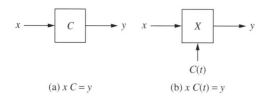

Figure 6.1. Coefficient block: (a) Constant coefficient and (b) time-varying coefficient, i.e., a multiplier.

operation represented by the block and its input and output variables depicted as signals.

The arrow-directed input and output signal lines are intended to represent only signal flow; they are not necessarily intended to represent only physical connections (although they sometimes do so). Note that the multiplier function ($\times$) is needed when the coefficient C is time varying, either as a known or as an unknown function of time. Similarly, a divider function ($\div$) is occasionally needed when x is to be divided by a time-varying coefficient C.

6.2.2 Summation Blocks

When a system variable is equal to the sum of two or more other system variables, the relationship is depicted by means of a small circle with the inputs and outputs arranged as shown in Fig. 6.2. This figure illustrates the three most commonly encountered forms of this block diagram.

6.2.3 Integration and Differentiation Blocks

When a system variable is the time integral of another system variable, the integration block is used to describe this functional relationship, as shown in Fig. 6.3(a). This figure also shows the initial value of the output at time $t = 0$ being summed with the integrator output, along with the corresponding mathematical equation. In reality, the initial conditions are not explicitly included in the block diagram but are implicit with the integrator block. The inverse operation, the function of differentiation, is represented by means of a differentiator block, as shown in Fig. 6.3(b). Because true differentiation is not physically achievable for stepwise changes, simulation diagrams usually do not use differentiator blocks. It seems ironic that derivatives with respect to time are commonly used in differential equations to describe dynamic systems,

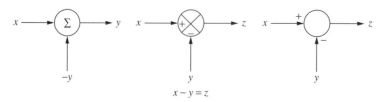

Figure 6.2. Three commonly used versions of the summation block diagram.

Example   145

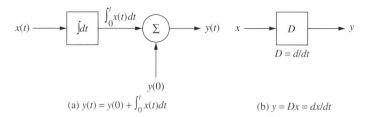

$$(a)\ y(t) = y(0) + \int_0^t x(t)dt \qquad\qquad\qquad (b)\ y = Dx = dx/dt$$

Figure 6.3. Block diagram symbols for integration and differentiation with respect to time.

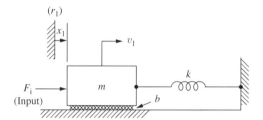

Figure 6.4. Simple mass–spring–damper system.

whereas in point of fact the dynamic behavior is more truly described physically in terms of time integrations.[2]

6.2.4 Drawing Complete Diagrams from Describing Equations

Usually one develops a typical simulation block diagram by starting with the input signal and using successive integrators combined with other functional blocks to arrive at the output. This approach will always work for describing physically realizable models of dynamic systems. If it is impossible to model a set of system equations without resorting to the use of one or more differentiators, it can be assumed that these equations represent a physically unrealizable system; moreover, the system input–output equation will have one or more right-hand-side (input) terms involving derivatives that are of higher order than the highest derivative term of the left-hand (output) side. The following example is used to illustrate the development of simulation block diagrams for physically realizable system models.

EXAMPLE 6.1

Develop the simulation block diagram for the mass–spring–damper system shown schematically in Fig. 6.4.

---

[2] It may be noted that achieving true time differentiation by physical means would be equivalent to predicting the future; this is implicitly involved in a system input–output differential equation in which the highest derivative of the input variable is larger than that of the output variable. Such an equation also represents a situation that is impossible to achieve physically when step changes occur to the input variable.

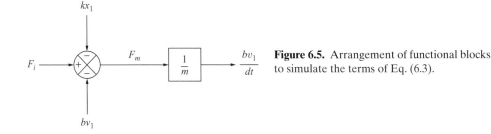

Figure 6.5. Arrangement of functional blocks to simulate the terms of Eq. (6.3).

Applying Newton's second law to mass m yields

$$F_i - F_k - F_b = m\frac{dv_1}{dt}. \tag{6.2}$$

Equation (6.2) may be combined with the elemental equations for the spring and damper forces and rearranged to provide the expression for the derivative of the velocity with respect to time:

$$\frac{dv_1}{dt} = \frac{1}{m}(F_i - kx_1 - bv_1). \tag{6.3}$$

In a sense, the block diagram to model this equation is built from "the inside out." The terms in the parentheses represent a summation of three different terms (calling for a summing block), the result of which is the equivalent force F_m experienced by the mass. This force is then multiplied by a constant gain value ($1/m$). The result of that multiplication will be a quantity equal to the time derivative of the velocity. Figure 6.5 shows a portion of a block diagram that represents Eq. (6.3).

The block diagram as shown in Fig. 6.5 is incomplete, and we need to develop both the velocity and displacement because they are required by the damping and spring force inputs to the summing junction. Clearly integration blocks are required for integrating the acceleration to yield velocity and then integrating the velocity to yield the displacement. To fully specify both integrations, the initial conditions for both variables (velocity and displacement) must be specified, although most block diagrams do not explicitly indicate this. Figure 6.6 shows the completed block diagram that schematically represents the typical spring–mass–damper problem.

The dynamic response of this system is readily traced from the input F_i through the summer, the coefficient $1/m$, the first integrator, and then the second integrator. Note the two feedback paths that go through the coefficient blocks b and k and close loops

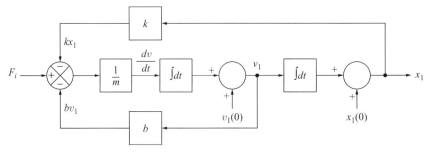

Figure 6.6. Simulation block diagram for mass–spring–damper system of Example 6.1.

around the integrator blocks. These loops are typical of physical systems; indeed they are essential for the results of the integrations to be bounded over time. The significance of such feedback connections in dynamic systems will be discussed at length in future chapters.

6.3 **BUILDING A SIMULATION**

6.3.1 Structure of the Simulation: The Block Diagram

As discussed in Section 5.1, the process of integration lies at the core of simulation, and hence the integrator blocks form the central basis of the simulation block diagram. In Section 6.2 it was pointed out that the building of the simulation block diagram begins by bringing in the correct number of integrator blocks (one for each state variable) and using the block library to "build" the correct state-variable functions. The actual mechanics of assembling the block diagram, and hence the simulation, can vary depending on the computer package chosen. Appendices 3 and 4 of this text contain brief tutorials for the MATLAB and Simulink simulation environment.

Regardless of which simulation package is used, the underlying structure of the simulation is the same. In general, each state variable requires an integrator block. The initial conditions for each state variable can be set within the block diagram or they can be tied to the program workspace, where they can be set externally, before each simulation run. The outputs of the integrators are the state variables themselves. Those outputs (which can be thought of as signals on electrical lines) are then used to "build" the derivatives (inputs to each integrator). For example, consider the spring–mass–damper system from Example 6.1. When the displacement and velocity of the mass are selected as the state variables, the state-variable equations for the system are

$$\dot{q}_1 = q_2,$$
$$\dot{q}_2 = \frac{1}{m}(F_i - b\,q_2 - k\,q_1). \tag{6.4}$$

The parameters b and k would be implemented as gains or constant blocks connected to the outputs of the integrators for the first and second states, respectively. The outputs of the constant blocks can then be summed together (with the appropriate signs) by use of a summation block. The results of the summation are then multiplied by $1/m$ by use of an additional gain block, which is then connected to the input of the first integrator.

Figure 6.7 shows the simulation block diagram from Simulink modeling the simple spring–mass–damper problem of Example 6.1. It is useful to compare the Simulink block diagram with the block diagram in Fig. 6.6, which was originally drawn for this system. The similarity between the two figures is a strong argument for the popularity and utility of these packages.

Note that the integrator blocks are single-input, single-output blocks and that the titles under the blocks can be edited by the user. These titles should be descriptive and pertinent to the application. Convention dictates that the integrator blocks be labeled

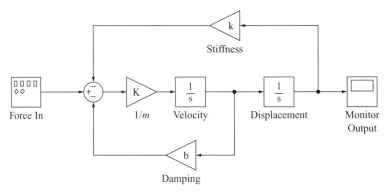

Figure 6.7. Simulink window showing the block diagram that represents the simulation structure for the spring–mass–damper system.

according to the output of the integration process. Note also that the integrator blocks are labeled "1/s." Students who have been exposed to Laplace transform methods will recognize 1/s as the reciprocal of the Laplace operator that represents integration in the time domain (see Appendix 2). Although it is acceptable and common to use such notation in the description of linear, time-invariant systems, it is *not* a general notation. Laplace transform methods are not applicable to nonlinear systems, and it is in the analysis of nonlinear systems that computer simulations have the most to offer. It would be preferable that integrator blocks be labeled with the integration sign followed by "*dt*," as shown in Section 6.2.

6.3.2 Model Parameters

Once the block diagram has been completed (and saved to disk), numerical values for each parameter must be set. In Simulink, one does this by double-clicking on each block and typing the value in the appropriate window. Alternatively, the user may enter the names of variables in these spaces. The variables can then be set within the MATLAB workspace. This technique results in considerably improved flexibility, particularly for parametric studies, as demonstrated in Subsection 6.5.4.

6.3.3 Initial Conditions

One can set, in a manner similar to setting parameter values, the initial conditions for the problem by double-clicking on the integrator blocks and entering numerical values. Alternatively, variable names can be entered in the integration blocks. The initial variables can then be set within the MATLAB environment.

6.3.4 Inputs

In the previous sections, the structure of the simulation was defined by the block diagram, model parameter values, and initial conditions. These three elements comprise the minimal amount of information required for implementing a simulation. How-ever, if external inputs are not defined, only the free or homogeneous component of the solution will be found through the simulation. As discussed in the introductory section of this chapter, one of the major reasons engineers turn to simulations is to find the response of a system to inputs and forcing functions of arbitrary shape.

Simulation packages offer a variety of means of generating complex and arbitrary inputs to excite the system. Simple inputs, such as step, sine, and square waves, are readily available as predefined blocks. Other inputs, such as a rectangular pulse, can be formed by use of two step-input blocks and a summing junction. The first step input is set to take place at the beginning of the desired pulse. The second step input is then set for the end of the pulse and summed to the first step with a minus sign. The net effect is a rectangular pulse with adjustable magnitude and width. Problem 6.1 at the end of this chapter examines this particular input more thoroughly.

Finally, completely arbitrary inputs can be formed from data stored in the pro-gram workspace or in a disk file. This gives the user great flexibility in that the inputs can be the results of other programs, data taken from the "real world," or artificially synthesized data.

6.3.5 Outputs

The major purpose for using a computer simulation is to study the response of a sys-tem to inputs and initial conditions. Implicit in this procedure is the identification of one or more system variables as output variables. Simulation packages allow various methods for the examination of output variables within the simulation. In particular, Simulink offers a wide variety of output blocks, including a "scope" block that acts like an oscilloscope directly measuring and displaying the designated variable. The time history of one or more variables can be stored to the workspace for later pro-cessing or directly to disk for processing by other applications. Appendix 4 presents some pertinent examples.

6.3.6 Simulation Solution Control

Up to this point, discussion has focused on those components of the simulation that represent characteristics of the system under study. Also important is a set of param-eters that govern the way in which the simulation package solves the differential equations. Most simulation packages offer the user a choice of integration algorithms (see Chap. 5) and parameters that affect the integration step size. Variable-step-size algorithms are very powerful and are generally considered the standard approach. As discussed in Chap. 5, the user specifies maximum and minimum step sizes and an error tolerance that directs the software in selecting a step size within the given range. The tighter (smaller) the tolerance, the smaller the step size. For a vast major-ity of applications, the following five parameters define the run-time control of the simulation.

1. *Integration algorithm*: Chapter 5 discusses the origins of various integration algo-rithms in detail. Most algorithms implement an adaptive-step-size approach that

constantly adjusts the integration time step within the constraints set by the remaining solution control parameters. Also available are algorithms specifically designed for stiff systems (ones with widely varying time scales). Section 5.6 provides some discussion of the problems posed by stiff systems.

2. *Initial and final time:* Nearly all simulations assume a starting time of $t = 0$. The final time parameter sets the condition for stopping the computer simulation.

3. *Minimum step size:* This parameter sets a lower limit on the time step used by the adaptive-step-size algorithm. This is usually set by the user on the basis of a variety of factors, including computational efficiency. The user should be prepared to wait as long as it takes for the computer to complete the simulation run at the minimum time step.

4. *Maximum step size:* This parameter sets the upper bound on the adaptive-step-size algorithm. Any adaptive-step-size algorithm can be converted to a fixed-step-size algorithm if the minimum step size and the maximum step size are set to the same value.

5. *Error tolerance:* As described in Section 5.4, adaptive-step-size algorithms adjust the integration step size up or down by comparing the difference between the values of the output for two attempted integration steps with an error tolerance. If the relative error between the two attempts is less than the tolerance, the step size remains the same, or is allowed to grow slightly. If not, the step size is reduced and the process is repeated. For example, if the error tolerance is set to 1.0×10^{-4}, the algorithm decreases the step size until the error is less than one part in 10,000.

Although it is impossible to make any generalizations about the most appropriate values of these parameters for all simulations, the following values of run-time control parameters will serve as a good starting point for most mechanical systems. These parameter values usually get the simulation going on the right track, and they can be adjusted up or down depending on the requirements of the application:

Integration method:	Fourth- or fifth-order Runge–Kutta
Minimum step size:	0.001 s
Maximum step size:	1.0 s
Error tolerance:	0.001

6.4   STUDYING A SYSTEM WITH A SIMULATION

Although this chapter, along with the software documentation that describes the particular simulation package, can convey a great deal of information about the mechanics of putting together a simulation, there are other, more subtle issues that arise in computer-based simulations. For example, a second-order linear system like the one shown in Example 6.1 and considered in the previous section can be described by the input–output differential equation

$$m \frac{d^2x}{dt^2} + b \frac{dx}{dt} + kx = F_i(t). \tag{6.5}$$

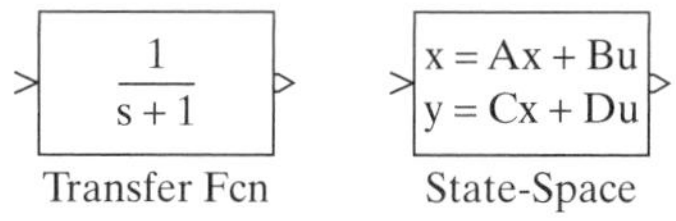

$$\boxed{\dfrac{1}{s+1}}$$
Transfer Fcn

$$\boxed{\begin{array}{l} x = Ax + Bu \\ y = Cx + Du \end{array}}$$
State-Space

Figure 6.8. (a) Simulink transfer function block, (b) Simulink state-space block.

It can also be represented in state-space matrix form:

$$\begin{bmatrix} \dot{x} \\ \dot{v} \end{bmatrix} = \begin{bmatrix} 0 & 1 \\ -\dfrac{k}{m} & -\dfrac{b}{m} \end{bmatrix} \begin{bmatrix} x \\ v \end{bmatrix} + \begin{bmatrix} 0 \\ \dfrac{1}{m} \end{bmatrix} F_i(t). \tag{6.6}$$

Simulink offers single blocks that implement either equation in very compact and efficient form. Equation (6.5) is easily represented as the "transfer function block" shown in Fig. 6.8(a). Similarly, the state-space formulation of Eq. (6.6) lends itself to the "state-space block" shown in Fig. 6.8(b).

In both cases, one enters the system parameters by double-clicking on the block and filling in the blanks. For the transfer function block, the parameters that are entered are the coefficients of the polynomials that make up the numerator and denominator of the transfer function.[3] The state-space block requires the parameters in the form of the system matrices, **A**, **B**, **C**, and **D**, as described in Section 3.3.

There is an important trade-off in choosing the level of detail that will be represented in the block diagram. On one hand, using a highly structured block such as the state-space block to represent a spring–mass–damper system brings about a very compact block diagram. On the other hand, use of this block requires more analytical work on the part of the user, putting the equations into state-space form, such as Eq. (6.4). It also requires that the system be linear, and, most important, it does not allow the user to introduce other effects into the system if desired. Subsection 6.4.3 explains this point in greater detail.

6.4.1 Monitoring Indices of Performance

Another major reason for studying simulation is that it provides a flexible means of evaluating a system with respect to various indices of performance. Again, although the system under study may be linear, many indicators of system performance (e.g., power consumption) are nonlinear.

For example, suppose one wishes to study the energy dissipated in the damper in the linear spring–mass–damper system shown in Fig. 6.1. The total energy dissipated can be expressed as

$$E_D = \int_0^t (F_{\text{damper}} \cdot v) d\tau = \int_0^t b v^2 d\tau. \tag{6.7}$$

Although this is a nonlinear function of the system states, and one could determine it analytically by solving for the velocity and then evaluating the integral, this

---

[3] The concept of a transfer function is presented in detail in Chap. 11.

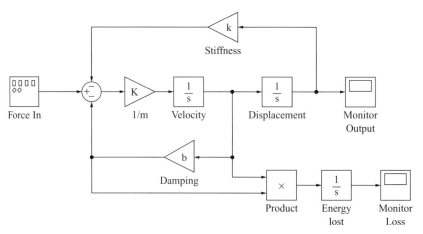

Figure 6.9. Coefficient block: (a) Constant coefficient and (b) time-varying coefficient, i.e. a multiplier.

function is very easily incorporated into the simulation by inclusion of the appropriate blocks. Figure 6.9 shows the results. If the system is allowed to respond to an initial velocity of 2 m/s, with no input, the energy loss can be seen in the plot shown in Fig. 6.10.

6.4.2 Parametric Studies: Engineering Design

One of the most powerful applications of computer simulation is its use as a design tool. With a well-crafted simulation, the engineer can investigate the implications of various design changes without the costly step of prototype construction and testing. One must be careful, however, in this approach. In order that the results of a simulation be significant, it must contain sufficient detail to reflect all of the implications of the proposed design changes. In addition, the simulation must be properly validated by means of a well-designed experimental protocol. An experienced engineer will be aware of these trade-offs and will use a simulation of sufficient detail to help define the general limitations and constraints of the design problem. In the end, the results of most engineering design efforts remain uncertain until the actual

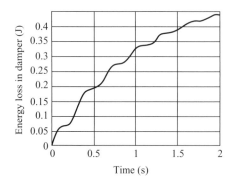

Figure 6.10. Time history of the energy dissipated in the damper, as computed in the simulation whose structure is shown in Fig. 6.3.

prototype is constructed and tested in its intended application. Nevertheless, the computer simulation, particularly one constructed by means of the block-diagram-oriented packages, offers a very valuable tool in evaluating the dynamic performance of systems.

Once the general structure of the design is determined, the design problem becomes one of choosing the best physical and geometric parameters that define the system. The computer simulation can examine the dynamic response of the system over a wide range of the parameter values, giving the engineer a great deal of information and guidance in choosing the parameter values. Analyses that examine the dynamic response of a system over a range of parameter values are often referred to as parametric studies. Figure 4.13 shows a series of step responses of a second-order system as the damping ratio changes. This is a well-known parametric study for second-order linear systems.

6.4.3 Nonlinear Systems

Up to this point, attention has focused on the use of simulation to solve equations similar to those solved analytically in Chap. 4. The real power of simulations lies in the solution of nonlinear and complex systems that are not solvable by analytical means. In this section, many common nonlinearities encountered in the study of mechanical systems are explored. In the next and final section of this chapter, a case study that illustrates many of these concepts is presented.

Friction. When mechanical systems are modeled, the treatment of friction is inevitable. Unfortunately, the most common treatment of friction – viscous friction, as seen in the well-known spring–mass–damper problem – is usually quite incorrect. In the vast majority of mechanical systems, the friction phenomenon is more accurately modeled by the "dry-friction" model treated in engineering mechanics. In engineering mechanics, friction is characterized as a force that opposes motion, the maximum value of which is normal force times the coefficient of friction. Unfortunately, dynamic models seldom include the normal forces, because they do not directly contribute to the dynamic response that is under study. For dynamic systems, dry friction is usually assumed to have a constant value, the sign of which is set to oppose motion. Therefore, like viscous friction, it is a function of velocity. Unlike viscous friction, the characteristic function is flat. Figure 6.11 shows the two characteristic curves.

Fortunately, most simulation packages offer easily implemented blocks that model this particular nonlinearity. In Simulink, the "discontinuities" block library, which is included in the package, allows considerable flexibility in modeling this phenomenon. One approach would be to use the "sign block" (which returns 1 or -1, depending on the sign of the block input) coupled with a gain block to bring about the dry-friction characteristic shown in Fig. 6.11. Another approach would be to use the "Coulomb & Viscous Friction Block," which models the algebraic sum of the linear and dry friction components. To demonstrate the effect of this nonlinearity, the example that was first introduced in Fig. 6.6 is modified by replacement of the

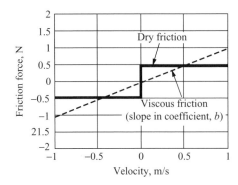

Figure 6.11. Characteristic curves for viscous (linear) friction and dry friction, which is also known as Coulomb friction.

linear gain block, which represented the viscous-friction element, with a combined Coulomb and viscous-friction block described by

$$F_{\mathrm{NLD}} = F_0\,\mathrm{sign}(v) + bv$$

where F_0 is the Coulomb friction force. The resultant block diagram is shown in Fig. 6.12.

Figure 6.13 compares the step response (magnitude = 2.0, start time = 0.1 s) for the preceding system with linear friction and with the combined Coulomb friction and viscous-friction block. The combined friction was set to have a linear damping coefficient of 1.0 N s/m and an offset at zero of 0.5 N. Note that the simulation that represents the effects of both linear and dry friction is markedly different from the response with the linear friction alone. This results partially from the fact that the nonlinear model contains an additional damping mechanism (the dry friction), so one would expect the transient response to die out sooner. On the other hand, the nature of the response, particularly its relatively quick approach to steady state, is indicative of the nonlinear nature of the problem.

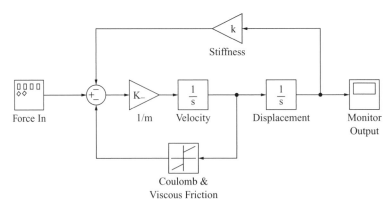

Figure 6.12. Block diagram showing the simulation structure that has been modified to include a Coulomb friction block. This block represents a velocity-dependent force that is the algebraic sum of a linear friction and a nonlinear dry-friction force.

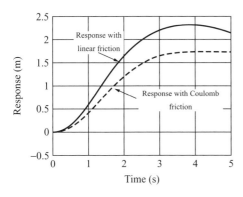

Figure 6.13. Comparison of spring–mass–damper system step responses with linear and nonlinear friction included.

Configuration dependence. Configuration dependence refers to systems in which the structure and/or parameters of the model change depending on the values of the state variables. A very common example of this type of system is the robotic manipulator. The equations of motion that describe the behavior of robots are very nonlinear, containing many trigonometric functions of state variables in inertial terms as well as products and squares of state variables. This particular nonlinearity gives rise to most of the difficulties encountered in controlling robotic systems.

Another common example of configuration dependence can be seen in systems containing two or more bodies that are in intermittent contact, such as a mechanical clutch. A system of this type presents some interesting challenges, which will be investigated by means of a case study in the next section.

Other common nonlinearities. In addition to the sign and Coulomb friction blocks just described, Simulink has a rich collection of other nonlinear general-purpose blocks, a small sample of which are subsequently described. Figure 6.14 shows 10 blocks that model nonlinear phenomena commonly encountered in the modeling of engineering systems. In the following discussions of the blocks, the input variable is u and the output is y.

Saturation. The saturation block models a common nonlinear phenomenon. Physical examples include motors that have torque limits beyond which they cannot operate or amplifiers that can produce voltages only within specified limits. Mathematically, the saturation block implements the following piecewise linear function,

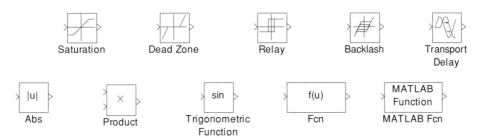

Figure 6.14. Sample of other common nonlinear blocks available in Simulink.

where u is the input to the saturation block and y is the output:

$$y = \begin{cases} y_{max} & \text{for} \quad u \geq u_{upper} \\ u & \text{for} \quad u_{lower} < u < u_{upper}. \\ y_{min} & \text{for} \quad u \leq u_{lower} \end{cases} \tag{6.8}$$

Dead zone. Many systems do not respond to inputs with values very near zero. The input must exceed some threshold value before response can be monitored. This phenomenon is often associated with some fundamental physical event such as dry friction, and it may be modeled with the dead-zone block. The mathematical definition is

$$y = \begin{cases} u - u_{upper} & \text{for} \quad u \geq u_{upper} \\ 0 & \text{for} \quad u_{lower} < u < u_{upper}. \\ u - u_{lower} & \text{for} \quad u \leq u_{lower} \end{cases} \tag{6.9}$$

Relay. Also encountered in the process industry, as well as in many very common applications, is the relay controller. This block models the common on–off controller module used in everything from home heating and air conditioning to chemical process control systems. In most cases, there are two set points defined on the input variable, one at which the system is turned on and another at which it is turned off. Chapter 14 includes a discussion of on–off control systems.

Backlash. Similar in form to the dead-zone block is the backlash block. Whereas the dead-zone block defines a zone of zero response when the input is in the vicinity of zero, the backlash block defines a zone of zero change in response when the rate of change of the input is near zero. The most common physical phenomenon that follows this pattern is gear backlash.

Transport delay. Transport delay, also known as "dead time," is commonly encountered in the process industry. It is also a phenomenon seen in computer-based control systems (see Chap. 16). The analytical definition is shown in this equation, where the duration of the delay is t_d:

$$y(t) = u(t - t_d). \tag{6.10}$$

Absolute value. This block takes the absolute value of the input signal:

$$y = |u|. \tag{6.11}$$

Product. This block was used earlier in this chapter to multiply the damper force by the velocity to produce the power dissipated in the damper. Note that this nonlinear block, which multiplies two variables together, should not be confused with the linear gain block, which is used to multiply a variable by a gain constant, which is a parameter.

Trigonometric function. As its name implies, you can implement a typical trigonometric function, as well as some inverse trigonometric functions, by using this block. This may be particularly useful for some robotic or vehicle applications.

Fcn. Simulink has several blocks that implement user-defined functions. The Fcn and the MATLAB Fcn (see next subsection) are two of the most common and powerful blocks. The `Fcn` block enables the user to implement any single-line function by using MATLAB syntax. In other words, if you can write a single line of MATLAB code, operating on the input to the block (represented as u), it can be implemented in this block.

`MATLAB Fcn.` A more powerful version of the `Fcn` block, this block is used to call an m-file function. Appendix 3 discusses m-files and functions implemented therein. A very common use of the MATLAB `Fcn` block is to implement complex nonlinear state equations when the mathematical descriptions are readily available.

6.5 SIMULATION CASE STUDY: MECHANICAL SNUBBER

Mechanical snubbers are common devices used to absorb large amounts of energy through relatively short displacements. They can also be difficult to simulate because they are active only when they are in contact with the system they are designed to protect. To properly model such a system, one needs to simulate both the unsnubbed system and the system while the snubber is engaged. In general, this requires the solution of at least two different sets of equations and switching between them when certain conditions are met. Figure 6.15 shows a highly idealized schematic of a snubber application. A mass m has acquired a certain amount of energy and at $t = t_0$ is moving at velocity v_0 at an initial displacement of x_0, which is greater than x_{snub}. The mass is suspended on an ideal spring k_1 and has no means of dissipating energy without the snubber. The drawing shows that once the displacement of the mass falls below the value of x_{snub} the mass is in contact with the snubber, which is represented as a spring and damper in parallel. During the time when x is less than or equal to x_{snub}, the snubber is in contact with the mass and can exert a force F_{snub} on it. Eventually, the mass slows to zero velocity and begins to move upward under the influence of the springs. While moving with a positive velocity, the damper within the snubber may dominate the response and allow the mass to lose contact with the snubber. Another way of thinking about this is to note that the snubber cannot pull on the mass, but can only push it. Clearly this system cannot be represented by a simple set of linear differential equations. However, a computer simulation, which can monitor the nature of the contact between the mass and the damper, is an appropriate tool to use in studying this system.

6.5.1 Modeling the System

The first step in any simulation process is to write the correct equations describing the behavior of the system. Even though there are three energy-storing elements in the system, only two state equations are required for describing its behavior. The reason behind this discrepancy is that both springs are active only when the snubber is engaged. During that time, the springs experience the same relative displacement and therefore are not independent. A single state variable describes their energy storage.

Regardless of whether or not the snubber is in contact with the mass, the first state equation is the same:

$$\dot{x} = v. \tag{6.12}$$

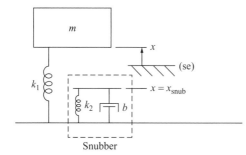

Figure 6.15. Schematic representation of a mass–snubber system.

The second equation can be stated in general form as

$$\dot{v} = \frac{1}{m}(F_{k1} + F_{snub}),\tag{6.13}$$

where

$$F_{k1} = -k_1 x,\tag{6.14}$$

$$F_{snub} = \begin{cases} 0 & \text{for } x > x_{snub} \\ \max[0, -k_2(x - x_{snub}) - b\,v] & \text{for } x \le x_{snub} \end{cases}.\tag{6.15}$$

It is interesting to note that the individual elements of the system, and all the equations that describe their behavior, are linear but that the system itself is nonlinear because the equations that describe the system change depending on the values of one of the states (in this case, the displacement.)

There are several ways to simulate this particular kind of nonlinearity. In the next subsections, different approaches are explored.

6.5.2 Block-Diagram Approach to Simulation Structure

Earlier in this chapter, a sampling of the nonlinear blocks available in Simulink was presented. All of the operations required for implementing the conditions of Eqs. (6.15) are available in simple block elements.

Specifically, the "switch" block accomplishes the condition testing shown in Eqs. (6.15). The switch block, shown in Fig. 6.16, has three inputs and one output. This block alternately routes the first or third (top or bottom) input to the output, depending on the value of the middle input line relative to some internally set threshold (the details of the switching condition are set in a dialog box when the switch block is double-clicked). If the switching condition is met, the first input (top signal) is routed to the output. On the other hand, if the condition is not met, the block switches over and routes the bottom signal.

This simulation will require two such switch blocks. The first ensures that the snubber force is applied only if the simulation finds that the mass is engaged with the snubber $(x < x_{snub})$. The second ensures that, if the snubber is engaged, it will not generate a negative force. Figure 6.17 shows the complete block diagram for

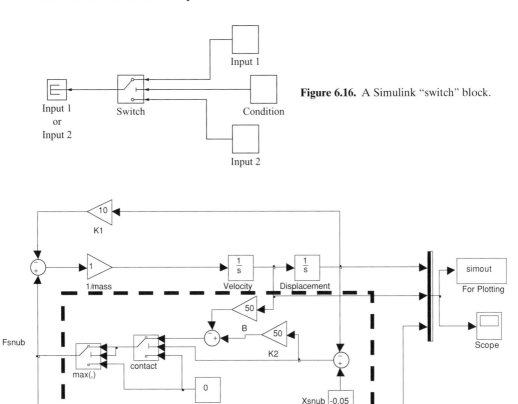

Figure 6.16. A Simulink "switch" block.

Figure 6.17. Simulink block diagram for the snubber system shown in Fig. 6.15.

a simulation of the snubber system. The portion of the diagram that models the snubber performance is highlighted by the dashed box (see Table 6.1 for values of parameters).

Figure 6.18 shows the displacement of the mass, starting at 0.05 m, and an initial velocity of −0.5 m/s. Close examination of the velocity and displacement curves reveals the times when the switches were used. At approximately 0.2 s (point A on the graph), the mass comes into contact with the snubber and the velocity moves sharply toward zero. As the snubber dissipates the energy of the mass, it slows the motion and reverses its direction. At approximately 0.27 s (point B), the mass is moving away from the snubber and the snubber cannot keep up, so F_{snub} falls to zero, even though the mass is still within the range of the snubber until 0.43 s (point C). These changes in the snubber force are more clearly indicated by the plot of F_{snub} in Fig. 6.19.

6.5.3 Alternative Approach to Configuration Dependence

The previous section demonstrated how block diagram tools can be used to model the mechanical snubber system, which has very pronounced nonlinearities. In this section, an alternative approach is presented that illustrates the flexibility of the

Table 6.1. Model parameters, initial conditions, and simulation control parameters for the model shown in Fig. 6.17

	Symbol	Value
Model parameters	m	1 kg
	k_1	10 N/m
	k_2	50 N/m
	b	50 N s/m
	x_{snub}	−0.05 m
Initial conditions	$x(0)$	0.05 m
	$v(0)$	−0.5 m/s
Simulation parameters	Integration method	ODE 45 (Dormand–Price)
	Minimum step	0.00001 s
	Maximum step	0.01 s
	Error tolerance (relative)	1×10^{-6}
	Error tolerance (absolute)	1×10^{-6}
	Final time	1.5 s

MATLAB environment. In this approach, the mathematical description of Eqs. (6.14) and (6.15) is modeled directly in a user-defined function.

As discussed in Subsection 6.4.3, Simulink has special blocks that allow users to write their own functions as script files that are then called from the simulation. In the case of the snubber, one of these functions is used to compute the force exerted by the snubber on the mass, F_{snub}, as a function of the displacement and the velocity of the mass. Table 6.2 shows the MATLAB script that implements the function fsnub(u), where **u** is the vector of displacement and velocity. Figure 6.20 shows the Simulink block diagram utilizing this approach to simulation.

It is left to the student to verify that the two simulations (Figs. 6.20 and 6.17) lead to the same solutions of the model equations.

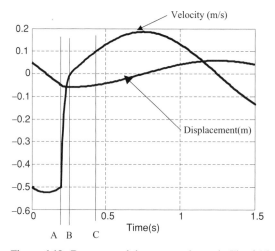

Figure 6.18. Response of the system shown in Fig. 6.15.

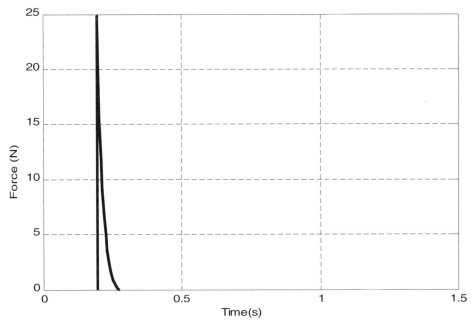

Figure 6.19. Plot of snubber force vs. time showing the onset of snubber contact and the point at which the computed snubber force is no longer positive (and hence set to zero).

6.5.4 Parametric Study – Running Simulation from a Script

In this section, a parametric study of the snubber shown in the previous subsection is performed. The snubber is defined by two parameters: its stiffness and its damping constant. The performance of the snubber may be indicated by any one of a number of variables that could be computed from the system response. Because snubbers are often used to protect an object from coming into contact with another one (as a doorstop protects the woodwork behind an open door) the maximum deflection of the snubber will be chosen as the index of performance. If the simulation produces

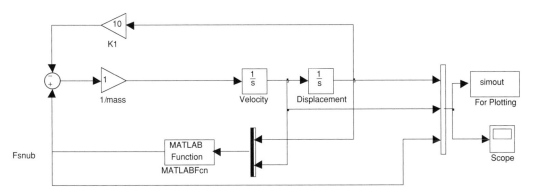

Figure 6.20. Simulink model using the MATLAB function block to compute F_{snub}.

Table 6.2. MATLAB function script to compute snubber force

```
function [force] = fsnub(u)
%
% FSNUB computes the nonlinear force for snubber simulation
%
% Computes the snubber force as a function of
% displacement and velocity
%
% u(1)  =  displacement                              u(2)  =  velocity
%
xsnub  =  -0.05;                                     % Location of snubber
k2  =  50.0;                                          % Spring constant (N/m)
 b  =  50.0;                                          % Damping (Ns/m)
%
if u(1) > xsnub
   force  =  0.0;                                     % not in contact
else
   force  =  k2*(xsnub-u(1))-b*u(2);
end;
%
% Snubber cannot exert a negative force
%
if force < 0
force  =  0;
end;
```

a column matrix, **X**, in which it stores the displacement of the mass over time, then the maximum deflection of the snubber can be found with the following expression:

$$x_{\max} = x_{\text{snub}} - \min(\mathbf{X}). \qquad (6.16)$$

In other words, find the minimum value of the displacement in the solution, then subtract it from the location of the snubber to find the maximum displacement. It will be assumed that values of stiffness from 1 to 50 N/m are representative of materials that may be used for the snubber. Likewise, values of damping from 1 to 50 N s/m will be investigated. To span those ranges, two row matrices are defined to contain the possible values that will be simulated:

$$k_{\text{all}} = \begin{bmatrix} 1 & 10 & 20 & 30 & 40 & 50 \end{bmatrix},$$
$$b_{\text{all}} = \begin{bmatrix} 1 & 10 & 20 & 30 & 40 & 50 \end{bmatrix}.$$

One approach to performing this parametric study is to go to the Simulink model, open the appropriate gain blocks by double-clicking, enter the appropriate stiffness and damping values, run the simulation, and analyze the results. This process would have to be repeated for each of the 36 combinations of representative parameter values in order to develop a complete picture of the system.

Fortunately, one can automate this approach of tediously rerunning the simulation for each combination of parameters by writing a script file that sets parameter values, runs the simulation, and stores the results. This approach requires the following features.

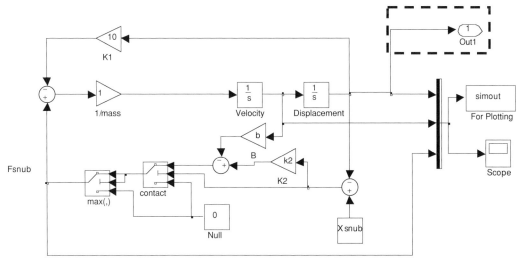

Figure 6.21. Simulink block diagram of the snubber simulation with the output block inserted on the displacement signal (highlighted in dashed block).

- The model parameters that are to be varied (in this case, k_2 and b) must be represented by workspace variables, not numerical values, in the Simulink model. In other words, the gain blocks that represent k_2 and b must refer to workspace variables (good names for these would be k2 and b), not the numbers that were used when the simulation was first run.
- The physical variable that contains the information of interest on system performance should be routed to a Simulink "out" block. This block defines the output of the simulation that will be returned to the calling script.

Figure 6.21 shows the simple modification of the simulation to incorporate these changes.

Table 6.3 shows the MATLAB script that performs the parametric study using the simulation represented in Fig. 6.21.

The last statement in the script file plots the maximum displacement for each combination of parameters vs. both the k_2 and b values by using a three-dimensional mesh plot, shown in Fig. 6.22

Figure 6.22 highlights the power of an automated parametric study. Not only does the plot quickly indicate those conditions under which the displacement is very large, but it also defines the interaction between the two parameters. For example, it can be seen that the response is not at all sensitive to differences in damping when k_2 is large. This is indicated by the fact that the surface is flat for all values of b when k_2 is 50 N/m. On the other hand, when k_2 is small, there is a noticeable influence of b on the response. Likewise, the response is more sensitive to changes in k_2 when b is small than when b is large, although those trends are not as significant. In conclusion, parametric studies offer a valuable tool for engineering analysis and design. As long as the desired performance features can be reduced to one or two performance indices, those indices can be computed for a wide range of parameter values and

Table 6.3. MATLAB script file for the parametric study using the `sim` command to call a Simulink simulation

```
%
% M-file to to perform parametric study of snubber
%
X0   = 0.05;
Xsnub= -0.05;
%
maxd=zeros(6,6);                        % matrix to store max displacements
                                        % for each run
kall=[1   10   20   30   40   50];      % values of stiffness to be used
ball=[1   10   20   30   40   50];      % values of damping to be used
%
for i=1:6;
   for j=1:6;
   b=ball(j);
   k2=kall(i);
   [T,Y]=sim('Chapt6_CS_0',2);          % run the simulation for 2 seconds
   maxd(i,j)= Xsnub- min(Y(:,1));       % find max displacement
   end
end
meshz(kall,ball,maxd);                  % plot results
```

combinations of those values and an overall picture of system performance can be viewed through a single plot.

6.6   SYNOPSIS

The use of block-diagram-oriented computer applications for the simulation of dynamic systems was presented. It was noted that, although a wide variety of commercial packages are available for the rapid construction and execution of simulations, they share certain common elements. In all cases, the differential equations are represented by means of a block diagram, constructed from a library of blocks representing many linear and nonlinear elements. Numerical values for the parameters that characterize each element must be provided. Initial conditions for the problem must also be defined by the user. For systems that respond to external inputs, the inputs must be defined and the user must select which variables from the block diagram are to be plotted or stored for later analysis. Finally, the user must direct the simulation package in its numerical solution of the equations by selecting the integration method and parameters that control the integration step size.

Several examples were presented in the Simulink–MATLAB environment, and it was shown that simulations can be used for a variety of engineering analyses.

Finally, the ability to extend the block-diagram-oriented packages by user-defined functions and scripts was discussed. This flexibility is crucial for a truly general-purpose simulation package.

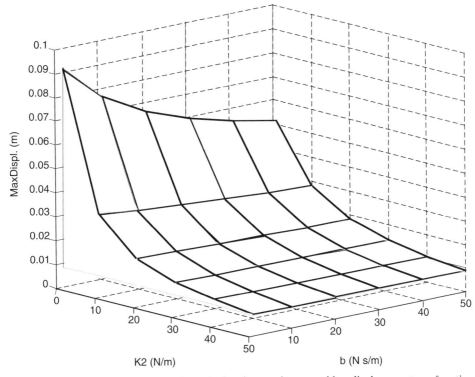

Figure 6.22. Results of the parametric study showing maximum snubber displacement as a function of the stiffness and damping of the snubber.

PROBLEMS

6.1 Construct a simulation of the spring–mass–damper problem described in Example 6.1. Using two "step" function blocks and a summing junction, construct a rectangular pulse input (see the theoretical discussion in Section 4.3). Execute several simulation runs, each time making the pulse width narrower (shorter duration) and the pulse magnitude larger. Keep the integral of the pulse constant. Continue this process until the response no longer exhibits any appreciable changes for narrower pulses. What is the common name of the response you computed?

6.2 Modify the spring–mass–damper model shown in Fig. 6.12 so that only dry (Coulomb) friction of unity magnitude is modeled. Comparing the response to a system with pure linear (viscous) damping of the same magnitude, what conclusions can you draw about the energy dissipating effects of dry vs. linear friction?

6.3 Modify the spring–mass–damper model shown in Fig. 6.12 to incorporate a more sophisticated friction model, as shown in Fig. P6.3. This model attempts to represent the "breakaway" friction that causes a higher force to start the body from zero velocity.

The friction force can be defined analytically:

$$F_f = \begin{cases} 6.25v^2 - 5v + 2 & for \quad v > 0 \\ -6.25v^2 - 5v - 2 & for \quad v \leq 0 \end{cases}. \tag{6.17}$$

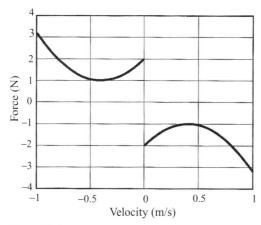

Figure P6.3.

Examine the response of the modified system model to step inputs and sinusoidal inputs.

6.4 Modify the spring–mass–damper model shown in Fig. 6.12 to include a nonlinear "stiffening" spring. Use the following relationship for the spring force:

$$F_{spring} = x^3.$$

Examine the responses of this system to step and sinusoidal inputs and compare them with those of the linear model. In particular, focus on steps and sinusoidal inputs of various amplitudes.

6.5 Construct a simulation model of the vehicle suspension system described in Problem 2.7. Use a random-noise function (available from the simulation package block library) for the wheel vertical position x. Examine the responses of the system to various amplitudes of input. How can this model be modified to detect the suspension "bottoming out" (when the spring compresses to its solid height)?

6.6 Construct a simulation model of the torsional system with nonlinear damping described in Problem 2.8. The system parameters are $J_1 = 0.01$ N m s^2/rad, $J_2 = 0.1$ N m s^2/rad, $K = 20$ N/rad. The input torque, T_s, undergoes a step change from 0 to 0.2 N m at time $t = 0$. Find the step response of the system for values of C ranging from 0.5 to 5.0 N m s/rad.

6.7 Building on the case study in Section 6.5, perform a parametric simulation study on the snubber to find the maximum velocity that the mass can have on coming into contact with the snubber and not have the snubber experience a displacement greater than 0.025 m.

6.8 Starting with the parameterized version of a second-order system model [Eq. (4.63)], construct a simulation to compute the step response of this system and use it to perform a parametric study for various values of the damping ratio (ζ). Plot the resulting step responses on a single graph and compare them with Fig. 4.13.

6.9 Construct a simulation of the nonlinear system that was linearized in Example 2.8. Use the simulation to compute the response of the system to a step change in force from $F = 0.1$ N to $F = 0.2$ N. Repeat the process for a number of step changes, all starting at 0.1 N, but rising to 0.4, 0.5, 0.8, 1.0, and 5.0 N. Compare the nature of the combined responses and compare them with the responses predicted by the linearized model (which was derived in the example). What do you conclude about the applicability of the linearized model?

6.10 Using the "switch" blocks shown in the snubber case study (Section 6.5), build a simulation of a bouncing basketball. Assume one-dimensional motion (only in the vertical direction) and that the ball has a mass of 600 g. Using both the simulation and careful field observations, find appropriate values of stiffness and damping that characterize the ball while it is bouncing off the floor. (Alternatively, use a soccer ball with a mass of 430 g.)

6.11 Construct a simulation of the system described in Problem 4.14. Show the step response of the system to a sudden change in torque. Use the simulation to find the value of damping b that will lead to a critically damped response. Compare this result with the answer for part (d) in Problem 4.14.

6.12 Construct a simulation of the system described in Problem 4.15, using the parameter values indicated there. Simulate the step response of the system for those values and for cases in which the shaft is twice as stiff and half as stiff. Plot the three step responses on the same graph.

7

Electrical Systems

LEARNING OBJECTIVES FOR THIS CHAPTER

7–1 To recognize the A-, T- and D-type elements of electrical systems.

7–2 To develop the mathematical equations that model the dynamic behavior of *RLC* electrical circuits.

7–3 Develop the equations that describe the dynamic behavior of *RLC* circuits with time-varying capacitors and inductors.

7–4 Analyze simple operational-amplifier circuits.

7.1 INTRODUCTION

The A-type, T-type, and D-type elements used in modeling electrical systems, which correspond to the mass, spring, and damper elements discussed in Chap. 2, are the capacitor, inductor, and resistor elements.

A capacitor, the electrical A-type element, stores energy in the electric field induced in an insulating medium between a closely spaced pair of conducting elements, usually plates of metal, when opposite charges are applied to the plates. Capacitance is a measure of the ability of a capacitor to accept charge and hence its ability to store energy. It occurs naturally between the conductors of a coaxial cable, between closely spaced parallel cables, and in closely packed coils of wire. In these cases the capacitance is distributed, along with resistance and inductance, along the line, and the analysis of such situations is beyond the scope of this text. However, in some cases the resistance and inductance are negligible, making it possible to use a lumped-capacitance model. More frequently, specially designed off-the-shelf capacitors are used that have negligible inductance and series resistance. In addition, the dielectric material between the plates is such a good insulator that the parallel leakage resistance between the plates is essentially infinite; thus it takes a very long time for a charge to leak away internally.

An inductor, the electrical T-type element, stores energy in the magnetic field surrounding a conductor or a set of conductors carrying electric current (i.e., flow of charge). Inductance is a measure of the ability of an inductor to store magnetic energy when a current flows through it. Like capacitance, inductance occurs naturally in coaxial cables, long transmission lines, and coils of wire. Here it is usually distributed along with capacitance and resistance, requiring a level of modeling and analysis that is beyond the scope of this text. Often, however, the case is such that the capacitance and resistance are negligible and a lumped-inductor model will suffice. Even if resistance is distributed along with the inductance, a series lumped-inductor, lumped-resistor model will be suitable.

A resistor, the electrical D-type element, dissipates energy, resulting in heat transfer to the environment equal to the electric energy supplied at its terminals. Resistance is a measure of how much voltage is required to driving one ampere (1 amp) of current through a resistor. It occurs naturally in all materials (except superconducting materials), including wires and metal structural elements. When resistance occurs in wires intended for conducting electricity, it is usually an unwanted phenomenon. In many other cases, resistance is necessary to accomplish circuit requirements, and a wide variety of resistors are available as off-the-shelf components, each type specially designed to provide needed characteristics. Carbon in some form is often used as the resistive medium, but coils of wire often serve this purpose, especially when a great deal of heat must be dissipated to the environment. Incandescent light bulbs are sometimes used as resistors, but they are very nonlinear.

This chapter deals only with lumped models made up of combinations of ideal electrical elements, which constitute a great majority of the electrical control systems encountered by engineers and scientists.

7.2 DIAGRAMS, SYMBOLS, AND CIRCUIT LAWS

All three of the basic types of circuit elements discussed in Section 7.1 are two-terminal elements, as shown schematically in Fig. 7.1. The electric potential at each terminal is measured by its voltage with respect to ground or some local reference potential, such as a machine frame or chassis. The rate of flow of electrical charge through the element in coulombs per second is measured in terms of amperes. The elemental equation usually takes the form

$$e_A = f_1(i_A) \quad \text{or} \quad i_A = f_2(e_A), \tag{7.1}$$

where e_A is the voltage across element A between terminals 1 and 2:

$$e_A = e_{12} = e_{1g} - e_{2g}. \tag{7.2}$$

The symbol e_{1g} represents the voltage between terminal 1 and ground, and e_{2g} represents the voltage between terminal 2 and ground.

In addition, there are two types of ideal source elements used for driving circuits, shown in Fig. 7.2. The ideal voltage source e_s is capable of delivering the designated voltage regardless of the amount of current i being drawn from it. The ideal current source i_s is capable of delivering the designated current, regardless of the voltage required for driving its load.

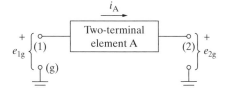

Figure 7.1. Circuit diagram of a two-terminal electrical element.

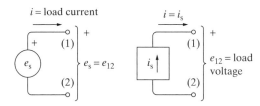

Figure 7.2. Circuit diagrams of voltage and current sources.

The two basic circuit laws needed to describe the interconnections between the elements of a circuit are known as Kirchhoff's voltage law and Kirchhoff's current law. The voltage law says that the sum of the voltage drops around a loop must equal zero [Fig. 7.3(a)]. A corollary to the voltage law is that the total voltage drop across a series of elements is the sum of the individual voltage drops across each of the elements in series [Fig. 7.3(b)].

The current law states that the sum of the currents at a node (junction of two or more elements) must be zero. This law is illustrated in Fig. 7.4.

7.3 ELEMENTAL DIAGRAMS, EQUATIONS, AND ENERGY STORAGE

7.3.1 Capacitors

The circuit diagram of an ideal capacitor is shown in Fig. 7.5. The elemental equation in terms of the stored charge q_C of a capacitor is

$$q_C = C(e_{1g} - e_{2g}) = Ce_C, \tag{7.3}$$

where e_C is the voltage across the capacitor, $e_C = e_{1g} - e_{2g} = e_{12}$.

In terms of current i_C (rate of flow of charge q_C), the elemental equation is

$$i_C = C\frac{de_C}{dt}. \tag{7.4}$$

Note the similarity to the elemental equation for an ideal mass given in Chap. 2. Because i_C is a through (T) variable and e_C is an across (A) variable, the capacitor C is said to be analogous to the mass m. Thus the state variable for a capacitor is its voltage e_C. However, the capacitor is different in one respect: Both of its terminals can "float above ground" – that is, neither must be grounded – whereas the velocity for a mass must always be referred to a nonaccelerating reference such as the earth (ground).

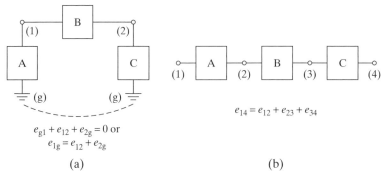

(a)   (b)

Figure 7.3. Illustration of Kirchhoff's voltage law for elements in a loop and for several elements in series.

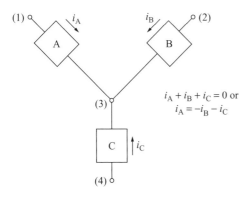

$$i_A + i_B + i_C = 0 \text{ or}$$
$$i_A = -i_B - i_C$$

Figure 7.4. Illustration of Kirchhoff's current law.

The comments made in Chapter 2 about the response of a mass m to an input force apply here to the response of a capacitor C to an input current – it takes time for a finite current to change the charge stored in, and hence the voltage across the terminals of, a capacitor.

The energy stored in a capacitor resides mainly in the very closely confined static field between the plates of the element, and it is often referred to as static field energy. The dielectric material between the plates is a very good insulator, so that very, very little current can flow (i.e., leak) directly from one plate to the other. The current i_C simply represents the rate of flow of charge to and from the plates where the charge is stored.

The stored electric-field energy is given by

$$\mathscr{E}_e = \frac{C}{2}e_C^2. \tag{7.5}$$

Here again it is apparent that imposing a sudden change in the voltage across a capacitor would not be realistic, because this would mean suddenly changing the stored energy, which would require an infinite current, i.e., an infinite power source.

7.3.2 Inductors

The circuit diagram of an ideal inductor is shown in Fig. 7.6. The elemental equation in terms of the flux linkage λ of the coil is

$$\lambda = Li_L. \tag{7.6}$$

In terms of the voltage across the inductor, $e_L = e_{1g} - e_{2g} = e_{12} = d\lambda/dt$, the elemental equation is

$$e_L = L\frac{di_L}{dt}. \tag{7.7}$$

$i_C = dq_C/dt$

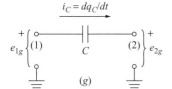

Figure 7.5. Circuit diagram of an ideal capacitor.

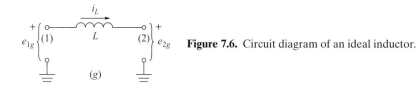

Figure 7.6. Circuit diagram of an ideal inductor.

Note the similarity to the elemental equations for an ideal spring. Because i_L is a T variable and e_L is an A variable, an inductor is said to be analogous to a spring, with L corresponding to $1/k$. Thus the state variable for an inductor is its current i_L.

The energy stored in an inductor is stored in the magnetic field surrounding its conductors, and is known as magnetic-field energy. The stored magnetic-field energy is given by

$$\mathcal{E}_m = \frac{L}{2} i_L^2. \tag{7.8}$$

It can be seen from Eq. (7.8) that to try to make a sudden change in the current through an inductor would not be realistic. Such a change would involve a sudden change in the stored energy, which would in turn require an infinite voltage and hence an infinite power source.

The magnetic field, which is induced by the current i_L flowing in many turns of a densely packed coil, is intensified if a ferromagnetic metal core is installed within the coil. Air core inductors are usually linear, but contain the inherent resistance of the coil wire, which may not be negligible. This parasitic resistance is modeled by an ideal resistor in series with the inductance of the coil.

When the core is a ferromagnetic material, the inductance for a given coil current is much greater and the ability of the inductor to store energy is much greater, but a ferromagnetic core also introduces nonlinearity and hysteresis in the flux linkage versus current relationship so that Eq. (7.6) becomes

$$\lambda = f_{NL}(i_{NLI}) \tag{7.9}$$

Compared with the variety and quantities of capacitors and resistors available as off-the-shelf elements, the probability of finding off-the-shelf inductors for specific applications is quite small. Thus inductors are usually custom designed and manufactured for each specific application.

7.3.3 Transformers

When two coils of wire are installed very close to each other so that they share the same core without flux leakage, an electric transformer results, as shown schematically in Fig. 7.7. Because a transformer is a four-terminal element, two elemental equations are needed to describe it. The elemental equation is[1]

$$e_{12} = n e_{34},$$
$$e_o = n e_i, \tag{7.10}$$

---

[1] The ideal relations used here are applicable only when time-varying voltages and currents, such as ac signals, occur. See G. Rizzoni, *Principles and Applications of Electrical Engineering*, 5th ed. (McGraw-Hill, New York, 2005), p. 309.

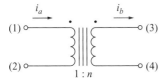

Figure 7.7. Circuit diagram of an ideal transformer.

n = ratio of turns between (3) and (4)
    to turns between (1) and (2)

where n is the ratio of the number of turns between terminals (3) and (4) to the number of turns between (1) and (2), and

$$i_b = \frac{1}{n}i_a. \tag{7.11}$$

The dots appearing over each coil indicate that the direction of each winding is such that n is positive; in other words, e_{12} has the same sign as e_{34} and i_a has the same sign as i_b.

Ideal transformers do not store energy and are frequently used to couple circuits dynamically. Combining Eqs. (7.10) and (7.11) reveals the equality of energy influx and efflux:

$$e_{34}i_b = ne_{12}\frac{1}{n}i_a. \tag{7.12}$$

7.3.4 Resistors

The circuit diagram of an ideal resistor is shown in Fig. 7.8. The elemental equation for an ideal resistor can be written either as

$$e_{12} = Ri_R \tag{7.13}$$

or as

$$i_R = \frac{1}{R}e_R, \tag{7.14}$$

where e_R is the voltage across the resistor, $e_R = e_{1g} - e_{2g} = e_{12}$. Note that the voltage across and the current through a resistor are related "instantaneously" to each other, because there is no energy storage, and no derivative of either e_R or i_R is involved in these equations.

The manner in which a source voltage in a circuit consisting of several resistors in series is distributed among the resistors is described mathematically by a very useful formula known as a voltage-divider rule. Consider the circuit with two resistors in series, R_1 and R_2, shown in Fig. 7.9.

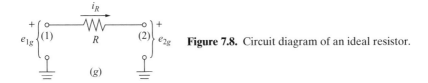

Figure 7.8. Circuit diagram of an ideal resistor.

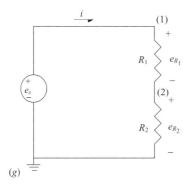

Figure 7.9. Circuit with two resistors in series.

The elemental equations for the two resistors are

$$e_{R1} = R_1 i, \tag{7.15}$$

$$e_{R2} = R_2 i, \tag{7.16}$$

where $e_{Ri} = e_{12}$ and $e_{R2} = e_{2g}$.

The current in the circuit is

$$i = \frac{e_s}{R_1 + R_2}. \tag{7.17}$$

Substituting the expression for i into Eqs. (7.15) and (7.16) gives

$$e_{R1} = \frac{R_1}{R_1 + R_2} e_s, \tag{7.18}$$

$$e_{R2} = \frac{R_2}{R_1 + R_2} e_s. \tag{7.19}$$

The results obtained in Eqs. (7.18) and (7.19) are examples of the voltage-divider formula, which for a circuit with n resistors in series takes the following mathematical form:

$$e_{Ri} = \frac{R_i}{R_1 + R_2 + \cdots + R_n} e_s, \quad i = 1, 2, \ldots, n. \tag{7.20}$$

Expressed verbally, the voltage-divider formula states that, in a circuit consisting of n resistors in series, the voltage across an ith resistor is equal to the source voltage times the ratio of the ith resistor over the sum of all resistors in series.

Although many resistors are carefully designed and manufactured to be linear, many others are inherently or intentionally nonlinear. The circuit diagram to be used here for a nonlinear resistor (NLR) is shown in Fig. 7.10.

The elemental equation for such a nonlinear resistor is written as either

$$e_R = f_{NL}(i_{NLR}) \quad \text{or} \quad i_{NLR} = f_{NL}^{-1}(e_R). \tag{7.21}$$

When Eq. (7.21) is linearized, for small perturbations about an operating point,

$$\hat{e}_R = R_{inc}\hat{i}_{NLR}, \tag{7.22}$$

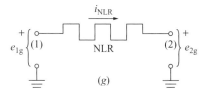

Figure 7.10. Circuit diagram of a NLR.

where the incremental resistance R_{inc} is given by

$$R_{inc} = \frac{df_{NL}}{di_{NLR}}\bigg|_{\bar{i}_{NLR}}. \qquad (7.23)$$

7.4   ANALYSIS OF SYSTEMS OF INTERACTING ELECTRICAL ELEMENTS

Electrical systems are usually referred to as electric circuits, and the techniques of analyzing electric circuits are very similar to those used in Chap. 2 for analyzing mechanical systems: Simply write the equation for each element and use the appropriate connecting laws (Kirchhoff's laws here) to obtain a complete set of n equations involving the n unknowns needed to describe the system completely; then combine (if everything is linear) to eliminate the unwanted variables to obtain a single input–output differential equation relating the desired output to the given input(s). If only a set of state-variable equations is needed, it is usually most convenient to "build" a state-variable equation around each energy-storage element, in which case nonlinearities are easily incorporated.

EXAMPLE 7.1

Find the input–output differential equation relating e_o to e_s for the simple RLC circuit shown in Fig. 7.11.

SOLUTION

For the source,

$$e_s = e_{1g}. \qquad (7.24)$$

For the resistor R_1,

$$i_{R1} = \frac{1}{R_1}(e_{1g} - e_{2g}). \qquad (7.25)$$

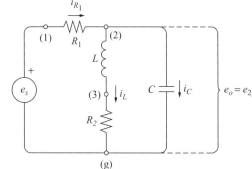

Figure 7.11. Circuit diagram of an RLC circuit.

For the inductor L,

$$e_{2g} - e_{3g} = \frac{di_L}{dt}. \tag{7.26}$$

For the resistor R_2,

$$e_{3g} = R_2 i_L. \tag{7.27}$$

For the capacitor C,

$$i_c = C \frac{de_{2g}}{dt}. \tag{7.28}$$

At node 2,

$$i_{R1} = i_L + i_C. \tag{7.29}$$

Equations (7.24)–(7.29) comprise a set of six equations involving six unknown variables: e_{1g}, e_{2g}, e_{3g}, i_{R1}, i_L, and i_C. (Note that using i_L to describe the current through both L and R_2 satisfies Kirchhoff's law at node 3 and eliminates one variable and one equation.)

The node method[2] may be applied to node 2 to eliminate the unwanted variables e_{1g}, e_{3g}, i_{R1}, i_L, and i_C. Substituting for i_{R1} from Eq. (7.25) and for i_C from Eq. (7.28) into Eq. (7.29) yields

$$\frac{1}{R_1}(e_{1g} - e_{2g}) = i_L + C \frac{de_{2g}}{dt}. \tag{7.30}$$

Then, using Eq. (7.24) for e_{1g} and rearranging Eq. (7.30) yields

$$i_L = \frac{1}{R_1} e_s - \frac{1}{R_1} e_{2g} - C \frac{de_{2g}}{dt}. \tag{7.31}$$

Differentiating Eq. (7.31) with respect to time gives

$$\frac{di_L}{dt} = \left(\frac{1}{R_1}\right) \frac{de_s}{dt} - \left(\frac{1}{R_1}\right) \frac{de_{2g}}{dt} - C \frac{d^2 e_{2g}}{dt^2}. \tag{7.32}$$

Combine Eqs. (7.26) and (7.27) to solve for e_{2g}:

$$e_{2g} = L \frac{di_L}{dt} + R_2 i_L. \tag{7.33}$$

Substitution from Eqs. (7.31) and (7.32) into Eq. (7.33) then eliminates i_L:

$$e_{2g} = \left(\frac{L}{R_1}\right) \left(\frac{de_s}{dt} - \frac{de_{2g}}{dt}\right) - LC \frac{d^2 e_{2g}}{dt^2} + \left(\frac{R_2}{R_1}\right)(e_s - e_{2g}) - R_2 C \frac{de_{2g}}{dt}. \tag{7.34}$$

Collecting terms and noting that e_o is the same as e_{2g}, we have

$$LC \frac{d^2 e_o}{dt^2} + \left(\frac{L}{R_1} + R_2 C\right) \frac{de_o}{dt} + \left(1 + \frac{R_2}{R_1}\right) e_o = \frac{R_2}{R_1} e_s + \frac{L}{R_1} \frac{de_s}{dt}. \tag{7.35}$$

---

[2] The node method starts by satisfying Kirchhoff's current law at each node and then uses the elemental equation for each branch connecting to each node.

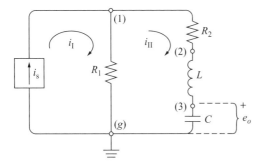

Figure 7.12. A different R, L, C circuit, driven by a current source.

EXAMPLE 7.2

Develop the input–output differential equation relating e_o to i_s for the circuit shown in Fig. 7.12.

SOLUTION

Here the loop method[3] is used and the elemental equations are developed as needed for each loop. Like the use of the node method, the use of the loop method helps to eliminate quickly the unwanted variables. The two loops, I and II, are chosen as shown in Fig. 7.12, carrying the loop currents i_I and i_{II}. Loop I is needed only to note that $i_I = i_s$. Using Kirchhoff's voltage law for loop II yields

$$R_2 i_{II} + L \frac{di_{II}}{dt} + \left(\frac{1}{C}\right) \int i_{II} dt + R_1 (i_{II} - i_s) = 0. \tag{7.36}$$

Because $i_{II} = C de_o/dt$, substituting for i_{II} in Eq. (7.36) yields

$$(R_1 + R_2) C \frac{de_o}{dt} + LC \frac{d^2 e_o}{dt^2} + e_o = R_1 i_s. \tag{7.37}$$

Rearranging, we have

$$LC \frac{d^2 e_o}{dt^2} + (R_1 + R_2) C \frac{de_o}{dt} + e_o = R_1 i_s. \tag{7.38}$$

EXAMPLE 7.3

(a) Write the state-variable equations based on the energy-storage elements L and C for the circuit shown in Fig. 7.13(a).
(b) Linearize these equations for small perturbations of all variables.
(c) Combine to eliminate the unwanted variable and obtain the input–output system differential equation relating the output $\hat{e}_{2g}$ to the input $\hat{e}_s(t)$.
(d) Draw the simulation block diagram for the linearized system.

SOLUTION

(a) In general, the state-variable for a capacitor is its voltage e_{2g}, and the state variable for an inductor is its current i_L. Thus, for the inductor L,

---

[3] The loop method starts by satisfying Kirchhoff's voltage law for each independent loop and then uses the elemental equations for each part of each loop.

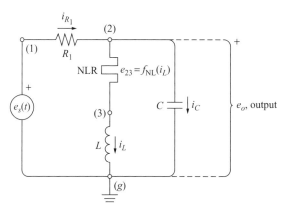

Figure 7.13(a). Electric circuit with NLR.

$$\frac{di_L}{dt} = \frac{1}{L}(e_{2g} - e_{23})$$

or

$$\frac{di_L}{dt} = \frac{-1}{L} f_{\mathrm{NL}}(i_L) + \frac{1}{L} e_{2g}, \qquad (7.39)$$

and for the capacitor C,

$$\frac{de_{2g}}{dt} = \frac{1}{C}(i_{R1} - i_L)$$

or

$$\frac{de_{2g}}{dt} = \frac{1}{C}\left(\frac{e_s - e_{2g}}{R_1} - i_L\right).$$

Rearranging yields

$$\frac{de_{2g}}{dt} = \left(\frac{-1}{C}\right)i_L + \left(\frac{-1}{R_1 C}\right)e_{2g} + \left(\frac{1}{R_1 C}\right)e_s. \qquad (7.40)$$

Note that Eqs. (7.39) and (7.40) comprise a set of nonlinear state-variable equations for the circuit.

Because of the nonlinear function $f_{\mathrm{NL}}(i_L)$, Eqs. (7.39) and (7.40) cannot be combined to eliminate i_L until they are linearized for small perturbations of all variables.

(b) After linearizing, Eqs. (7.39) and (7.40) become

$$\frac{d\hat{i}_L}{dt} = \left(\frac{-1}{L}\right)\frac{de_{23}}{di_L}\Big|_{\bar{i}_L}\hat{i}_L + \left(\frac{1}{L}\right)\hat{e}_{2g}, \qquad (7.41)$$

$$\frac{d\hat{e}_{2g}}{dt} = \left(\frac{-1}{C}\right)\hat{i}_L + \left(\frac{-1}{R_1 C}\right)\hat{e}_{2g} + \left(\frac{1}{R_1 C}\right)\hat{e}_{2g}. \qquad (7.42)$$

(c) Equations (7.41) and (7.42) may now be combined as follows to eliminate $\hat{i}_L$. Solving Eq. (7.42) for $\hat{i}_L$ yields

$$\hat{i}_L = -C\frac{d\hat{e}_{2g}}{dt} - \left(\frac{1}{R_1}\right)\hat{e}_{2g} + \left(\frac{1}{R_1}\right)\hat{e}_s.$$

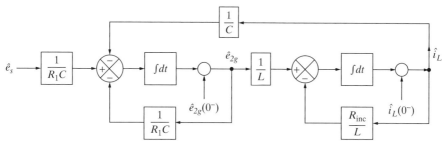

Figure 7.13(b). Simulation block diagram for the linearized system.

Substituting the expression for $\hat{i}_L$ in Eq. (7.41) and using $R_{inc} = \left.\frac{de_{2g}}{di_L}\right|_{\bar{i}_L}$ yields

$$-C\frac{d^2\hat{e}_{2g}}{dt^2} - \frac{1}{R_1}\frac{d\hat{e}_{2g}}{dt} + \frac{1}{R_1}\frac{d\hat{e}_s}{dt} = -\frac{R_{inc}}{L}\left(-C\frac{d\hat{e}_{2g}}{dt} - \frac{\hat{e}_{2g}}{R_1} - \frac{\hat{e}_s}{R_1}\right) + \frac{\hat{e}_{2g}}{L}.$$

Collecting terms, we have

$$C\frac{d^2\hat{e}_{2g}}{dt^2} + \left(\frac{L + R_1 R_{inc}C}{R_1 L}\right)\frac{d\hat{e}_{2g}}{dt} + \left(\frac{R_{inc} + R_1}{R_1 L}\right)\hat{e}_{2g} = \frac{1}{R_1}\frac{d\hat{e}_s}{dt} + \frac{R_{inc}}{LR_1}\hat{e}_s. \quad (7.43)$$

(d) The simulation block diagram for the linearized system appears in Fig. 7.13(b).

7.5   OPERATIONAL AMPLIFIERS

An operational amplifier (op-amp) is an electrical amplifier with a very high voltage gain, of the order of 10^6 V/V. It is constructed as an integrated circuit or an interconnection of many (30–40, or so) components, primarily transistors and resistors, fabricated on a piece of silicon. Op-amps have found an enormous number of versatile applications in engineering systems that perform processing of electrical signals. There are hundreds of different types of op-amps on the market today, and most of them can be purchased for less than a dollar. Although a complete discussion of the topic is clearly beyond the scope of this text, this section briefly reviews the principles of operation and several most common applications of op-amps that are likely to be of interest to an engineer dealing with instrumentation, measurement, or control system applications.[4]

The schematic circuit diagram of an op-amp is presented in Fig. 7.14. It shows two inputs, a noninverting input (+) and an inverting input (−), and one output. Figure 7.15 shows a top view of a typical op-amp package, such as the LM 741, which has eight terminals distributed along two sides of the package.

Because of their very high gain, op-amps are almost never used to simply amplify the voltage applied to the two input terminals shown in Fig. 7.14. In this configuration, the output signal swings over its full range and into saturation (near supply voltage) for difference in input signals of a very small magnitude, of the order of 10 μV. This is illustrated in Fig. 7.16, which shows a typical input–output voltage characteristic of

---

[4] For a more complete treatment of op-amps and their applications, see P. Horowitz and W. Hill, *The Art of Electronics*, 2nd ed. (Cambridge University Press, Cambridge, 1999), pp. 175–261.

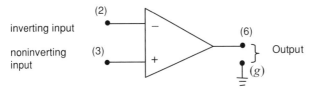

Figure 7.14. Circuit diagram of an op-amp.

an op-amp. A linear range of the output voltage extends from 0 V to approximately 2 to 3 V below and above the supply voltage. If the supply voltage of ± 15 V is used, the output voltage of an op-amp varies in a linear fashion over a range of approximately -12 to $+12$ V. Within this range, the relationship between the output, $e_o = e_{6g}$, and input voltage, e_{23}, is linear, i.e.,

$$e_{6g} = k_a e_{23}, \qquad (7.44)$$

where k_a is the gain of the amplifier. For k_a equal to 10^6 V/V, the corresponding linear range of the input voltage is approximately from -12 to $+12$ μV, which is far below the magnitude of typical signals encountered in engineering systems.

Therefore, in almost all practical applications, op-amps are used in a negative-feedback configuration, which not only lowers the voltage gain but also improves considerably their other performance characteristics. Negative feedback is the main topic of Chaps. 13 and 14, but for the purposes of the present discussion it will suffice to say that it is a process whereby a signal related to the system output is transferred back to the input of the system in such a way that it reduces the net input signal applied to the system. As the output signal tends to increase, so does the reduction of the input signal. As a result, an op-amp's voltage gain is reduced to a level at which it can operate effectively, without being driven into saturation when subjected to typical signals encountered in engineering systems.

There are two overarching assumptions that are generally accepted in the analysis of op-amps with negative feedback:

Assumption 1: Because the op-amp gain is so high, when the output voltage is within its normal operating range, the voltage between the input terminals is so small that it can be assumed to be zero.

Assumption 2: The current drawn by an op-amp input is zero because its input resistance is very, very large.

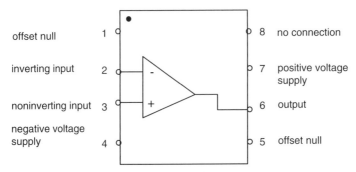

Figure 7.15. Top view of the op-amp package.

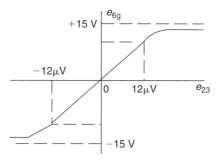

Figure 7.16. Output voltage versus input voltage characteristic of an op-amp.

Several of the most common op-amp circuits are now analyzed based on these two assumptions.

7.5.1 Inverting Amplifier

The inverting amplifier is shown in Fig. 7.17. In this configuration, the noninverting input terminal of the op-amp is at ground and thus, from assumption 1, terminal 2 is at ground also. In fact, terminal 2 is sometimes called a virtual ground. This implies that the voltage across R_1 is e_i and the voltage across R_2 is e_o. From the Kirchhoff's current law equation at the inverting input,

$$i_1 + i_2 = i_{in}. \tag{7.45}$$

Because of assumption 2,

$$i_{in} = 0. \tag{7.46}$$

Hence, from Eq. (7.45),

$$i_1 = -i_2, \tag{7.47}$$

or, equally,

$$\frac{e_i}{R_1} = -\frac{e_o}{R_2}. \tag{7.48}$$

The expression for the output voltage is

$$e_o = -\frac{R_2}{R_1} e_i. \tag{7.49}$$

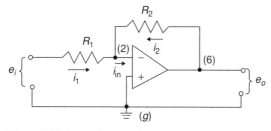

Figure 7.17. Inverting amplifier.

Expressed verbally, the output voltage of the inverting amplifier is equal to the inverted input voltage times the amplifier gain (R_2/R_1). One undesirable characteristic of an inverting amplifier is that its input resistance, which is equal to R_1, can be relatively small, especially when high voltage gain is required. Therefore excessive current may be drawn from whatever device is connected to R_1.

7.5.2 Noninverting Amplifier

The diagram of a noninverting amplifier is shown in Fig. 7.18. From assumption 1,

$$e_{2g} = e_{3g} = e_i. \tag{7.50}$$

However, e_{2g}, can also be determined from a voltage-divider formula:

$$e_{2g} = \frac{R_1}{R_1 + R_2} e_{6g}. \tag{7.51}$$

Recognizing that e_{6g} is the same as the output voltage e_o and using Eq. (7.50) gives

$$e_i = \frac{R_1}{R_1 + R_2} e_o. \tag{7.52}$$

Hence the output voltage of the noninverting amplifier is

$$e_o = \left(1 + \frac{R_2}{R_1}\right) e_i. \tag{7.53}$$

The noninverting amplifier has a huge input resistance of the order of 10^{12} Ω and a very small output resistance of a fraction of an ohm.

7.5.3 Voltage Follower

A voltage follower is a noninverting amplifier with R_1 infinite and R_2 equal to zero. The circuit diagram of a voltage follower is shown in Fig. 7.19. Because this is a special case of a noninverting amplifier, Eq. (7.53) still applies but with the specified values

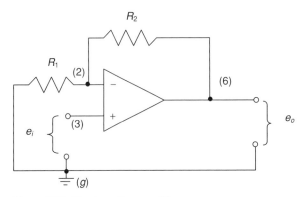

Figure 7.18. Noninverting amplifier.

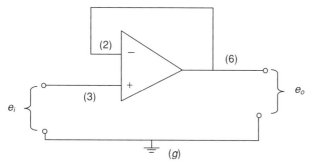

Figure 7.19. Voltage follower.

of R_1 and R_2, the output voltage in this case is

$$e_o = e_i. \tag{7.54}$$

Like the noninverting amplifier, the voltage follower has a very large input resistance and a very small output resistance, which makes it an ideal unit to isolate two parts of an electrical system. For this reason, the voltage follower is sometimes called a buffer.

7.5.4 Summing Amplifier

A summing amplifier is an inverting amplifier with multiple inputs. Figure 7.20 shows a diagram of a summing amplifier with two inputs. Following the same steps as in Subsection 7.5.1, here we find that the Kirchhoff's current law equation at terminal 2 is

$$i_1 + i_2 + i_3 = i_{\text{in}}. \tag{7.55}$$

Using Ohm's law for each current on the left-hand side and applying assumption 2 yields

$$\frac{e_{i1}}{R_1} + \frac{e_{i2}}{R_2} + \frac{e_o}{R_3} = 0. \tag{7.56}$$

If all resistors are selected the same, $R_1 = R_2 = R_3$, the output voltage is an inverted sum of the input voltages:

$$e_o = -(e_{i1} + e_{i2}). \tag{7.57}$$

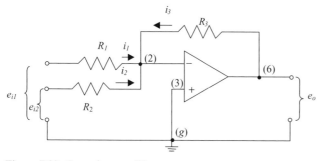

Figure 7.20. Summing amplifier.

7.5.5 Differential Amplifier

The circuit diagram of a differential amplifier is shown in Fig. 7.21. From assumption 2, Kirchhoff's current law equation at terminal 2 is

$$i_1 + i_2 = 0, \tag{7.58}$$

where the two currents are

$$i_1 = \frac{e_{i1} - e_{2g}}{R_1}, \tag{7.59}$$

$$i_2 = \frac{e_o - e_{2g}}{R_2}. \tag{7.60}$$

Substituting the expressions for currents into Eq. (7.58) and solving for e_{2g} yields

$$e_{2g} = \left(\frac{R_2}{R_1 + R_2} \right) e_{i1} + \left(\frac{R_1}{R_1 + R_2} \right) e_o. \tag{7.61}$$

When the voltage-divider expression is used, the voltage at terminal 3 is

$$e_{3g} = \left(\frac{R_2}{R_1 + R_2} \right) e_{i2}. \tag{7.62}$$

Based on assumption 1, the voltages defined by Eqs. (7.61) and (7.62) are the same, $e_{2g} = e_{3g}$, which gives

$$\left(\frac{R_2}{R_1 + R_2} \right) e_{i1} + \left(\frac{R_1}{R_1 + R_2} \right) e_o = \left(\frac{R_2}{R_1 + R_2} \right) e_{i2}, \tag{7.63}$$

and hence,

$$e_o = \left(\frac{R_2}{R_1} \right) (e_{i2} - e_{i1}). \tag{7.64}$$

If both R_1 and R_2 are selected the same, the output voltage is equal to the difference between the input voltages,

$$e_o = e_{i2} - e_{i1}. \tag{7.65}$$

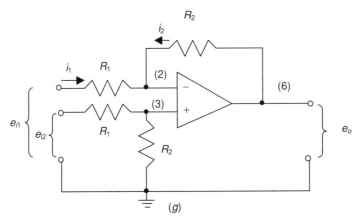

Figure 7.21. Differential amplifier.

7.5.6 Active Filter

In the applications discussed so far, the only electrical components (other than op-amps) used in the circuits were resistors. Because resistors are energy-dissipating elements and are not capable of storing energy, all the circuits act as static systems with various static relationships between input and output voltage signals. By including energy-storing elements, such as capacitors, in the op-amp circuits, dynamic systems can be designed to perform a wide range of dynamic operations such as integration, differentiation, or filtering with desired frequency characteristics. All those dynamic op-amp circuits, generally referred to as active filters, find a great variety of applications in signal processing and in control systems. The analysis of the most common active filters is performed in a manner similar to that used in Subsections 7.5.1–7.5.5, and it is illustrated in the following example of an integrator.

EXAMPLE 7.4

Derive a mathematical expression for the output voltage in the circuit shown in Fig. 7.22.

The circuit shown in Fig. 7.22 is similar to the inverting amplifier circuit shown in Fig. 7.17, except that in this active filter a capacitor is used instead of resistance R_2 in the feedback path. To follow the steps used in the analysis of an inverting amplifier in Subsection 7.5.1, start with Kirchhoff's current law equation at terminal 2,

$$i_R = -i_C, \tag{7.66}$$

where the two currents can be expressed as

$$i_R = \frac{e_i}{R}, \tag{7.67}$$

$$i_C = C\frac{de_o}{dt}. \tag{7.68}$$

Substituting the expressions for currents into Eq. (7.66) gives

$$\frac{e_i}{R} = -C\frac{de_o}{dt}, \tag{7.69}$$

and hence the output voltage is

$$e_o = -\frac{1}{RC}\int_0^t e_i\,dt + e_o(0). \tag{7.70}$$

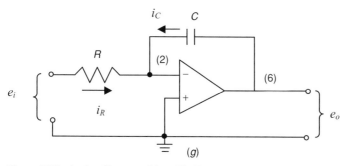

Figure 7.22. Active filter considered in Example 7.4.

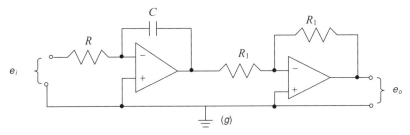

Figure 7.23. Noninverting integrator.

Thus the output voltage is proportional to the integral of the input voltage with a gain of $-(1/RC)$. To change the gain to a positive number, so that the output signal is positive when an integral of the input signal is positive, the output of this circuit can be connected to an inverting amplifier with $R_1 = R_2$, which results in a voltage gain of -1. A complete circuit diagram is shown in Fig. 7.23. With the initial condition set to zero, the output voltage from this circuit is

$$e_o = \frac{1}{RC} \int_0^t e_i \, dt. \tag{7.71}$$

A simulation block diagram for this filter acting as a noninverting integrator is shown in Fig. 7.24.

7.6 LINEAR TIME-VARYING ELECTRICAL ELEMENTS

Many of the instrument and control systems used in engineering use electrical elements that are time varying. The most common is the variable resistor, in which the resistance is caused to vary with time.

The elemental equation of this time-varying resistor is

$$e_R = e_{1g} - e_{2g} = e_{12} = R(t)i_R. \tag{7.72}$$

If such a resistor is supplied with a constant current, its voltage drop will have a variation with time that is proportional to the variation in R. Conversely, if the resistor is supplied with a constant voltage drop, it will have a current that varies inversely with the varying of R.

The capacitor may also be a time-varying element, having the time-varying elemental equations

$$q_c = C(t)e_C, \tag{7.73}$$

and, because $i_C = dq_C/dt$,

$$i_C = C(t)\frac{de_C}{dt} + e_C\frac{dC}{dt}. \tag{7.74}$$

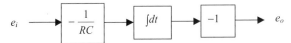

Figure 7.24. Simulation block diagram for noninverting integrator.

Similarly, the elemental equations for a time-varying inductor are

$$\lambda = L(t)i_L,$$

or

$$e_L = \frac{d\lambda}{dt}, \tag{7.75}$$

and, because $e_L = d\lambda/dt$,

$$e_L = L(t)\frac{di_L}{dt} + i_L\frac{dL}{dt}. \tag{7.76}$$

In general, the state variable for a time-varying capacitor is its charge q_C, and the state variable for a time-varying inductor is its flux linkage λ.

EXAMPLE 7.5

A time-varying inductor in series with a voltage source and a fixed resistor is being used as part of a moving-metal detector. The circuit diagram for this system is shown in Fig. 7.25. Derive the state-variable equations for the circuit.

SOLUTION

Use Kirchhoff's voltage law to write

$$e_s = e_L + Ri_L. \tag{7.77}$$

Substituting from Eq. (7.76) for e_L in Eq. (7.77) gives

$$e_s = L(t)\frac{di_L}{dt} + i_L\frac{dL}{dt} + Ri_L. \tag{7.78}$$

If i_L is selected as the state variable, the state-variable equation takes the following form:

$$\frac{di_L}{dt} = -\frac{1}{L(t)}\left(R + \frac{dL}{dt}\right)i_L + \frac{1}{L(t)}e_s. \tag{7.79}$$

This equation would be very difficult to solve because of the presence of the derivative of inductance. The state model of the circuit can be greatly simplified if a flux linkage λ is used as the state variable instead of current i_L. Substituting from Eq. (7.75) for e_L and i_L yields

$$e_s = \frac{d\lambda}{dt} + \frac{R}{L(t)}\lambda. \tag{7.80}$$

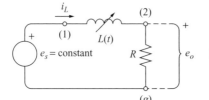

Figure 7.25. Time-varying inductor and resistor circuit.

Hence the state-variable equation in a standard form is

$$\frac{d\lambda}{dt} = -\frac{R}{L(t)}\lambda + e_s.\tag{7.81}$$

The output equation relating e_o to the state variable λ is

$$e_o = \frac{R}{L(t)}\lambda.\tag{7.82}$$

This example clearly demonstrates the benefits of selecting flux linkage as the state variable in circuits involving time-varying inductors.

7.7   SYNOPSIS

The basic physical characteristics of the linear, lumped-parameter electrical elements were discussed together with their mechanical system analogs: type A, electrical capacitor and mechanical mass; type T, electrical inductor and mechanical spring; type D, electrical resistor and mechanical damper. Thus capacitance C is analogous to mass m, inductance L is analogous to inverse stiffness $1/k$, and resistance R is analogous to inverse damping $1/b$. Diagrams and describing equations were presented for the two-terminal elements, including ideal voltage and ideal current sources. Kirchhoff's voltage and current laws were provided to serve as a means of describing the interactions occurring between the two-terminal elements in a system of interconnected elements (i.e., electrical circuits).

The dynamic analysis of electric circuits was then illustrated by means of several examples chosen to demonstrate different approaches to finding the input–output differential equation for a given circuit. In each case the procedure began with writing the elemental equations for all of the system elements and writing the required interconnection (Kirchhoff's law) equations to form a necessary and sufficient set of n equations relating n unknown variables to the system input(s) and time. Then, to eliminate the unwanted unknown variables, one of four different methods was used in each example: (a) step-by-step manipulation of the n equations to reduce the number of unwanted variables and reduce the number of equations one or two at a time; (b) the node method, which eliminates many or all of the unwanted T-type variables (currents) in a few steps; (c) the loop method, which eliminates many or all of the unwanted A-type variables (voltages) in a few steps; or (d) developing the set of state-variable equations based on each of the energy-storage elements, which eliminates all unwanted variables except the state variables, so that only the voltages across capacitors and the currents through inductors remain as unknowns. In cases (b), (c), and (d), the elimination of the remaining unwanted variables involves the algebraic manipulation of a reduced set of only m equations containing only m unknowns to arrive at the desired input–output differential equation containing the one remaining unknown and its derivatives on the left-hand side. For the case in which the system is to be simulated by computer, method (d) is preferred because it leads directly to the equations in proper form for programming on the computer, and the final reduction to one variable is needed only for the purposes of producing a check solution when needed. *Note that a properly constituted set of state-variable equations must not have any derivative terms on the right-hand sides of any of its equations.*

Analyses of electrical systems were carried out, including a circuit with a nonlinear resistor. The small-perturbations technique was used to derive a linearized model of the nonlinear circuit. In deriving state models for circuits containing time-varying capacitors and inductors, charge and flux linkage should be used as the state variables to avoid problems with the presence of derivatives of capacitance or inductance in the state variable equations.

Operational amplifiers were introduced with a brief discussion of their general performance characteristics. The discussion was illustrated with examples of several most commonly used op-amp circuits.

PROBLEMS

7.1 The source in the circuit shown in Fig. P7.1 undergoes a step change so that e_s suddenly changes from 0 to 10 V at $t = 0$. Before the step change occurs, all variables are constant, i.e., e_s has been zero for a long time.

(a) Find $e_{32}(0^-)$ and $e_{32}(0^+)$.

(b) Find $e_{1g}(0^-)$ and $e_{1g}(0^+)$.

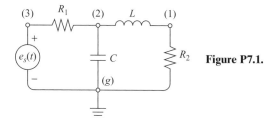

Figure P7.1.

7.2 The circuit shown in Fig. P7.2 is subjected to a step change in e_s from 5.0 to 7.0 V at $t = 0$. Before the step change occurs, all variables are constant, i.e., e_s has been 5.0 V for a very long time.

(a) Find $i_L(0^-)$, $i_L(0^+)$, and $i_L(\infty)$.

(b) Find $e_{3g}(0^-)$, $e_{3g}(0^+)$, and $e_{3g}(\infty)$.

(c) Find $e_{2g}(0^-)$, $e_{2g}(0^+)$, and $e_{2g}(\infty)$.

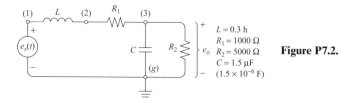

$$L = 0.3 \text{ h}$$
$$R_1 = 1000 \ \Omega$$
$$R_2 = 5000 \ \Omega$$
$$C = 1.5 \ \mu\text{F}$$
$$(1.5 \times 10^{-6} \text{ F})$$

Figure P7.2.

7.3 You have been asked if you can determine the capacitance of an unknown capacitor handed to you by a colleague. This capacitor is not an electrolytic device, so you do not have to worry about its polarity. Not having a capacitance meter, you decide to use a battery that is close at hand, together with a decade resistor box, to run a simple test that will enable you to determine the capacitance. After the capacitor has been charged to 12.5 V from the battery and then disconnected from the battery, a high value of resistance, 100,000 Ω, is connected across the terminals of the capacitor, resulting in its slow discharge. The values of its voltage at successive intervals of time are recorded below.

Time, s	Capacitor voltage, e_{1g}, V
0	12.5
10	9.3
20	7.1
30	5.3
40	3.9
50	2.9
60	2.2

(a) Write the system differential equation for the capacitor voltage e_{1g} during the discharge interval: What is the input to this system? (*Hint:* What happens at $t = 0$?)

(b) Determine your estimated value for the capacitance C.

7.4 This is a continuation of Problem 7.3. You have rearranged the components R and C as shown in Fig. P7.4 and run another test by suddenly closing the switch and recording the capacitor voltage at successive increments of time. The results appear in the accompanying table.

Time, s	e_{1g}, V
0	2.2
10	4.3
20	5.9
30	7.2
40	8.3
50	9.1
60	9.6

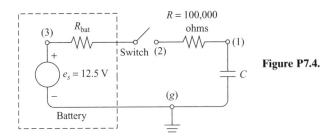

Figure P7.4.

Use the value of capacitance found in Problem 7.3(b) to determine an estimated value for the internal resistance of the battery, R_{bat}.

7.5 (a) Develop the system differential equation relating the output voltage e_{3g} to the input voltage $e_s = e_{2g}$ for the circuit shown in Fig. P7.5. (Use only the symbols in the figure.)

(b) The input voltage, which has been zero for a very long time, is suddenly increased to 5.0 V at $t = 0$. Find $e_{3g}(0^-)$, $e_{3g}(0^+)$, and $e_{3g}(\infty)$.

(c) Find the system time constant, and sketch the response of e_{3g} versus time t.

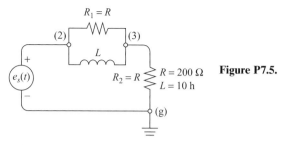

Figure P7.5.

7.6 An electric circuit is being driven with a voltage source e_s as shown in Fig. P7.6, and the output of primary interest is the capacitor voltage $e_o = e_{3g}$.

(a) Derive the system differential equation relating e_o to e_s.

(b) Find the undamped natural frequency and damping ratio for this system.

(c) Find $i_L(0^+)$, $e_o(0^+)$, and de_o/dt at $t = 0^+$, assuming that e_s has been equal to zero for a very long time and then suddenly changes to 10 V at $t = 0$.

(d) Sketch the response of e_o versus time, indicating clearly the period of oscillation (if applicable) and the final steady-state value of e_o.

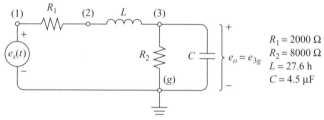

Figure P7.6.

7.7 (a) Develop a complete set of state-variable equations for the circuit shown in Fig. P7.7, using e_{3g} and i_L as the state variables.

(b) Draw the simulation block diagram for this system, using a separate block for each independent parameter and showing all system variables.

(c) Consider e_{3g} to be the output variable of primary interest and develop the input–output system differential equation for this system.

(d) Having been equal to 5 V for a very long time, e_s is suddenly decreased to 2 V at time $t = 0$. Find $i_L(0^+)$, $i_C(0^+)$, $e_{3g}(0^+)$, and de_{3g}/dt at $t = 0^+$.

(e) Sketch the response of e_{3g} versus time, given the following system parameters: $R_1 = 80\ \Omega$, $R_2 = 320\ \Omega$, $L = 2$ h, $C = 50\ \mu$F, $R_3 = 5\ \Omega$.

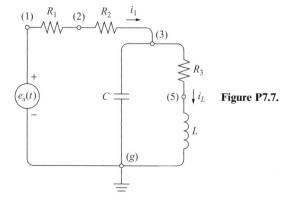

Figure P7.7.

7.8 You have been given a field-controlled dc motor for use in a research project, but the values of the field resistance and field inductance are unknown. By conducting a simple test of the field with an initially charged capacitor of known capacitance, you should find it possible to estimate the values of R_f and L_f for this motor winding. The schematic diagram in Fig. P7.8 shows such a test arrangement, the test being initiated by closing the switch at $t = 0$.

(a) Assuming a series R–L model for the field winding, draw the lumped-parameter circuit diagram for this system. (Why would a parallel R–L model for the field winding not be correct?)

(b) Derive the system differential equation for the voltage e_{21} after the switch is suddenly closed. (Note that the input for this system is a sudden change in the system at $t = 0$; after $t = 0$, this is a homogeneous system with no input variable.)

(c) After the switch is closed, the voltage e_{21} is displayed versus time on an oscilloscope. A damped oscillation, having a period of 0.18 s and a DR of 0.8 per cycle, is observed. Determine estimated values of R_f and L_f, based on your knowledge about the response of second-order systems.

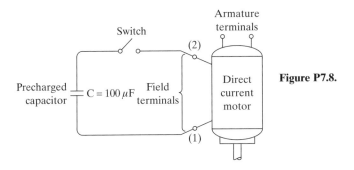

Figure P7.8.

7.9 The nonlinear circuit shown in Fig. P7.9 is to be subjected to an input voltage consisting of a constant normal operating value $\bar{e}_s$ and an incremental portion $\hat{e}_s$, which changes with time.

The system parameters are $R_1 = 400\,\Omega$, $L = 3.0\,\text{h}$, $C = 5\,\mu F$, and, for the NLR, $e_{23} = 3.0 \times 10^6 (i_{\text{NLR}})^3$. The normal operating value of the source voltage $\bar{e}_s(t)$ is 5.0 V.

(a) Find the normal operating-point value of the output voltage $\bar{e}_{2g}$ and the incremental resistance R'_{inc} of the NLR.

(b) Develop the nonlinear state-variable equations for the circuit, using e_{12} and i_{NLR} as the state variables, and linearize them for small perturbations from the normal operating point.

(c) Find the natural frequency ω_n and the damping ratio ζ of the linearized system, and carry out the analytical solution for e_{2g} when e_s suddenly increases by 0.5 V at $t = 0$. Sketch $e_{2g}(t)$.

7.10 The nonlinear electric circuit shown in Fig. P7.10 is subjected to a time-varying current source i_s, and its output is the voltage e_{2g}. The nonlinear resistor obeys the relation $e_{23} = 1.0 \times 10^7 (i_{\text{NLR}})^3$.

(a) Write the set of nonlinear state-variable equations using i_{NLR} and e_{2g} as the state variables.

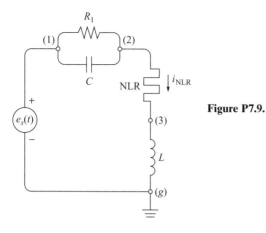

Figure P7.9.

(b) Given the following normal operating-point data, find the values of the resistance R_1 and the normal operating current i_{NLR}:

$$\bar{e}_{12} = 1.2\mathrm{V} \quad \bar{e}_{23} = 3.75\,\mathrm{V}.$$

(c) Develop the set of linearized state-variable equations for this system.

(d) Draw the simulation block diagram for this linearized system, using a separate block for each system parameter and showing all system variables.

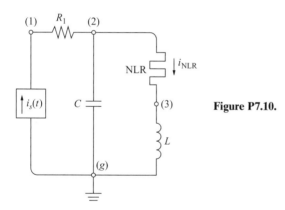

Figure P7.10.

7.11 The nonlinear electric circuit shown in Fig. P7.11 is subjected to an input voltage e_s, which consists of a constant normal operating component $\bar{e}_s = 5.0\,\mathrm{V}$ and a time-varying component $\hat{e}_s(t)$. The NLR is a square-law device such that $e_{23} = K|i_{\mathrm{NLR}}|i_{\mathrm{NLR}}$.

(a) Develop the nonlinear state-variable equations for the circuit using e_{3g} and i_L as the state variables and using symbols for the system parameters.

(b) Determine the incremental resistance $R_{\mathrm{inc}} = (de_{23}/di_{\mathrm{NLR}})|i_{\mathrm{NLR}}$ (i.e., the linearized resistance) for the NLR. Use symbols only.

(c) Write the linearized state-variable equations for the system, using symbols only.

(d) Draw the simulation block diagram for the linearized system, using a separate block for each parameter and showing all system variables.

(e) Derive the input–output system differential equation for the linearized system, with $e_o = e_{3g}$ as the output.

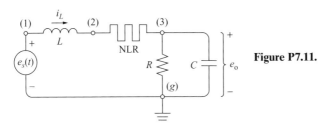

Figure P7.11.

7.12 In the variable-reactance transducer circuit shown in Fig. P7.12, the air gap varies with time, producing a corresponding variation of inductance with time given by

$$L(t) = 1.0 + 0.1 \sin 5t.$$

The input voltage $e_s = 6.0\,\text{V}$ is constant, and the output signal is $e_{2g} = \bar{e}_{2g} + \hat{e}_{2g}(t)$. The values of the resistances are $R_1 = 1.2 \times 10^3\,\Omega$ and $R_2 = 2.4 \times 10^3\,\Omega$.

(a) Derive the state-variable equation for this circuit using magnetic flux linkage λ as the state variable.

(b) Determine the normal operating conditions (i.e., the values of $\bar{e}_{2g}$ and $\bar{i}_L$) for this system and linearize the state-variable equation for small perturbations about the normal operating point.

(c) Write the differential equation relating $\hat{e}_{2g}$ to $L(t)$.

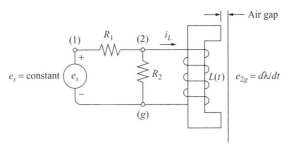

Figure P7.12.

7.13 In the circuit shown in Fig. P7.13, the value of the capacitance varies with time about its mean value $\bar{C}$ and the voltage source e_s is constant. In other words,

$$C = \bar{C} + \hat{C}(t).$$

(a) Develop the state-variable equation for this system, using charge q_C on the capacitor as the state variable.

(b) Linearize this state-variable equation for small perturbations about the normal operating point.

(c) Write the linearized system differential equation relating $\hat{e}_{2g}$ to $\hat{C}(t)$.

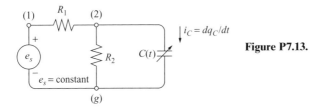

Figure P7.13.

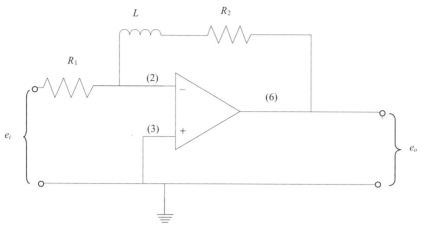

Figure P7.14.

7.14 The circuit shown in Fig. P7.14 is often used to obtain desired dynamic characteristics for use in automatic control systems. Make the commonly accepted assumptions about op-amps (as discussed in Section 7.5) and derive the input–output equation for the system.

7.15 The circuit shown in Fig. P7.15 is a second-order low-pass Butterworth filter used widely in data acquisition systems to filter out high-frequency noise signals that contaminate the measuring signal.

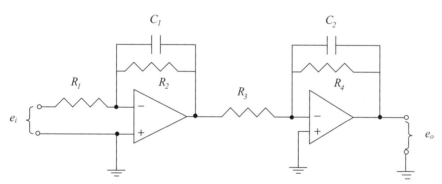

Figure P7.15. Second-order Butterworth filter.

Make the necessary simplifying assumptions and derive the state-variable equations for the filter. Obtain the input–output equation relating output voltage e_o to input voltage e_i. Assuming that the system is overdamped, find the expressions for the time constants and for the steady-state gain of the system.

7.16 A thermistor is a semiconducting temperature sensor whose operation is based on the temperature-dependent electrical resistance. A typical characteristic of the thermistor relating output voltage V_t to temperature T is shown in Fig. P7.16(a).

This form of the characteristic, which represents the sensor's calibration curve, is very inconvenient for signal recording and processing purposes. The desirable form of the calibration curve is shown in Fig. P7.16(b).

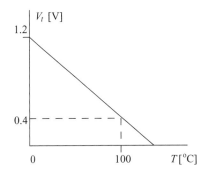

Figure P7.16(a). Characteristic of a thermistor sensor.

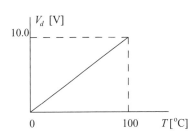

Figure P7.16(b). Desired calibration characteristic of thermistor sensor.

The thermistor actual calibration curve shown in Fig. P7.16(a) can be modified to the desired form, shown in Fig. P7.16(b), by means of a simple op-amp circuit shown in Fig. P7.16(c).

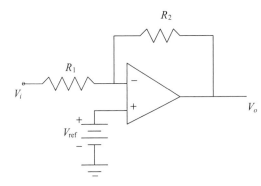

Figure P7.16(c). Op-amp thermistor calibration circuit.

(a) Write the expression for the output voltage V_o in terms of inputs V_i and V_{ref}. Note that the circuit is a combination of an inverting amplifier (V_o versus V_i) and a noninverting amplifier (V_o versus V_{ref}).

(b) Write the mathematical equation for the actual calibration curve shown in Fig. 3(a).

(c) Write the mathematical equation for the desired calibration curve shown in Fig. 3(b).

(d) Assuming $R_1 = 1\,k\Omega$, find the values of R_2 and V_{ref} in the op-amp circuit to modify the original calibration curve of the thermistor into the desired form.

7.17 The coil in the magnetic levitation apparatus described in Problem 2.15 is driven by the electric circuit shown in Fig. P7.17. Including the dynamics of the ball as derived in that problem, add the appropriate state equation(s) to include the electrical dynamics introduced by this circuit. The additional parameter values are as follows:

$$R_1 \qquad\qquad 10 \; \Omega$$
$$R_s \qquad\qquad 1 \; \Omega$$
$$L \qquad\qquad 0.4124 \; H$$

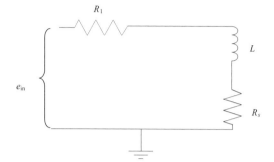

Figure P7.17. Schematic of circuit modeling the magnetic levitation coil.

(a) Do you think the additional dynamics introduced by the circuit make significant impact on the overall system dynamics? Explain and justify your answer.

(b) The second resistor in the circuit is an actual discrete resistor, specifically included by the engineers who designed this apparatus, and they specified that it be 1.000 Ω to a very high degree of accuracy. Can you say why this has been included in the system? (*Hint*: In designing a workable control system for the apparatus, it is necessary that the instantaneous current in the coil be measured.)

8

Thermal Systems

LEARNING OBJECTIVES FOR THIS CHAPTER

8–1 To recognize the A- and D-type elements of thermal systems.

8–2 To identify and model the three fundamental modes of heat transfer.

8–3 To use the energy-balance method to develop models of lumped-parameter thermal systems.

8.1 INTRODUCTION

Fundamentals of mathematical methods used today to model thermal systems were developed centuries ago by such great mathematicians and scientists as Laplace, Fourier, Poisson, and Stefan. The analytical solution of the equations describing the basic mechanisms of heat transfer – conduction, convection, and radiation– was always considered to be an extremely challenging mathematical problem. The study of energy transfer in thermal systems continues to be an important topic in engineering because it forms the basis of analysis of energy efficiency for indoor environmental controls, industrial processes, and all forms of energy transformation.

As described in Chap. 1, temperature is an A-type variable, determining the amount of energy stored in a thermal capacitance, the A-type energy-storing element corresponding to a mass in mechanical systems or a capacitor in electrical systems. All matter has thermal capacitance (which is proportional to mass), and energy is stored as internal energy because of its temperature. The T-type variable in thermal systems is heat flow rate Q_h; however, as was pointed out in Section 1.2, there is no T-type element in thermal systems that would be capable of storing energy as a result of heat flow rate. The D-type element in thermal systems is the thermal resistor and will be defined as the resistance to heat transfer.

Temperature is usually dependent on spatial as well as temporal coordinates. As a result, the dynamics of thermal systems has to be described by partial differential equations. Moreover, radiation and convection are by their very nature nonlinear, further complicating the solutions. Few problems of practical interest described by nonlinear partial differential equations have been solved analytically, and the numerical solution usually requires an extensive study to be carried out with sophisticated finite-element and computational fluid dynamics software.

In this chapter, methods are presented to model the dynamic behavior of thermal systems that can be seen as lumped-parameter systems. In addition, techniques to approximate spatial variations of temperature by use of lumped approximations are also presented. First, the equations describing the basic mechanisms of heat transfer by conduction, convection, and radiation are reviewed in Section 8.2. In Section 8.3, lumped models of thermal systems are introduced. Application of the lumped models leads to approximate solutions, and great care must be exercised in interpreting the results produced by this method. In many cases, however, especially in the early stages of system analysis, the lumped models are very useful because of their simplicity and easy solution methods.

8.2   BASIC MECHANISMS OF HEAT TRANSFER

In this section the classical mathematical equations describing heat transfer by conduction, convection, and radiation are reviewed. This part of the material is presented in a rather condensed form because it is assumed that the reader is generally familiar with it.

8.2.1 Conduction

The net rate of heat transfer by conduction across the boundaries of a unit volume is equal to the rate of heat accumulation within the unit volume, which is mathematically expressed by the Laplace equation

$$\frac{\partial^2 T}{\partial x^2} + \frac{\partial^2 T}{\partial y^2} + \frac{\partial^2 T}{\partial z^2} = \frac{1}{\alpha}\frac{\partial T}{\partial t}, \tag{8.1}$$

where T is temperature, α is thermal diffusivity, and x, y, and z are the Cartesian space coordinates. The thermal diffusivity can be expressed as $\alpha = k/\rho c_p$, where k is thermal conductivity, ρ is density, and c_p is the specific heat of the material.

At steady state – that is, when temperature remains constant in time – the Laplace equation takes the form

$$\frac{\partial^2 T}{\partial x^2} + \frac{\partial^2 T}{\partial y^2} + \frac{\partial^2 T}{\partial z^2} = 0. \tag{8.2}$$

For one-dimensional heat conduction in the x direction, the rate of heat flow is determined by the Fourier equation

$$Q_{hk} = -kA\frac{dT}{dx}, \tag{8.3}$$

where A is an area of heat transfer normal to x. Integrating with respect to x yields

$$\int_0^L Q_{hk}dx = -\int_{T_1}^{T_2} kAdT, \tag{8.4}$$

and hence the rate of heat transfer is

$$Q_{hk} = -\left(\frac{A}{L}\right)\int_{T_1}^{T_2} kdT. \tag{8.5}$$

The one-dimensional steady-state heat conduction is depicted in Fig. 8.1. Note that the assumption of the temperature gradients in the y and z directions being equal to

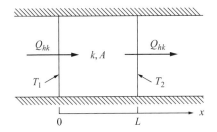

Figure 8.1. One-dimensional steady-state heat conduction.

zero implies that there is no heat loss through the sides, which are said to be perfectly insulated. At steady state, the temperature is constant in time and thus there is no energy storage.

If the thermal conductivity of the material does not depend on temperature, the rate of heat transfer can be expressed as

$$Q_{hk} = \left(\frac{kA}{L}\right)(T_1 - T_2). \tag{8.6}$$

In Eq. (8.6), T_1 is the temperature at $x = 0$ and T_2 is the temperature at $x = L$. In general, if the value of the rate of heat transfer Q_{hk} is known, the value of temperature at any location between 0 and L, $0 < x < L$, can be calculated, substituting $T(x)$ for T_2 in Eq. (8.6):

$$T(x) = T_1 - \left(\frac{x}{Ak}\right)Q_{hk}. \tag{8.7}$$

Note that Q_{hk} is positive if heat is transferred in the direction of x and negative if heat is flowing in the direction opposite to x.

8.2.2 Convection

Convective heat transfer is usually associated with the transfer of mass in a boundary layer of a fluid over a fixed wall. In the system shown in Fig. 8.2, the rate of heat transfer by convection Q_{hc} between a solid wall and a fluid flowing over it is given by

$$Q_{hc} = h_c A(T_w - T_f), \tag{8.8}$$

where h_c is a convective heat transfer coefficient, A is an area of heat transfer, and T_w and T_f represent wall and fluid temperatures, respectively.

From the distribution of the fluid velocity within the boundary layer shown in Fig. 8.2, it can be seen that the velocity is zero on the surface of the wall and thus no convective heat transfer can take place there. However, although heat is transferred across the wall–fluid boundary by conduction, it is carried away from there by convection with the flowing fluid. The rate at which the fluid carries heat from the wall surface is thus determined by the heat convection equation [Eq. (8.8)]. If the flow of fluid is caused by its density gradient in the gravity field, the heat convection is said to be free. The density gradient usually occurs as a result of temperature gradient. Forced convection, on the other hand, takes place when the flow of fluid is forced by an external energy source, such as a pump or a blower.

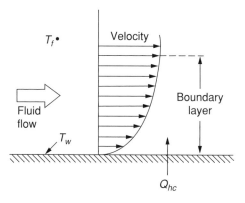

Figure 8.2. Convective heal transfer between a fluid and a wall.

Determination of the value of the convective heat transfer coefficient for an actual thermal system presents a challenging task. In many practical cases, empirical models are developed that express a convective heat transfer coefficient as a function of other system variables for specified operating conditions. The applicability of such empirical formulas is usually limited to a narrow class of problems. Moreover, the mathematical forms used in convective heat transfer correlations are often nonlinear, which leads to nonlinear system-model equations, which are difficult to solve analytically.

8.2.3 Radiation

The rate of heat transfer by radiation between two separated bodies having temperatures T_1 and T_2 is determined by the Stefan–Boltzmann law,

$$Q_{hr} = \sigma F_E F_A A (T_1^4 - T_2^4), \tag{8.9}$$

where $\sigma = 5.667 \times 10^{-8} \, \mathrm{W/m^2 \, K^4}$ (the Stefan–Boltzmann constant), F_E is effective emissivity, F_A is the shape factor, and A is the heat transfer area.

The effective emissivity F_E accounts for the deviation of the radiating systems from blackbodies. For instance, the effective emissivity of parallel planes having emissivities ϵ_1 and ϵ_2 is

$$F_E = \frac{1}{\dfrac{1}{\epsilon_1} + \dfrac{1}{\epsilon_2} - 1}. \tag{8.10}$$

The values of the shape factor F_A range from 0 to 1 and represent the fraction of the radiative energy emitted by one body that reaches the other body. In a case in which all radiation emitted by one body reaches the other body – for example, in heat transfer by radiation between two parallel plane surfaces – the shape factor is equal to 1.

The nonlinearity of Eq. (8.9) constitutes a major difficulty in developing and solving mathematical model equations for thermal systems in which radiative heat transfer takes place. Determining values of model parameters also presents a challenging task, especially when radiation of gases is involved.

8.3   LUMPED MODELS OF THERMAL SYSTEMS

Mathematical models of thermal systems are usually derived from the basic energy-balance equations that follow the general form

$$
\begin{pmatrix} \text{rate of energy} \\ \text{stored} \\ \text{within system} \end{pmatrix} = \begin{pmatrix} \text{heat flow} \\ \text{rate} \\ \text{into system} \end{pmatrix} - \begin{pmatrix} \text{heat flow} \\ \text{rate} \\ \text{out of system} \end{pmatrix}
$$

$$
+ \begin{pmatrix} \text{rate of heat} \\ \text{generated} \\ \text{within system} \end{pmatrix} + \begin{pmatrix} \text{rate of work} \\ \text{done} \\ \text{on system} \end{pmatrix}. \qquad (8.11)
$$

For a stationary system composed of a material of density ρ, specific heat c_p, and constant volume V, the energy-balance equation takes the form

$$
\rho c_p V \frac{dT}{dt} = Q_{hin}(t) - Q_{hout}(t) + Q_{hgen} + \frac{dW}{dt}. \qquad (8.12)
$$

Equation (8.12) can be used only if the temperature distribution in the system, or in a part of the system, is uniform – that is, when the temperature is independent of spatial coordinates, $T(x, y, z, t) = T(t)$. The assumption about the uniformity of the temperature distribution also implies that the system physical properties, such as density and specific heat, are constant within the system boundaries.

Two basic parameters used in lumped models of thermal systems are thermal capacitance and thermal resistance.

The thermal capacitance of a thermal system of density ρ, specific heat c_p, and volume V is

$$
C_h = \rho c_p V. \qquad (8.13)
$$

The rate of energy storage in a system of thermal capacitance C_h is

$$
Q_{hstored} = C_h \frac{dT}{dt}. \qquad (8.14)
$$

Note that a thermal capacitance is an A-type element because its stored energy is associated with an A-type variable, T. Physically, thermal capacitance represents a systems ability to store thermal energy. It also provides a measure of the effect of energy storage on the system temperature. If the system thermal capacitance is large, the rate of temperature change owing to heat influx is relatively low. On the other hand, when the system thermal capacitance is small, the temperature increases more rapidly with the amount of energy stored in the system. For example, if the rate of thermal energy storage changes by ΔQ_h in a stepwise manner, the change in rate of change of temperature is, from Eq. (8.14),

$$
\frac{dT}{dt} = \frac{\Delta Q_h}{C_h}. \qquad (8.15)
$$

Another parameter used in lumped models of thermal systems is thermal resistance R_h. Thermal resistance to heat flow rate Q_h between two points having different temperatures, T_1 and T_2, is

$$R_h = \frac{T_1 - T_2}{Q_h}. \tag{8.16}$$

The mathematical expressions for thermal resistance are different for the three different mechanisms of heat transfer – conduction, convection, and radiation. The conductive thermal resistance R_{hk} can be obtained from Eqs. (8.6) and (8.16):

$$R_{hk} = \frac{L}{kA}. \tag{8.17}$$

From Eqs. (8.8) and (8.16), the convective thermal resistance is found to be

$$R_{hc} = \frac{1}{h_c A}. \tag{8.18}$$

The ratio of the conductive thermal resistance to the convective thermal resistance yields a dimensionless constant known as the Biot number[1] Bi:

$$\text{Bi} = \frac{R_{hk}}{R_{hc}}. \tag{8.19}$$

Using Eqs. (8.17) and (8.18), one can express the Biot number as

$$\text{Bi} = \frac{h_c L}{k}. \tag{8.20}$$

The value of the Biot number provides a measure of adequacy of lumped models to represent the dynamics of thermal systems. At the beginning of this section, lumped models were characterized as having a uniform temperature distribution, which occurs when an input heat flow (usually convective) encounters relatively low resistance within the system boundaries. The resistance to heat flow within the system is usually conductive in nature. Thermal systems represented by lumped models should therefore have relatively high thermal conductivities. It can thus be deduced that lumped models can be used when the ratio of convective thermal resistance to conductive thermal resistance is large or, equally, when the Biot number, which represents the reciprocal of that ratio, is small enough. The value of 0.1 is usually accepted as the threshold for the Biot number below which lumped models can be used to describe actual thermal systems with sufficient accuracy. If $\text{Bi} > 0.1$, distributed models involving partial differential equations are necessary for adequate representation of system dynamics.

To derive an expression for a radiative thermal resistance, the basic equation [Eq. (8.9)], describing heat transfer by radiation, has to be linearized. Applying the general linearization procedure based on Taylor series expansion to the function of two variables $f_{\text{NL}}(T_1, T_2)$ yields

$$f_{\text{NL}}(T_1, T_2) \approx f_{\text{NL}}(\overline{T}_1, \overline{T}_2) + \hat{T}_1 \frac{\partial f_{\text{NL}}}{\partial T_1}\bigg|_{\overline{T}_1, \overline{T}_2} + \hat{T}_2 \frac{\partial f_{\text{NL}}}{\partial T_2}\bigg|_{\overline{T}_1, \overline{T}_2}, \tag{8.21}$$

[1] J. P. Holman, *Heat Transfer*, 9th ed. (McGraw-Hill, New York, 2002), pp. 133–5.

where $\overline{T}_1$ and $\overline{T}_2$ represent the normal operating point. The linearized expression for radiative heat transfer is then

$$Q_{hr} \approx \overline{Q}_{hr} + b_1 \hat{T}_1 - b_2 \hat{T}_2, \tag{8.22}$$

where

$$\overline{Q}_{hr} = \sigma F_E F_A A (\overline{T}_1^4 - \overline{T}_2^4),$$
$$b_1 = 4\sigma F_E F_A A \overline{T}_1^3,$$
$$b_2 = 4\sigma F_E F_A A \overline{T}_2^3.$$

The form of expression (8.22) is not very convenient in modeling thermal systems because of its incompatibility with corresponding equations describing heat transfer by convection and conduction in lumped models. To achieve this compatibility, a different linearization procedure can be used. First, rewrite Eq. (8.9) in the following form:

$$Q_{hr} = \sigma F_E F_A A (T_1^2 + T_2^2)(T_1 + T_2)(T_1 - T_2) \tag{8.23}$$

Assuming that T_1 and T_2 in the first two parenthetical factors in Eq. (8.23) represent normal operating-point values, the linear approximation is

$$\hat{Q}_{hr} \approx \left(\frac{1}{R_{hr}}\right)\left[\hat{T}_1(t) - \hat{T}_2(t)\right], \tag{8.24}$$

where the radiative thermal resistance is

$$R_{hr} = \frac{1}{\sigma F_E F_A A \left(\overline{T}_1^2 + \overline{T}_2^2\right)\left(\overline{T}_1 + \overline{T}_2\right)}. \tag{8.25}$$

A lumped linear model of combined heat transfer by convection and radiation in parallel can be approximated by

$$\hat{Q}_h = \left(\frac{1}{R_{hc}} + \frac{1}{R_{hr}}\right)\left(\hat{T}_1 - \hat{T}_2\right). \tag{8.26}$$

The lumped-model parameters, thermal capacitance and thermal resistance, will now be used in specific examples.

EXAMPLE 8.1

As advances in microelectronic circuits have allowed continual miniaturization of digital circuits, the issue of cooling has become more important. If you open the PC on your desktop, you will find that the central processing unit (CPU) has a large black metal piece attached to it. This piece, the *heatsink*, is designed to conduct heat away from the processor (where high temperatures would lead to failure) and transfer that heat, by way of convection, to the air being circulated by one or more fans in the case of the computer. A significant portion of failures in electronic components are linked to failure in the system designed to cool the components.

In this example, a single vertical pin from one of these heatsinks is modeled and the equations that describe its dynamic behavior are solved in a Simulink model. Figure 8.3 shows the geometry of the pin, which is made of pure aluminum.

T_a

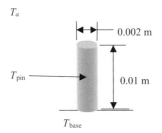

0.002 m

T_{pin}

0.01 m

Figure 8.3. Model of one pin in an electronic heatsink in Example 8.1.

T_{base}

The pin is initially at ambient temperature (25 °C). At time $t = 0$, the base is subjected to a temperature of 100 °C. Find the response of the temperature of the pin and the temperature profile along the length of the pin as it changes over time.

SOLUTION

The first step of the solution is to use the Biot number to see if the pin can be modeled as a single lump or if the problem must be broken down into many smaller elements. Equation 8.20 defines the Biot number, where L is the characteristic length for the problem, k is the thermal conductivity of aluminum, and h_c is the convective heat transfer coefficient for this problem.

The characteristic length is that dimension along which one is likely to see variations in temperature. In this case, the length (0.01 m), rather than the diameter, is the proper dimension to use. Thermal properties of aluminum are easily found in handbooks and are summarized in Table 8.1. As discussed in Subsection 8.2.2, the determination of convective heat transfer coefficients can be a very difficult task; however, for simple geometries, heat transfer textbooks provide reasonable ranges for these coefficients. For this geometry, it was decided that a value of 20 W/m² °C was a reasonable estimate.

Therefore the Biot number is

$$\mathrm{Bi} = \frac{h_c L}{k} = \frac{(20)(0.01)}{220} = 0.001,$$

which would indicate that a lumped model for the entire pin is adequate. The essential assumption to move forward from this point is that the entire pin is at a uniform temperature throughout the analysis. If part of the information you hoped to extract from this analysis is the temperature distribution along the pin, then the problem will still have to be broken down to smaller elements.

Table 8.1. Thermal properties of pure aluminum

Property	Value	Units
Density	2707	kg/m³
Thermal conductivity	220	W/m °C
Specific heat (c_p)	896	J/kg °C

Beginning with a simple single-lump model, we note the following energy balance:

$$\left(\begin{array}{l}\text{rate of heat stored} \\ \text{in the pin}\end{array}\right) = \left(\begin{array}{l}\text{rate of heat} \\ \text{conducted through base}\end{array}\right) - \left(\begin{array}{l}\text{rate of heat} \\ \text{convected to air}\end{array}\right),$$

(8.27)

from which the single differential equation can be written:

$$C_h \frac{dT_{\text{pin}}}{dt} = \frac{1}{R_{hk}}\left(T_{\text{base}} - T_{\text{pin}}\right) - \frac{1}{R_{hc}}\left(T_{\text{pin}} - T_a\right),$$

(8.28)

where

$$C_h = \rho c_p \mathcal{V} = (2707)(896)(3.14 \times 10^{-8}) = 0.0762 \text{ J/}^\circ\text{C},$$

$$R_{hk} = \frac{L}{kA_k} = \frac{0.005}{(220)(3.14 \times 10^{-6})} = 7.23 \,^\circ\text{C/W},$$

$$R_{hc} = \frac{1}{h_c A_c} = \frac{1}{(20)(6.28 \times 10^{-5})} = 796 \,^\circ\text{C/W}.$$

Note that the length used in the conductive term represents the length from the base of the pin to its center whereas the area A_k is the circular (cross-section) area of the base of the pin through which the conduction takes place. On the other hand, the area used to compute the convective resistance, A_c, is the surface area of the pin exposed to the air.

Substituting the numerical values and putting the equation in standard form leads to the following equation:

$$\frac{dT_{\text{pin}}}{dt} + 1.83\,T_{\text{pin}} = 1.81\,T_{\text{base}} + 0.017\,T_a.$$

(8.29)

Following the techniques discussed in Chap. 4, one can find the solution of this differential equation:

$$T_{\text{pin}}(t) = 100 - 75e^{-1.83t}.$$

(8.30)

Figure 8.4 shows the response of the temperature of the pin for 3 s following a sudden change in the base temperature.

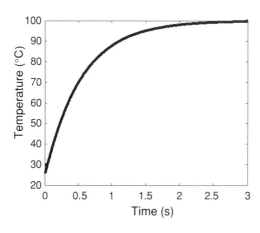

Figure 8.4. Response of pin temperature to sudden change in base temperature by use of a one-lump model.

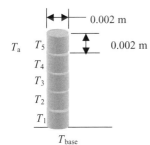

T_a T_5 0.002 m

T_4

T_3 **Figure 8.5.** Pin model using five equal-sized lumps.

T_2

T_1

0.002 m

T_{base}

Although the low value of the Biot number indicates that the time response computed with a single lumped model is relatively accurate and that the entire pin will be approximately at the same temperature, some temperature variation along the length of the pin will no doubt exist. If this variation is of interest (and it often is for heat transfer analysis), a more detailed approach is called for.

When the pin is divided into five equal segments along its length, the result is a stack of cylinders, as seen in Fig. 8.5.

Figure 8.6 shows a sketch of one of the pin elements, indicating the paths through which thermal energy is transmitted. As shown in the one-lump model, an energy balance for each element results in a first-order differential equation. Because there are five elements, this model is represented by five first-order equations. These equations are subsequently shown in state-space form.

$$\frac{d}{dt}T_1 = \frac{1}{C_h}\left[-\left(\frac{1}{R_{hk0}} + \frac{1}{R_{hk}} + \frac{1}{R_{hc}}\right)T_1 + \frac{1}{R_{hk}}T_2 + \frac{1}{R_{hk0}}T_{\text{base}} + \frac{1}{R_{hc}}T_a\right], \quad (8.31)$$

$$\frac{d}{dt}T_2 = \frac{1}{C_h}\left[\frac{1}{R_{hk}}T_1 - \left(\frac{2}{R_{hk}} + \frac{1}{R_{hc}}\right)T_2 + \frac{1}{R_{hk}}T_3 + \frac{1}{R_{hc}}T_a\right], \quad (8.32)$$

$$\frac{d}{dt}T_3 = \frac{1}{C_h}\left[\frac{1}{R_{hk}}T_2 - \left(\frac{2}{R_{hk}} + \frac{1}{R_{hc}}\right)T_3 + \frac{1}{R_{hk}}T_4 + \frac{1}{R_{hc}}T_a\right], \quad (8.33)$$

$$\frac{d}{dt}T_4 = \frac{1}{C_h}\left[\frac{1}{R_{hk}}T_3 - \left(\frac{2}{R_{hk}} + \frac{1}{R_{hc}}\right)T_4 + \frac{1}{R_{hk}}T_5 + \frac{1}{R_{hc}}T_a\right], \quad (8.34)$$

$$\frac{d}{dt}T_5 = \frac{1}{C_h}\left[\frac{1}{R_{hk}}T_4 - \left(\frac{1}{R_{hk}} + \frac{1}{R_{hc}} + \frac{1}{R_{hce}}\right)T_5 + \left(\frac{1}{R_{hc}} + \frac{1}{R_{hce}}\right)T_a\right], \quad (8.35)$$

Energy out by conduction
(or convection for top element)

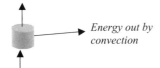

Energy out by   **Figure 8.6.** Energy balance for one element of the five-lump
convection   model.

Energy in by
conduction

where the thermal capacitance of an element C_h, conductive resistances R_{hk}, and convective resistances R_{hc}, are given by the following expressions:

$$C_h = \rho c_p V = (2707)(896)(6.28 \times 10^{-9}) = 0.0152 \frac{\text{J}}{{}^\circ\text{C}},$$

thermal capacitance of an element;

$$R_{hk0} = \frac{L_0}{kA_k} = \frac{0.001}{(220)(3.14 \times 10^{-6})} = 1.45 \frac{{}^\circ\text{C}}{\text{W}},$$

conductive resistance between element 1 and base;

$$R_{hk} = \frac{L}{kA_k} = \frac{0.002}{(220)(3.14 \times 10^{-6})} = 2.89 \frac{{}^\circ\text{C}}{\text{W}},$$

conductive resistance between adjacent elements;

$$R_{hc} = \frac{1}{h_c A_c} = \frac{1}{(20)(1.26 \times 10^{-5})} = 3968 \frac{{}^\circ\text{C}}{\text{W}},$$

convective resistance between element sides and the environment;

$$R_{hce} = \frac{1}{h_c A_{ce}} = \frac{1}{(20)(3.14 \times 10^{-6})} = 16,000 \frac{{}^\circ\text{C}}{\text{W}},$$

convective resistance between the top surface of element 5 and the environment.

Although conventional analytical methods can be used to solve these equations, computer methods are much more convenient. Figure 8.7 shows the Simulink model that solves a state-space linear model with two inputs. The parameters of the model are embedded in the state matrices. Because the main state matrix is a 5×5 matrix, it is best to write an m-file to set up the system matrices, thus allowing the user to more easily make modifications and experiment with the model. The m-file corresponding to this model is shown in Table 8.2.

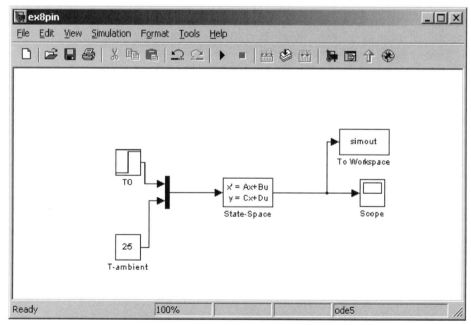

Figure 8.7. Simulink block diagram of pin simulation. The details of the model are embedded in the state-space matrices.

Table 8.2. M-file to set up state-space matrices for five-lump pin model

```
% Example 8.1
%
%Thermal properties for aluminum
k  = 220.0;
cp = 896.0;
rho = 2702.0;
%Geometric Parameters of the model
L  = 0.002;
L0 = 0.001;
Ak = 3.14E-6;
Ac = 1.26E-5;
Ace = 3.14E-6;
%
hc = 20.0;
% Compute Thermal Coefficients
Ch  = rho*cp*Ak*L;
Rhk = L/(k*Ak);
Rhk0 = L0/(k*Ak);
Rhc = 1/(hc*Ac);
Rhce = 1/(hc*Ace);
%
T0 = 25 * ones (5,1);%  Set initial conditions for pin temps
a = zeros (5,5);     %  system matrix (5 × 5)
b = zeros (5,2);     %  input matrix (5 × 2)
c = eye (5);         %  output matrix (5 × 5)
d = zeros (5,2);     %  transmission (5*2)
%
a (1,1) = -1/Ch*(1/Rhk0+1/Rhk+1/Rhc);
a (1,2) = 1/Ch*(1/Rhk);
%
a (2,1) = 1/Ch*(1/Rhk);
a (2,2) = -1/Ch*(2/Rhk+1/Rhc);
a (2,3) = 1/Ch*(1/Rhk);
%
a (3, 2) = 1/Ch*(1/Rhk);
a (3, 3) = -1/Ch*(2/Rhk+1/Rhc);
a (3, 4) = 1/Ch*(1/Rhk);
%
a (4, 3) = 1/Ch*(1/Rhk);
a (4, 4) = -1/Ch*(2/Rhk+1/Rhc);
a (4, 5) = 1/Ch*(1/Rhk);
%
a (5, 4) = 1/Ch*(1/Rhk);
a (5, 5) = -1/Ch*(1/Rhk+1/Rhc+1/Rhce);
%
b (1, 1) = 1/(Ch*Rhk0);
b (1, 2) = 1/(Ch*Rhc);
b (2, 2) = 1/(Ch*Rhc);
b (3, 2) = 1/(Ch*Rhc);
b (4, 2) = 1/(Ch*Rhc);
b (5, 2) = 1/Ch*(1/Rhc+1/Rhce);
```

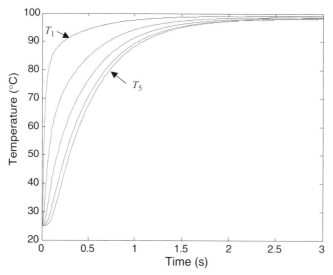

Figure 8.8. Temperature response of the five-lump model to step change in base temperature.

Figure 8.8 shows the temperature response of the five elements computed with this Simulink model.

The temperature response that is in the center of the five lines represents the temperature of the center element (T_3). This response can be compared directly with the single-lump response computed in Fig. 8.4. It is interesting to note that the responses are similar, but by no means identical, showing that the single-lump model does not completely represent the complexity of the pin.

Finally, Figure 8.9 represents the variation of temperature along the pin for various points in the time response. It is important to note that, even though the Biot number for the single-lump model indicated that the temperature could be assumed

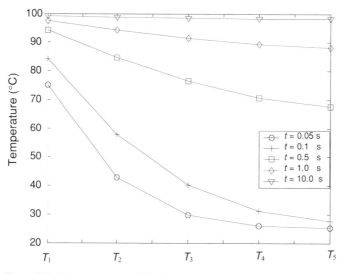

Figure 8.9. Temperature profile along the length of the pin for various times during the response.

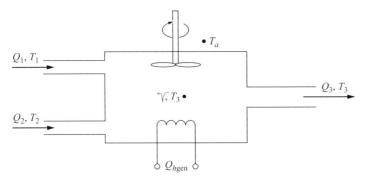

Figure 8.10. Blending systems.

to be uniform along the pin, that conclusion would not be valid for the transient portion of the response (less than 1 s).

EXAMPLE 8.2

Consider the blending system shown schematically in Fig. 8.10. Two identical liquids of different temperatures T_1 and T_2, flowing with different flow rates Q_1 and Q_2, are perfectly mixed in a blender of volume V. The mixture of liquids is also heated in the blender by an electric heater supplying heat at a constant rate, Q_{hgen}. There are heat losses in the system, and the coefficient of heat transfer between the blender and ambient air of temperature T_a is h_c. Although the mixing in the tank is assumed to be perfect, the work done by the mixer is negligible and the kinetic energies of the flows Q_1, Q_2, and Q_3 are very small. Derive a mathematical model of the blending process.

SOLUTION

The unsteady-flow energy-balance equation includes the following terms:

$$\begin{pmatrix} \text{rate} \\ \text{of} \\ \text{enthalpy} \end{pmatrix}_1 + \begin{pmatrix} \text{rate} \\ \text{of} \\ \text{enthalpy} \end{pmatrix}_2 + \begin{pmatrix} \text{rate} \\ \text{of} \\ \text{heat} \end{pmatrix}_{gen} = \begin{pmatrix} \text{rate} \\ \text{of} \\ \text{enthalpy} \end{pmatrix}_3$$

$$+ \begin{pmatrix} \text{rate} \\ \text{of} \\ \text{heat} \end{pmatrix}_{loss} + \begin{pmatrix} \text{rate} \\ \text{of} \\ \text{change} \\ \text{of} \\ \text{energy} \end{pmatrix}_{sto}.$$

The first two terms in this equation represent the rates of enthalpy supplied with the two incoming streams of liquids given by

$$Q_{h1} = \rho c_p Q_1 (T_1 - T_a),$$
$$Q_{h2} = \rho c_p Q_2 (T_2 - T_a).$$

The next term in the energy-balance equation represents the heat generated by the heater, Q_{hgen}. The heat carried away from the tank is represented by the first two terms on the right-hand side of the energy-balance equation. The rate of enthalpy carried with the outgoing stream of the mixture of the two input liquids is

$$Q_{h3} = \rho c_p Q_3 (T_3 - T_a) = \rho c_p (Q_1 + Q_2)(T_3 - T_a).$$

The rate at which heat is lost by the liquid through the sides of the tank to the ambient air is

$$Q_{hloss} = h_c A (T_3 - T_a).$$

Finally, the rate of change of energy stored in the liquid contained in the tank is

$$\frac{d\mathscr{E}}{dt} = \rho c_p \mathcal{V} \frac{dT_3}{dt}.$$

Substituting detailed mathematical expressions for the heat and enthalpy rates into the energy-balance equation yields

$$\rho c_p Q_1 (T_1 - T_a) + \rho c_p Q_2 (T_2 - T_a) + Q_{hgen}$$
$$= \rho c_p (Q_1 + Q_2)(T_3 - T_a) + h_c A (T_3 - T_a) + \rho c_p \mathcal{V} \frac{dT_3}{dt}. \qquad (8.36)$$

Equation (8.36) can be rearranged into the simpler form

$$\frac{dT_3}{dt} = -\frac{1}{\mathcal{V}} \left(\frac{h_c A}{\rho c_p} + Q_1 + Q_2 \right) T_3 + \frac{Q_1}{\mathcal{V}} T_1 + \frac{Q_2}{\mathcal{V}} T_2$$
$$+ \frac{1}{\rho c_p \mathcal{V}} Q_{hgen} + \frac{h_c A}{\rho c_p \mathcal{V}} T_a. \qquad (8.37)$$

Equation (8.37) represents a first-order multidimensional model with six potential input signals, Q_1, T_1, Q_2, T_2, Q_{hgen}, and T_a. The system time constant is

$$\tau = \frac{\rho c_p \mathcal{V}}{h_c A + \rho c_p Q_3}.$$

The model can be further simplified if the blender is assumed to be perfectly insulated, $h_c = 0$, and if there is no heat generation in the system, $Q_{hgen} = 0$. Under such conditions the system state equation becomes

$$\frac{dT_3}{dt} = -\left(\frac{1}{C} \right) \left(\frac{1}{R_1} + \frac{1}{R_2} \right) T_3 + \left(\frac{1}{R_1 C} \right) T_1 + \left(\frac{1}{R_2 C} \right) T_2,$$

where the lumped-model parameters C, R_1, and R_2, are defined as follows:

$$C = \rho c_p \mathcal{V},$$

$$R_1 = \frac{1}{\rho c_p Q_1},$$

$$R_2 = \frac{1}{\rho c_p Q_2}.$$

8.4   SYNOPSIS

Exact models of thermal systems usually involve nonlinear partial differential equations. Deriving closed-form analytical solutions for such problems is often impossible, and even obtaining valid computer models poses a very difficult task. Simplified lumped models of thermal systems were introduced in this chapter. Lumped models are very useful in the early stages of system analysis and also in verifying more complex computer models. It is always very important, when simplified models are used, to make sure that such models retain the basic dynamic characteristics of the original systems. The criterion of applicability of lumped models in thermal problems is provided by the Biot number, which is defined as the ratio of the conductive thermal resistance to the convective thermal resistance. Thermal systems can be represented by lumped models if the Biot number is small enough – usually less than 0.1. Basic elements of the lumped models, thermal capacitance and thermal resistance, are analogous to the corresponding elements in mechanical and electrical systems (although there is no thermal inductance). The state-model equations are derived around the thermal system elements associated with the A-type variable, temperature, and the T-type variable, heat flow rate. The same analytical and numerical methods of solution of state-variable equations as those described in earlier chapters can be used with thermal systems. Several examples of thermal systems were presented, including a computer heatsink and liquid blending process, to illustrate thermal energy storage.

PROBLEMS

8.1 A slab of cross-sectional area A and length $L = L_1 + L_2$ is made of two materials having different thermal conductivities k_1 and k_2, as shown in Fig. P8.1. The left-hand- and right-hand-side surfaces of the slab are at constant temperatures $T_1 = 200\,^\circ\text{C}$ and $T_2 = 20\,^\circ\text{C}$, whereas all other surfaces are perfectly insulated. The values of the system parameters are

$$k_1 = 0.05\,\text{W/m}\,^\circ\text{C}, \quad L_1 = 0.04\,\text{m},$$
$$k_2 = 0.7\,\text{W/m}\,^\circ\text{C}, \quad L_2 = 1.4\,\text{m},$$
$$A = 1\,\text{m}^2.$$

(a) Find the temperature at the interface of the two materials.

(b) Sketch the temperature distribution along the slab.

(c) Derive an expression for an equivalent thermal resistance of the entire slab in terms of the thermal resistances of the two parts.

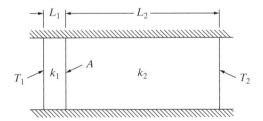

Figure P8.1. Thermal system considered in Problem 8.1.

8.2 The temperatures of the side surfaces of the composite slab shown in Fig. P8.2 are T_1 and T_2. The other surfaces are perfectly insulated. The cross-sectional areas of the two parts of the slab are A_1 and A_2, and their conductivities are k_1 and k_2, respectively. The length of the slab is L.

(a) Find an equivalent thermal resistance of the slab and express it in terms of the thermal resistances of the two parts.

(b) Sketch the steady-state temperature distribution along the slab.

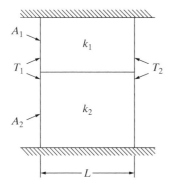

Figure P8.2. Thermal system considered in Problem 8.2.

8.3 A hollow cylinder is made of material of thermal conductivity k. The dimensions of the cylinder are as follows: inside diameter D_i, outside diameter D_o, and length L. The inside and outside surfaces are at constant temperatures T_i and T_o, respectively, whereas the top and bottom surfaces are both perfectly insulated. Find the expression for the lumped conductive thermal resistance of the cylinder for heat transfer in the radial direction only.

8.4 Consider again the cylinder in Problem 8.3, assuming that the inside diameter $D_i = 1$ m, the outside diameter $D_o = 2$ m, and the thermal conductivity of the material $k = 54$ W/m °C. The outside surface of the cylinder is now exposed to a stream of air at temperature T_a and velocity v_a. The inside surface remains at constant temperature T_i. The convective heat transfer coefficient between the outside surface of the cylinder and the stream of air is approximated by the expression $h_c = 2.24 v_a$. Determine the condition for the velocity of air under which the cylinder can be adequately represented by a lumped model.

8.5 A slab of material of density ρ and specific heat c_p, shown in Fig. P8.5, is perfectly insulated on all its sides except the top, which is in contact with fluid of temperature T_f in motion above the slab. The convective heat transfer coefficient between the fluid and the top side of the slab is h_c. The cross-sectional area of the slab is A, and its height is L.

(a) Determine the condition, in terms of the system parameters, under which the slab can be described by a lumped model.

(b) Assuming that the condition determined in part (a) holds, derive the state-variable model for the slab using its temperature T_s as the state variable and the temperature of the fluid T_f as the input variable.

(c) Write the expression for the system response $T_s(t)$ for a step change in the fluid temperature from an initial equilibrium condition $T_{f0} = T_{s0}$ to ΔT_f.

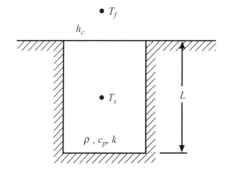

Figure P8.5. Thermal system considered in Problem 8.5.

8.6 A turkey is pulled out of the oven ready to be served at a dinner. It is initially at a uniform temperature of 180 °C. At the last minute, one of the invited guests calls to say that she will be a half hour late. Assuming that the turkey can be served only if its temperature is above 80 °C, can it be left at the room temperature of 20 °C until the late guest arrives 30 min later? To justify your answer, perform simple calculations using a lumped model of the turkey and the following estimates of the relevant parameters: mass of turkey, 4 kg; specific heat, 4200 J/kg deg; heat transfer surface area, 0.5 m²; and the convective heat transfer coefficient, 15 W/m² deg.

8.7 A thermocouple circuit is used to measure the temperature of a perfectly mixed liquid, as shown in Fig. P8.7. The hot junction of the thermocouple has the form of a small sphere of radius r_t. The density of the hot junction material is ρ_t, and the specific heat is c_t. The thermal capacitance of the thermocouple wire is negligible. The measuring voltage e_{21} is related to the hot junction temperature T_t by the equation $e_{21} = aT_t$.

(a) Derive a mathematical model for this system relating e_{21} to T_L.

(b) Sketch the response of e_{21} to a step change in the liquid temperature T_L.

(c) How long will it take for the measuring signal e_{21} to reach approximately 95 percent of the steady-state value after a step change of the liquid temperature? The thermocouple parameters are $\rho_t = 7800\ \text{kg/m}^3$, $c_t = 0.4\ \text{kJ/kg}\,^\circ\text{C}$, and $r_1 = 0.2$ mm. The convective heat transfer between the liquid and the hot junction is $h_c = 150\ \text{W/m}^2\,^\circ\text{C}$.

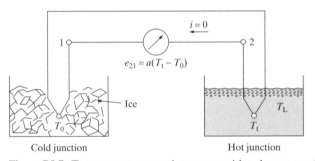

Cold junction   Hot junction

Figure P8.7. Temperature-measuring system with a thermocouple.

8.8 Figure P8.8 shows a simple model of an industrial furnace. A packing of temperature T_1 is being heated in the furnace by an electric heater supplying heat at the rate $Q_{\text{hi}}(t)$. The temperature inside the furnace is T_2, the walls are at temperature T_3, and ambient temperature is T_a. The thermal capacitances of the packing, the air inside the furnace, and the furnace walls are C_{h1}, C_{h2}, and C_{h3}, respectively. Derive state-variable equations for this system assuming that heat is transferred by convection only, with the convective heat transfer coefficients h_{c1} (air – packing), h_{c2} (air – inside walls), and h_{c3} (outside walls – ambient air).

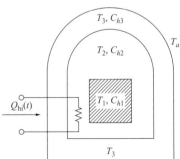

Figure P8.8. Simplified model of an industrial furnace.

8.9 The ceramic object shown in Figure P8.9 consists of two layers having different thermal capacitances C_{h1} and C_{h2}. The top layer, having temperature T_1, is exposed to thermal radiation from a heater of temperature T_r and effective emissivity F_E. The area exposed to radiation is A_1 and the shape factor is F_{A1}. Both layers exchange heat by convection with ambient air of temperature T_a through their sides of areas A_s. The convective heat transfer coefficient is h_c.

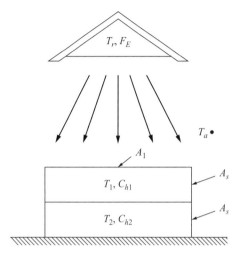

Figure P8.9. Radiative heating system considered in Problem 8.9.

Heat is also transferred between the two layers at the rate given by

$$Q_{h12} = \left(\frac{1}{R_{12}} \right) (T_1 - T_2),$$

where R_{12} represents the thermal resistance of the interface between the layers. Heat transfer through the bottom of the lower layer is negligible.

(a) Derive nonlinear state-variable equations for this system using temperatures T_1 and T_2 as the state variables.

(b) Obtain linearized state-model equations describing the system in the vicinity of the normal operating point given by $\overline{T}_r$, $\overline{T}_1$, $\overline{T}_2$.

8.10 A heat storage loop of a solar water-heating system is shown schematically in Fig. P8.10. The rate of solar energy incident per unit area of the collector is $S(t)$. The collector can be modeled by the Hottel–Whillier–Bliss equation, which gives the rate of heat absorbed by the collector, Q_{hcol}:

$$Q_{hcol}(t) = A_c F_R [S(t) - U_L (T_s - T_a)].$$

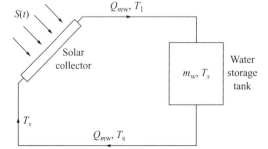

Figure P8.10. Heat storage loop in a solar water-heating system.

The values of the collector parameters are collector surface area, $A_c = 30\,\mathrm{m}^2$, heat removal factor, $F_R = 0.7$, heat loss coefficient, $U_L = 4.0\,\mathrm{W/m^2\,°C}$.

The rate of water flow in the loop is $Q_{mw} = 2160\,\mathrm{kg/h}$ and the specific heat of water is $c_w = 4180\,\mathrm{J/kg\,°C}$. The water storage tank has capacity $m_w = 1800\,\mathrm{kg}$ and is considered to be perfectly insulated. It is also assumed that there are no heat losses through the piping in the system. Derive a state-variable model for this system, assuming that the water in the storage tank is fully mixed.

8.11 Consider again the solar heating system shown in Fig. P8.10, but now without assuming that the storage tank is fully mixed. In fact, stratification of a storage tank leads to enhanced performance of a solar heating system. Develop a mathematical model for the system shown in Fig. P8.10, assuming that the water storage tank is made up of three isothermal segments of equal volume. Assume also that heat conduction between successive segments of water is negligible.

8.12 Figure P8.12 shows a segment of steel pipe with a layer of insulation. The insulation is surrounded by a protective aluminum layer. Assume that the temperature of the fluid in the pipe is determined by other elements of the system (and can be assumed to be an input for this analysis). Write the equations that describe the dynamic response to changes in the fluid temperature. Assume that both the pipe and the cladding are at uniform temperatures (lumped capacitances) and that the insulation has a temperature gradient along its radius, but that the thermal capacitance of the insulation is negligible. Finally, assume no significant heat transfer from the ends of the pipe and that the relationship between the radius and the thickness of the layers is such that the curvature effects are not significant.

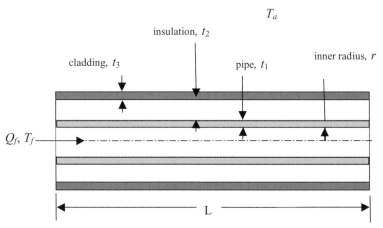

Figure P8.12. Cross section of pipe.

8.13 Refer to Fig. P8.12 and the appropriate assumptions described in Problem 8.12 and the system is initially at equilibrium with $T_f = 50\,°\mathrm{C}$. If T_f experiences a sudden change to $90\,°\mathrm{C}$, determine the final temperature of the aluminum cladding and how long it will take for that temperature to reach 95 percent of its final value.

Use $L = 1.0\,\mathrm{m}$, $r = 0.1\,\mathrm{m}$, and the following values for your analysis:

Property/parameter	Steel pipe	Insulation	Cladding
$\rho\ (\mathrm{kg/m^3})$	7810	25	2707
$k\ (\mathrm{W/m\,°C})$	43.0	0.04	204.0
$c_p\ (\mathrm{kJ/kg\,°C})$	0.473	0.70	0.896
thickness (m)	$t_1 = 0.005$	$t_2 = 0.05$	$t_3 = 0.002$
h_c (fluid to pipe)	$500\,\mathrm{W/m^2\,°C}$	h_c (cladding to air)	$20\,\mathrm{W/m^2\,°C}$

8.14 At a picnic, a watermelon initially cooled to 5°C is exposed to 30°C air. Assume that the temperature of the watermelon is uniform throughout. Assume the following parameter values for this problem:

c_p	Heat capacity	4200 J/kg deg
m	Mass of watermelon	4.0 kg
A	Surface area	0.5 m²
h	Convection coefficient	15 W/m² deg

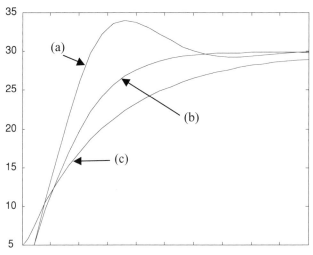

Figure P8.14. Three possible step responses for Problem 8.14.

(a) How long will it take before the watermelon warms up to 20.75°C?

(b) Which of the three curves shown in Fig. P8.14 most resembles the response predicted by this analysis?

9

Fluid Systems

9-1 To recognize the A-, T-, and D-type elements of fluid systems.

9-2 To use laws of continuity and compatibility to develop lumped-parameter models of hydraulic fluid systems.

9-3 To model lumped-parameter compressible (pneumatic) fluid systems by use of ideal gas assumptions.

9.1 INTRODUCTION

Corresponding to the mass, spring, and damper elements discussed in Chap. 2, the A-type, T-type, and D-type elements used in modeling fluid systems are the fluid capacitor, the fluid inertor, and the fluid resistor elements.

Capacitance occurs as a result of elasticity or compliance in the fluid or in the walls of the container. Although liquids such as water and oil are often considered to be incompressible in many fluid flow situations, these hydraulic fluids are sometimes compressible enough to produce fluid capacitance. In other cases, the walls of the chambers or passages containing the fluid have enough compliance, when the fluid pressure changes rapidly, to produce fluid capacitance. In long lines and passages, fluid capacitance is distributed along the line, together with inertance and resistance, and the analysis of such situations is beyond the scope of this text.[1] If both the resistance and inertance are negligible, however, it is possible to use a lumped-capacitance model of the line.

When fluid capacitance is wanted for energy-storage purposes, specially designed off-the-shelf capacitors with minimal resistance and inertance, called hydraulic accumulators, are used. Storage tanks and reservoirs also serve as fluid capacitors. The energy stored in a capacitor is potential energy and is related to the work required for increasing the pressure of the fluid filling the capacitor.

When the working fluid is a compressed gas such as air, fluid compliance is much more significant and must be accounted for even for small chambers and passages.

Inertance results from the density of the working fluid when the acceleration of the fluid in a line or passage requires a significant pressure gradient for producing the

---

[1] For the case of a lossless line with distributed inertance and capacitance, see J. F. Blackburn, G. Reethof, and J. L. Shearer, *Fluid Power Control* (MIT Press, Cambridge, MA, 1960), pp. 83–9, 137–43.

rate of change of flow rate involved. It occurs mainly in long lines and passages, but it can be significant even in relatively short passages when the rate of change of flow rate is great enough. The energy stored in an inertor is kinetic, and it is related to the work done by the pressure forces at the terminals of the element to increase the momentum of the flowing fluid. If resistance (see subsequent discussion) is distributed along with the inertance in a passage, a series lumped-inertance, lumped-resistance model will be suitable, as long as the capacitance is negligible.

Fluid resistance is encountered in small passages and usually is a result of the effects of fluid viscosity, which impedes the flow and requires that significant pressure gradients be used to produce the viscous shearing of the fluid as it moves past the walls of the passage. Fluid resistance often becomes significant in long lines; it is then modeled in series with lumped inertance for the line when the rate of change of flow rate is large. A fluid resistor dissipates energy in the fluid, resulting in an increase in fluid temperature.

When fluid resistance results from turbulent flow in a passage, or from flow through an orifice, the kinetic effects of predominant inertia forces in the fluid flow result in nonlinear characteristics that can sometimes be linearized successfully.

9.2   FLUID SYSTEM ELEMENTS

9.2.1 Fluid Capacitors

The symbolic diagram of a fluid capacitor is shown in Fig. 9.1. Note that the pressure in a fluid capacitor must be referred to a reference pressure P_r. When the reference pressure is that of the surrounding atmosphere, it is the gauge pressure; when the reference pressure is zero, i.e., a perfect vacuum, it is the absolute pressure. In this respect, a fluid capacitor is like a mass—that is, one of its across variables must be a reference.

The volume rate of flow Q_c is the through variable, even though no flow "comes out the other side," so to speak. The net flow into the capacitor is stored and corresponds to the process of charging a capacitor in electrical systems.

The elemental equation for an ideal fluid capacitor is

$$Q_c = C_f \frac{dP_{1r}}{dt}, \tag{9.1}$$

where C_f is the fluid capacitance. The simplest form of fluid capacitance arises when the compressibility of the fluid, constrained within a rigid vessel, experiences pressure fluctuations as the amount of fluid within it varies. This variation in pressure as a result

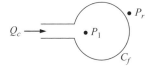

Figure 9.1. Symbolic diagram of a fluid capacitor.

of addition or subtraction of fluid is the basis of a characteristic of fluids known as the bulk modulus β, defined as

$$\beta = \frac{\Delta P}{\Delta V / V}. \tag{9.2}$$

To interpret this equation, imagine a rigid vessel of volume V containing fluid at an arbitrary pressure. Now add a small additional amount of fluid (ΔV) and measure the resulting change in pressure (ΔP). The ratio given in (9.2) defines the bulk modulus of that fluid. Therefore the fluid capacitance of a given volume of fluid in a rigid container (often referred to as entrapped volume) is given by

$$C_f = \frac{V}{\beta}. \tag{9.3}$$

When Eqs. (9.2) and (9.3) are combined, the fluid capacitance can also be expressed as

$$C_f = \frac{\Delta V}{\Delta P}. \tag{9.4}$$

Based on this definition, many useful relationships for common fluid capacitances can easily be derived. Several of them are given in Fig. 9.2. Note that, when one is dealing with compressed gases, the pressure in the capacitor must always be expressed as an absolute pressure; that is, the pressure must be referred to a perfect vacuum.

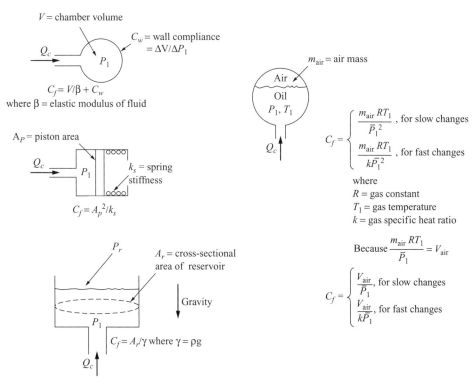

Figure 9.2. Typical fluid capacitors and their capacitances.

Q_1

$\bullet\, P_1$ ▰▰▰▰▰ $\bullet\, P_2$ **Figure 9.3.** Symbolic diagram of a fluid inertor.

I

An alternative form of the elemental equation that is sometimes more useful, especially for time-varying or nonlinear fluid capacitors, is given by

$$V_c = C_f P_{1r}, \tag{9.5}$$

where V_c is the volume of the net flow, i.e., the time integral of Q_c.

The potential energy stored in an ideal fluid capacitor is given by

$$\mathcal{E}_P = \frac{C_f}{2} P_{1r}^2. \tag{9.6}$$

Hence the fluid capacitor is an A-type element, storing energy that is proportional to the square of its across variable P_{1r}, and it would be unrealistic to try to change its pressure suddenly.

9.2.2 Fluid Inertors

The symbolic diagram of a fluid inertor is shown in Fig. 9.3. The elemental equation for an inertor is

$$P_{12} = I \frac{dQ_I}{dt}, \tag{9.7}$$

where I is the fluid inertance. For frictionless incompressible flow in a uniform passage having cross-sectional area A and length L, the inertance $I = (\rho/A)L$, where ρ is the mass density of the fluid. For passages having nonuniform area, it is necessary to integrate $[\rho/A(x)]dx$ over the length of the passage to determine I. The expression just given for I can be modified for a flow with a nonuniform velocity profile by application of the unsteady-flow momentum equation to a small element of the passage and integration across the passage area: The correction factor for a circular area with a parabolic velocity profile is 2.0; i.e., $I = (2\rho/A)L$. Because the nonuniform velocity profile usually results from viscosity effects, the accompanying fluid resistance would then be modeled as a series resistor.

The kinetic energy stored in an ideal inertor is given by

$$\mathcal{E}_K = \frac{I}{2} Q_I^2. \tag{9.8}$$

Hence the inertor is a T-type element, storing energy as a function of the square of its T-type variable; it would be unrealistic to try to change the flow rate suddenly through an inertor.

When enough fluid compressibility or wall compliance is also present, a lumped-parameter capacitance–inertance model might be suitable.[2] The justification for using this model would follow the lines discussed for modeling a mechanical spring in

---

[2] For a brief discussion of lumped-parameter models of transmission lines, see *Handbook of Fluid Dynamics*, edited by V. L. Streeter (McGraw-Hill, New York, 1961), pp. 21–24 through 21–28.

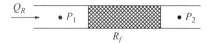

$$\underrightarrow{Q_R} \quad \bullet P_1 \quad \boxed{} \quad \bullet P_2$$

$$R_f$$

Figure 9.4. Symbolic diagram of a fluid resistor.

Chap. 2. Otherwise, a distributed-parameter model is needed, which is beyond the scope of this book.

9.2.3 Fluid Resistors

The symbolic diagram of a fluid resistor is shown in Fig. 9.4.

The elemental equation of an ideal fluid resistor is

$$P_{12} = R_f Q_R, \tag{9.9}$$

where R_f is the fluid resistance, a measure of the pressure drop required for forcing a unit of flow rate through the resistor. Alternatively, the elemental equation may be expressed by

$$Q_R = \frac{1}{R_f} P_{12}. \tag{9.10}$$

Figure 9.5 shows some typically encountered linear fluid resistors, together with available expressions for their resistance.

When the flow in a passage becomes turbulent or when the flow is through an orifice, a nonlinear power-law relationship is used to express the pressure drop as a function of flow rate:

$$P_{12} = C_R Q_{NLR}^n \tag{9.11}$$

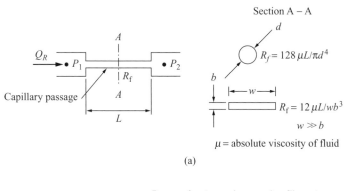

(a)

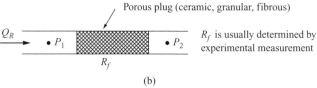

(b)

Figure 9.5. Linear fluid resistors: (a) capillary passages and (b) porous plugs.

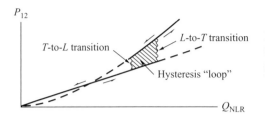

Figure 9.6. Steady-state pressure drop vs. flow rate characteristics for a fluid line undergoing laminar-to-turbulent-to-laminar transition.

where C_R is a flow constant and n is either approximately 7/4 (from the Moody diagram) for turbulent flow in long, straight, smooth-walled passages or 2 (from Bernoulli's equation) for flow through a sharp-edged orifice. The latter is a particularly useful relationship because it closely approximates the condition of flow through a valve. In those cases, the pressure-flow relationship is often given in this form:

$$Q_{\mathrm{NLR}} = A_\circ C_d \sqrt{\frac{2}{\rho} P_{12}}. \qquad (9.12)$$

For valve flow, A_o is the cross-sectional area of the valve port, C_d is the sharp-edged discharge coefficient, assumed to be 0.61 in the absence of additional information from the valve manufacturer, and ρ is the mass density of the fluid. Section 9.4 presents the use of this equation to model flow through a commercial hydraulic valve. For more demanding applications, for which more accurate models are needed, it is necessary to experimentally determine the nonlinear relationship between the pressure drop and the flow rate.

When a flow in a passage that undergoes transition from laminar to turbulent to laminar is to be modeled, hysteresis is likely to be present, as shown in Fig. 9.6. Careful measurements of the actual line should be made to determine the transition points, and so on.

In pneumatic systems, in which the fluid is very compressible, the Mach number of an orifice flow can easily exceed 0.2, and for cases involving such high local flow velocities, the compressible flow relations must be used or approximated.[3] The flow rate is then expressed in terms of mass or weight rate of flow because volume rate of flow of a given amount of fluid varies so greatly with local pressure. A modified form of fluid capacitance is then needed, unless the local volume rate of flow is computed at each capacitor. Section 9.5 discusses the application of these principles to pneumatic systems.

9.2.4 Fluid Sources

The ideal sources used in fluid system analysis are shown in Fig. 9.7. An ideal pressure source is capable of delivering the indicated pressure, regardless of the flow required by what it is driving, whereas an ideal flow source is capable of delivering the indicated flow rate, regardless of the pressure required for driving its load.

---

[3] For a discussion of compressible flow effects in orifices, see J. F. Blackburn *et al.*, *op. cit.*, pp. 61–80, 214–34.

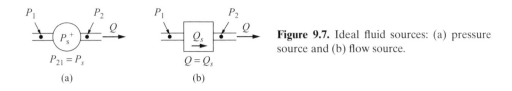

Figure 9.7. Ideal fluid sources: (a) pressure source and (b) flow source.

9.2.5 Interconnection Laws

The two fluid system interconnection laws, corresponding to Kirchhoff's laws for electrical systems, are the laws of continuity and compatibility. The continuity law says that the sum of the flow rates at a junction must be zero, and the compatibility law says that the sum of the pressure drops around a loop must be zero. These laws are illustrated in Fig. 9.8.

9.3 ANALYSIS OF FLUID SYSTEMS

The procedure followed in the analysis of fluid systems is similar to that followed earlier for mechanical and electrical systems: Write the elemental equations and the interconnection equations and then (a) combine, removing unwanted variables, to obtain the desired system input–output differential equation or (b) build a state-variable equation around each energy-storage element by combining to remove all but the state variables for the energy-storage elements. These state variables are (a) pressure for capacitor (A-type) elements and (b) flow rate for inertor (T-type) elements.

EXAMPLE 9.1

Find a set of state-variable equations and develop the input–output differential equation relating the output pressure P_{3r} to the input pressure P_s for the fluid system shown in Fig. 9.9.

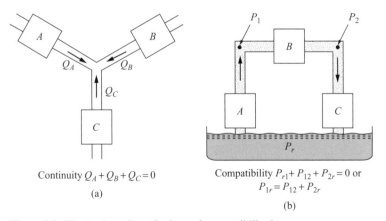

Continuity $Q_A + Q_B + Q_C = 0$

(a)

Compatibility $P_{r1} + P_{12} + P_{2r} = 0$ or
$$P_{1r} = P_{12} + P_{2r}$$

(b)

Figure 9.8. Illustration of continuity and compatibility laws.

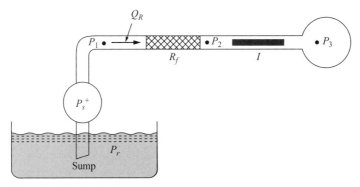

Figure 9.9. A simple fluid R, I, C system.

SOLUTION

The elemental equations are as follows: For the fluid resistor,

$$P_{12} = R_f Q_R. \tag{9.13}$$

For the inertor,

$$P_{23} = I \frac{dQ_R}{dt}. \tag{9.14}$$

For the fluid capacitor,

$$Q_R = C_f \frac{dP_{3r}}{dt}. \tag{9.15}$$

Continuity is satisfied by use of Q_R for Q_I and Q_c. To satisfy compatibility.

$$P_s = P_{1r} = P_{12} + P_{23} + P_{3r}. \tag{9.16}$$

Combining Eqs. (9.13), (9.14), and (9.16) to eliminate P_{12} and P_{23} yields

$$I \frac{dQ_R}{dt} = P_s - R_f Q_R - P_{3r}. \tag{9.17}$$

Rearranging Eq. (9.17) yields the first state-variable equation:

$$\frac{dQ_R}{dt} = -\frac{R_f}{I} Q_R - \frac{1}{I} P_{3r} + \frac{1}{I} P_s. \tag{9.18}$$

Rearranging Eq. (9.15) yields the second state-variable equation:

$$\frac{dP_{3r}}{dt} = \frac{1}{C_f} Q_R. \tag{9.19}$$

Combining Eqs. (9.18) and (9.19) to eliminate Q_R and multiplying all terms by I yields the input–output system differential equation:

$$C_f I \frac{d^2 P_{3r}}{dt^2} + R_f C_f \frac{dP_{3r}}{dt} + P_{3r} = P_s. \tag{9.20}$$

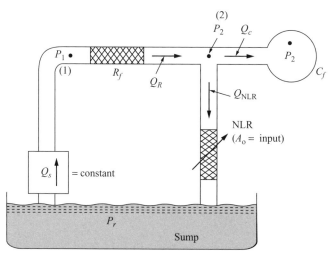

Figure 9.10. Simple fluid control system.

EXAMPLE 9.2

A variable-orifice NLR is being used to modulate the flow rate Q_{NLR} and control the pressure P_{2r} in the simple fluid control system shown in Fig. 9.10. The flow equation for the orifice is $Q_{\mathrm{NLR}} = C_1 A_o (P_{2r})^{0.5}$. Develop the input–output differential equation relating small changes in the output pressure P_{2r} to small changes in the orifice area A_o when the supply flow Q_s is constant.

SOLUTION

The elemental equations for small perturbations in all variables are as follows: For the linear resistor,

$$\hat{P}_{12} = R_f \hat{Q}_R. \tag{9.21}$$

For the fluid capacitor,

$$\hat{Q}_c = C_f \frac{d\hat{P}_{2r}}{dt}. \tag{9.22}$$

For the time-varying NLR,

$$\hat{Q}_{\mathrm{NLR}} = \left(\frac{C_1 \overline{A_o}}{2|\overline{P}_{2r}|^{0.5}} \right) \hat{P}_{2r} + C_1 |\overline{P}_{2r}|^{0.5} \hat{A}_o. \tag{9.23}$$

To satisfy continuity at (1),

$$\hat{Q}_R = \hat{Q}_s. \tag{9.24}$$

To satisfy continuity at (2),

$$\hat{Q}_R = \hat{Q}_C + \hat{Q}_{\mathrm{NLR}}. \tag{9.25}$$

The compatibility relation,

$$\hat{P}_{1r} = \hat{P}_{12} + \hat{P}_{2r}, \tag{9.26}$$

is not needed here because of a lack of interest in finding $\hat{P}_{1r}$.

Combining Eqs. (9.22)–(9.25) to eliminate $\hat{Q}_R$, $\hat{Q}_C$, and $\hat{Q}_{NLR}$ yields

$$\hat{Q}_s = C_f \frac{d\hat{P}_{2r}}{dt} + \left(\frac{C_1 \overline{A}_o}{2|\overline{P}_{2r}|^{0.5}} \right) \hat{P}_{2r} + C_1 |\overline{P}_{2r}|^{0.5} \hat{A}_o. \tag{9.27}$$

Because Q_s is constant, $\hat{Q}_s$ is zero, and the output terms on the right-hand side may be rearranged to yield the system input–output differential equation for small perturbations in all variables:

$$C_f \frac{d\hat{P}_{2r}}{dt} + \left(\frac{C_1 \overline{A}_o}{2|\overline{P}_{2r}|^{0.5}} \right) \hat{P}_{2r} = -C_1 |\overline{P}_{2r}|^{0.5} \hat{A}_o. \tag{9.28}$$

9.4 ELECTROHYDRAULIC SERVOACTUATOR

In this section, a common high-performance hydraulic system is described and modeled. Electrohydrualic servoactuators are found on aerospace and robotic systems for which accurate and fast responses are required in the presence of large and unpredictable external loads. A servoactuator is made up of a hydraulic cylinder (or rotary actuator) closely coupled to an electrohydraulic servovalve. Figure 9.11 shows a servoactuator in schematic form.

The key component in this system is the servovalve. Servovalves are specially designed flow control valves that are capable of modulating flow between zero and some rated value. This is in contrast to most hydraulic valves, which are designed to control only flow direction by moving an internal component, the valve spool, between two or three preferential positions.

Figure 9.12 shows a cutaway view of a valve spool and ports of a typical four-way hydraulic valve. Note that the internal component, the valve spool, is free to move in a cylindrical bore that has ports cut into it. As the spool moves to the right, it exposes the control ports (C_1 and C_2) to the supply pressure and return ports, respectively. As the spool moves to the left, the connections are reversed and fluid can flow from supply to C_2 and from C_1 to return. A servovalve has a complex mechanism that allows the spool to be rapidly and accurately positioned at any point between fully

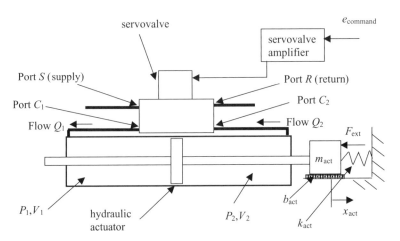

Figure 9.11. Electrohydraulic servoactuator.

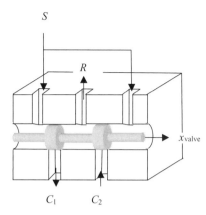

Figure 9.12. Cutaway view of a four-way valve.

closed and fully open. The mechanism that accomplishes this includes a stationary coil (stator), a movable ferrous component (armature), a flexible component, and flapper valves. A detailed analysis of servovalves lies beyond the scope of this text, but more information is readily available from companies that manufacture and supply the valves such as Moog, Inc., of Albany, NY.

Manufacturers' specifications for servovalves include the response of the valve spool, that is, how fast one can expect the valve spool to change position in response to changes in the current in the valve coil. For most applications, a second-order linear model adequately describes this response, and the parameters of that model (coefficients of the differential equation) can be derived from manufacturers' literature. The form of the model equations is

$$\dot{x}_{\text{valve}} = v_{\text{valve}}, \tag{9.29}$$

$$\dot{v}_{\text{valve}} = a_1 v_{\text{valve}} + a_0 x_{\text{valve}} + b_0 e_{\text{command}}, \tag{9.30}$$

where a_1, a_0, and b_0 are the model coefficients.

Equations (9.29) and (9.30) make up the first two state equations for the model of the servoactuator.

From Fig. 9.11, the remaining required states can be deduced. There are two energy-storing elements in the mechanical part of the system. One is the combined mass of the piston, piston rod, and the external load, m_{act}, and the other is the spring k_{act}. The appropriate state variables for these two elements are the position and the velocity of the actuator:

$$\dot{x}_{\text{act}} = v_{\text{act}}, \tag{9.31}$$

$$\dot{v}_{\text{act}} = \frac{1}{m_{\text{act}}}[(P_1 - P_2)A_{\text{pist}} - b_{\text{act}}v_{\text{act}} - k_{\text{act}}x_{\text{act}} - F_{\text{ext}}], \tag{9.32}$$

where k_{act}, b_{act}, and m_{act} represent the combined stiffness, damping, and mass associated with the motion of the actuator and its load. For example, b_{act} is associated with the velocity-dependent force encountered by the actuator as it moves. This force will come from both the piston itself (sliding inside the cylinder) and the external mechanism to which the actuator is attached. Any externally applied load force (considered a model input for this discussion) can be considered as F_{ext}.

On the hydraulic side of the system, the motive force for the actuator comes from pressures in the cylinder, which, in turn, result from flow entering and leaving the actuator by way of the valve. The entrapped volumes between the valve and the actuator piston can be modeled as fluid capacitances (see Subsection 9.2.1) and the pressures P_1 and P_2 are the remaining state variables representing the potential energy stored in the two hydraulic capacitors:

$$\dot{P_1} = \frac{1}{C_{f1}}(Q_1 - A_{\text{pist}} v_{\text{act}}), \tag{9.33}$$

$$\dot{P_2} = \frac{1}{C_{f2}}(A_{\text{pist}} v_{\text{act}} - Q_2), \tag{9.34}$$

where

$$C_{f1} = \frac{V_1(x_{\text{act}})}{\beta}, \tag{9.35}$$

$$C_{f2} = \frac{V_2(x_{\text{act}})}{\beta}, \tag{9.36}$$

and β is the bulk modulus of the fluid, defined in Subsection 9.2.1, and is a physical property that quantifies the relationship between the change in pressure and the change in volume.

Note that the definitions of the fluid capacitances indicate that the entrapped volumes (V_1 and V_2) are dependent on the instantaneous position of the piston within the actuator. Also note the sign assumptions for Q_1 and Q_2. Q_1 is defined as the flow out of valve port C_1 and into the left-hand side of the cylinder in Fig. 9.11. Q_2 is defined as the flow into the valve at port C_2 coming from the right-hand side of the cylinder. These definitions are arbitrary but must be consistent with the valve flow equations subsequently described.

Equations (9.29)–(9.34) are the six state-variable equations that describe the servoactuator. However, the model is incomplete because the flows through the valves (Q_1 and Q_2) are not described as functions of the state variables.

Subsection 9.2.3 describes the algebraic relationship between flow and pressure differential through many types of fluid resistance elements. In general, spool valves like the one shown in Fig. 9.12 are modeled as sharp-edged orifices with turbulent flow in which the flow is proportional to the square root of the pressure differential across the valve. For example, if the position of the valve spool is positive (to the right in Fig. 9.12), then Q_1 can be modeled as

$$Q_1 = C_d A_v \sqrt{\frac{2}{\rho}(P_s - P_1)}, \tag{9.37}$$

where C_d is the orifice coefficient, generally taken as 0.61 for valve applications and A_v is the cross-sectional area of the valve orifice, which is the area of the port exposed by the valve spool. Further examination of Fig. 9.12 shows that the orifice area is directly proportional to the valve position, x_{valve}. In general, the actual port is rectangular in shape, and this area is equal to the product of the valve spool position and the width of the valve port, w_{valve}. Equation (9.37) can now be rewritten in terms of x_{valve}:

$$Q_1 = C_d w_{valve} x_{valve} \sqrt{\frac{2}{\rho}(P_s - P_1)}. \tag{9.38}$$

Two other considerations remain before the model is complete. First, Eq. (9.38) cannot be evaluated if the pressure in the cylinder (P_1) is higher than the supply pressure. Although it may not be intuitively obvious, it is entirely possible that such a condition might exist, if only for a brief period, and the equations should be modified to allow for that possibility. Second, if the valve spool position is negative, Q_1 and Q_2 represent flows to and from different valve ports. The following conditional equations adequately account for these situations. Note that the formulation is somewhat simplified by the reasonable assumption that the pressure at the return port of the valve (normally connected directly to the reservoir) is zero.

If $x_{valve} > 0$,

$$Q_1 = C_d w_{valve} x_{valve} \, \text{sgn}(P_s - P_1) \sqrt{\frac{2}{\rho}|P_s - P_1|}, \tag{9.39}$$

$$Q_2 = C_d w_{valve} x_{valve} \sqrt{\frac{2}{\rho}(P_2)}. \tag{9.40}$$

If $x_{valve} < 0$,

$$Q_1 = C_d w_{valve} x_{valve} \sqrt{\frac{2}{\rho}(P_1)}, \tag{9.41}$$

$$Q_2 = C_d w_{valve} x_{valve} \, \text{sgn}(P_s - P_2) \sqrt{\frac{2}{\rho}|P_s - P_2|}. \tag{9.42}$$

In the following example, the six-state electrohydraulic servoactuator model is simulated by use of Simulink.

EXAMPLE 9.3

An electrohydaulic servoactuator is used in a military aircraft to precisely control the position of a flight control surface. The effective mass of the control surface and the actuation mechanism is 40 kg (including the moving part of the actuator), and the effective stiffness of the mechanism is 800 N/m. It has been estimated that the overall linear damping is about 6000 N s/m.

The hydraulic cylinder has a bore of 0.1 m, a rod diameter of 0.02 m, and a total stroke of 0.7 m. The valve is a high-performance servovalve rated at 5 gal/min, capable of responding to sinusoidal commands as high as 100 Hz without significant degradation.

Develop a simulation of this system by using Simulink. As a test of the model, simulate the response of the actuator to an input that commands the valve from fully closed, to fully open for 0.5 s, then fully closed again.

The model parameters are summarized in Tables 9.1 and 9.2.

Figure 9.13 shows a Simulink model of the servoactuator. Although the diagram is fairly complicated, it helps to focus on the integrator blocks. Note that there are six integrator blocks, two associated with the valve position (servovalve dynamics), two

Table 9.1. Servovalve and fluid parameters

Parameter	Value	Units
w_{valve}	0.0025	m
b_0	90	m/V s^2
a_0	360,000	1/s^2
a_1	1000	1/s
β_{fluid}	689.0	MPa
ρ_{fluid}	900	kg/m^3
C_d	0.61	
P_{supply}	20.7	MPa

Note: These parameters are consistent with an MTS 252.23 or Moog 760-X19 servo-valve/servoamplifier set up to respond to a voltage command of $+/-$ 10 V.

Table 9.2. Load/actuator parameters

Parameter	Value	Units
m_{act}	40	kg
b_{act}	6000	N s/m
k_{act}	800	N/m
Stroke	0.7	m
A_{pist}	0.0075	m^2

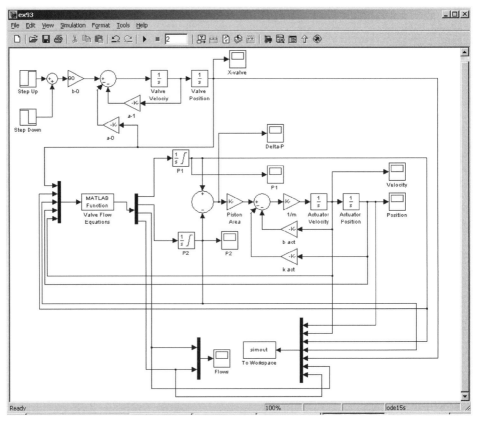

Figure 9.13. Simulink model of an electrohydraulic actuator.

Table 9.3. M-file function to compute valve flows

```
function pdots = vflow (u)
%
% function to compute valve flows and
% pressure derivatives
%
% u(1) = valve position
% u(2) = P1
% u(3) = P2
% u(4) = actuator position
% u(5) = actuator velocity
%
% parameters
%
Apist = 0.0075; % m^2
S = 0. 7;       % m
B = 6. 89E8;    % Pa
rho = 900.0;    % kg/m^3
Cd = 0.61;
wv = 0.0025;    % m
Ps = 2.07E7;    % Pa
%
if u(1) > = 0
    Q1 = Cd*wv*u(1) * sign(Ps-u(2)) * (2/rho*abs(Ps-u(2)))^0.5;
    Q2 = Cd*wv*u(1) * (2/rho*u(3))^0.5;
else
    Q2 = Cd*wv*u(1) * sign(Ps-u(3)) * (2/rho*abs(Ps-u(3)))^0.5;
    Q1 = Cd*wv*u(1) * (2/rho*u(2))^0.5;
end
%
Cf1 = 1/B*1.2*Apist * (S/2+u(4));
Cf2 = 1/B*1.2*Apist * (S/2-u(4));
%Cf1 = (Apist*S/2)/B;
%Cf2 = (Apist*S/2)/B;
%
pdots(1) = 1/Cf1 * (Q1 - Apist*u(5));
pdots(2) = 1/Cf2 * (Apist* u(5) - Q2);
pdots(3) = Q1;
pdots(4) = Q2;
```

associated with the actuator/load combination, and two associated with the pressures on either side of the actuator position. Note that the latter two blocks (P_1 and P_2) are somewhat different in that they are internally limited by the software so that they cannot go negative (this is done by use of the dialog box associated with the integrators).

Finally, note that the equations associated with the valve flow [(9.39)–(9.42)] as well as the computation of the pressure derivatives are embedded in a function call. Table 9.3 lists the MATLAB code associated with this function.

Simulation Results In this model, very fast dynamics associated with the fluid capacitances are combined with the relatively slow load dynamics, giving rise to a system

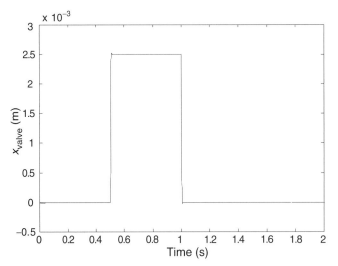

Figure 9.14. Valve spool position response.

that is rather stiff (see Section 5.6). Therefore the ODE15S algorithm is used in Simulink to compute the response.

As described in the beginning of the example, the input to the simulation is a time-varying voltage that is at 0 V at the initial time, rises to 10 V at 0.5 s, and returns to 0 V at 1.0 s. According to our knowledge of the system, we would expect the servovalve to open to its fully open position (corresponding to a 10-V command) at 0.5 s and return to the closed position at 1.0 s. Figure 9.14 shows the valve spool response from the simulation that confirms this expectation.

Figure 9.15 shows the flow rates through the valve for the system. Note that the flow rate into the left-side chamber responds nearly instantly when the valve opens,

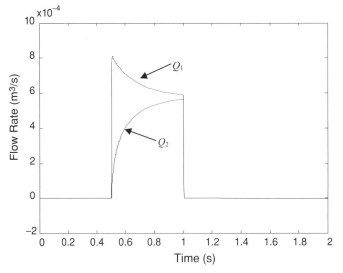

Figure 9.15. Valve flow rates Q_1 and Q_2.

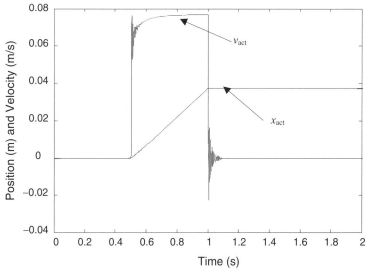

Figure 9.16. Velocity and position of servoactuator load.

but the return side flow is slow to respond because this flow arises as the load piston "sweeps" the low-pressure fluid out of the actuator.

Finally, the motion of the actuator is shown in Fig. 9.16. Note that the velocity shows considerable oscillation at the points of valve opening and closure, in spite of the fact that the damping coefficient is relatively high. This is typical with electro-hydraulic servosystems because the dynamics associated with the actuator pressures and valve flows is usually highly underdamped.

9.5 PNEUMATIC SYSTEMS

Up to this point, this chapter has dealt with fluid systems in which the effects of fluid compressibility are small, but not necessarily negligible. Thus the volume occupied by a given mass of fluid has been nearly constant throughout the system in which it was being used, regardless of the pressure changes that occurred as the fluid mass moved from one part of the system to another. For the compressibility effects in the fluid to be small enough, the fluid has had to be a liquid that was not near its boiling point – in other words, a hydraulic fluid such as water or oil at ambient temperature; or, if gaseous – in other words, a pneumatic fluid – the gross changes in pressure have had to be small enough so that the density changes were only of the order of a few percent.

Thus the discussions and methods used so far have been limited to what are commonly known as hydraulic systems. In this section we deal with modifications of the describing equations needed to cope with pneumatic systems, in which fluid compressibility plays a much greater role than in hydraulic systems.

Of foremost concern in the analysis of pneumatic systems is the need to satisfy the continuity requirement at connecting points between elements (in other words, at node points). This makes it necessary to use mass or weight rate of flow as the

T variable instead of volume rate of flow.[4] Also, it is necessary to use different or modified elemental equations to incorporate the greater effects of compressibility on the flow relations for the pneumatic system elements.

Here we use as the T variable the weight rate of flow $Q_w = \gamma v_{ave} A$ instead of volume rate of flow $Q = v_{ave} A$, where γ is the local weight density of the fluid and v_{ave} is the mean velocity in a flow passage of cross-sectional area A. Similar procedures may be followed if one wishes to use mass rate of flow $Q_m = \rho v_{ave} A$, where ρ is the local mass density of the fluid, or to use the standard volume rate of flow $Q_s = (P/T)(T_s/P_s)v_{ave} A$, where P and T are the local pressure and temperature of the fluid and P_s and T_s are the standard pressure and temperature conditions to be used.

First, we modify the elemental equation for a fluid capacitor [Eq. (9.1)] by multiplying both sides by the local weight density of the fluid γ:

$$Q_{wC} = C_{fw}\frac{dP_{1r}}{dt}, \qquad (9.43)$$

where $Q_{wC} = \gamma Q_C$ and $C_{fw} = \gamma C_f$.

The most commonly encountered form of pneumatic capacitor is a chamber or passage filled with the pneumatic fluid (usually air, but other gases, such as nitrogen, oxygen, helium, etc., are used as pneumatic working fluids). Thus the expressions shown for the gaseous part of a gas-charged hydraulic accumulator in Fig. 9.2 are directly available for modification.

In the modifications that follow, it is assumed that the gaseous fluid behaves as a perfect gas (in other words, it is not so highly compressed so as to be near its liquid state) and obeys the perfect-gas law $P = \rho RT = (\gamma/g)RT$, where R is its perfect-gas constant and g is the acceleration that is due to gravity.

Thus the expression for the modified fluid capacitance of a gas-filled chamber of constant volume is

$$C_{fw} = \begin{cases} \gamma_1 \mathcal{V}/P_1 = g\mathcal{V}/(RT_1), & \text{for very "slow" changes in } P_1 \\ \gamma_1 \mathcal{V}/kP_1 = g\mathcal{V}/(kRT_1), & \text{for very "fast" changes in } P_1 \end{cases}, \qquad (9.44)$$

where $\mathcal{V}$ is the total volume of the gas-filled chamber, P_1 is the absolute pressure of the chamber gas, the specific heat ratio $k = c_p/c_v$ for the chamber gas and T_1 is the absolute temperature of the chamber gas. The term "slow" here denotes changes that take place slowly enough for heat transfer from the surroundings to keep the chamber gas temperature constant – in other words, the changes in the chamber are isothermal. The term "fast" here denotes changes that occur so rapidly that heat transfer from the surroundings is negligible – in other words, the changes in the chamber are adiabatic. For cases in which the rapidity of change is intermediate, a polytropic coefficient n, where $1.0 < n < 1.4$, may be used instead of k for the adiabatic case – that is, the equation for "fast" changes.

When the capacitor consists of a cylinder with a spring-restrained piston of area A_p (see Fig. 9.2) or a bellows of area A_p and spring stiffness k_s enclosing a storage

---

[4] Another alternative is to use an equivalent standard volume rate of flow in which the mass of the fluid is expressed in terms of the volume that it would occupy if it were at some standard pressure and temperature such as ambient air at sea level.

chamber, and the fluid is a compressible gas, the capacitance consists of two parts: (a) a part resulting from the compressibility of the gas in the chamber and (b) a part resulting from changes in volume V and energy storage in the spring k_s:

$$C_f = \begin{cases} \dfrac{gV(t)}{RT_1} + \dfrac{gP_1(t)A_P^2}{RT_1 k_s}, & \text{for very slow changes in } P_1 \\[3mm] \dfrac{gV(t)}{kRT_1} + \dfrac{gP_1(t)A_P^2}{RT_1 k_s}, & \text{for very fast changes in } P_1 \end{cases} \qquad (9.45)$$

Note that P_1 must be expressed as an absolute pressure and that P_1 and V are now functions of time. The absolute temperature T_1 also varies slightly with time, but this variation superposed on its relatively large absolute value usually represents a negligible effect. Using absolute zero pressure (a perfect vacuum) as the reference pressure (for example, P_r) eliminates the problem of having some pressures being absolute pressures and some being relative (or gauge) pressures.

The equation for an ideal pneumatic inertor having a uniform-flow velocity profile across its cross-sectional area is obtained by modification of Eq. (9.7) so that $Q_{wI} = \gamma Q_1$ and $I_w = I/\gamma = L/(gA)$:

$$P_{12} = I_w \frac{dQ_{wI}}{dt}, \qquad (9.46)$$

where $\gamma = g(P_1 + P_2)/RT$ is the average weight density of the gas in the passage.

Note that a gas-filled passage is more likely to need to be modeled as a chamber-type pneumatic capacitor having inflow and outflow at its ends than as an inertor; or it may even need to be modeled as a long transmission line having distributed capacitance and inertance, requiring partial differential equations, which is beyond the scope of this text.[5]

The capillary or porous plug type of hydraulic resistor that was modeled earlier as a linear resistor with resistance R_f becomes nonlinear with a compressible gas flowing through it, having the nonlinear elemental equation

$$Q_{wNLR} = K_w \left(P_1^2 - P_2^2 \right), \qquad (9.47)$$

where K_w is a conductance factor that needs to be evaluated experimentally for best results. (This nonlinear model is limited to operation of the device with flow velocities of Mach numbers less than 0.2, because at higher Mach numbers the operation begins to resemble that of an orifice.)

For pneumatic flow through sharp-edged orifices, the flow rate is related to the ratio of the downstream pressure to the upstream pressure by the classical equation

---

[5] For discussions of transmission line models, see J. Watton, *Fluid Power Systems* (Prentice-Hall, New York, 1989), pp. 224–48; J. F. Blackburn *et al.*, *op. cit.*, pp. 81–8; and V. L. Streeter, ed., *op. cit.*, pp. 21–20 through 21–22.

for isentropic flow of a perfect gas through a converging nozzle,[6] corrected for the *vena contracta* effect by use of a discharge coefficient C_d:

$$\frac{Q_{w\text{NLR}}}{C_d A_o} = \begin{cases} \dfrac{C_2 P_u}{(T_u)^{0.5}}, & \text{for } 0 < \dfrac{P_d}{P_u} < PR_{\text{crit}}, \text{ "choked flow"} \\[3ex] \dfrac{C_2 P_u}{(T_u)^{0.5} C_3} \left(\dfrac{P_d}{P_u}\right)^{1/k} \left[1 - \left(\dfrac{P_d}{P_u}\right)^{k-1/k}\right]^{0.5}, \\[3ex] & \text{for } PR_{\text{crit}} < \dfrac{P_d}{P_u} < 1.0, \text{ "unchoked flow, "} \end{cases} \qquad (9.48)$$

where

$$C_d = \text{discharge coefficient, approximately 0.85 for air,}$$
$$A_o = \text{orifice area, in}^2 \text{ (m}^2\text{),}$$
$$Q_{w\text{NLR}} = \text{the weight rate of flow, lb/s (N/s),}$$
$$C_2 = g\sqrt{\frac{k}{R}\left(\frac{k+1}{2}\right)^{k+1/k-1}} = 0.532(°R)^{0.5}/s \text{ for air}$$
$$[0.410(K)^{0.5}/s],$$
$$C_3 = \sqrt{\frac{2\left(\frac{k+1}{2}\right)^{k+1/k-1}}{k-1}} = 3.872 \text{ for air,}$$
$$PR_{\text{crit}} = \left(\frac{2}{k+1}\right)^{k/(k-1)} = 0.528 \text{ for air,}$$
$$P_u = \text{upstream pressure, lb/in}^2 \text{ (N/m}^2\text{),}$$
$$P_d = \text{downstream pressure, lb/in}^2 \text{ (N/m}^2\text{),}$$
$$T_u = \text{upstream temperature, }°R \text{ (K),}$$
$$R = \text{the perfect-gas constant} = 2.48 \times 10^5 \text{ in}^2/(s^2 °R)$$
$$[268 \text{ m}^2/(s^2 \text{ K})],$$
$$k = \text{the specific heat ratio } \frac{c_p}{c_v} = 1.4 \text{ for air.}$$

A very close approximation, requiring much less computing effort, can be obtained by use of the following relationship for gases such as air having $k = 1.4$:

$$\frac{Q_{w\text{NLR}}}{C_d A_o} = \begin{cases} \dfrac{C_2 P_u}{(T_u)^{0.5}}, & \text{for } 0 < \dfrac{P_d}{P_u} < 0.5 \\[3ex] \dfrac{2C_2 P_u}{(T_u)^{0.5}}\left[\dfrac{P_d}{P_u}\left(1 - \dfrac{P_d}{P_u}\right)\right]^{0.5}, & \text{for } 0.5 < \dfrac{P_d}{P_u} < 1.0 \end{cases} \qquad (9.49)$$

This approximation leads to values of weight flow rate for air that are consistently lower than the values obtained by use of the ideal isentropic flow equation [Eq. (9.40)], but that never depart by more than 3% from the ideal values.

When $P_1 > P_2$ the flow is from (1) to (2), and $P_d = P_2$ and $P_u = P_1$. However, when $P_2 > P_1$, the flow is reversed, and $P_d = P_1$ and $P_u = P_2$.

The graph for $Q_{w\text{NLR}}$ vs. P_1/P_2 for the complete range of P_1/P_2 from zero to a very large value is plotted in Fig. 9.17(a). The curve shown in Fig. 9.17(b) is the square-law graph of the volume flow rate Q vs. $(P_1 - P_2)$ for ideal hydraulic orifice flow, included for comparison purposes.

---

[6] J. F. Blackburn *et al.*, *op. cit.*, pp. 214–23.

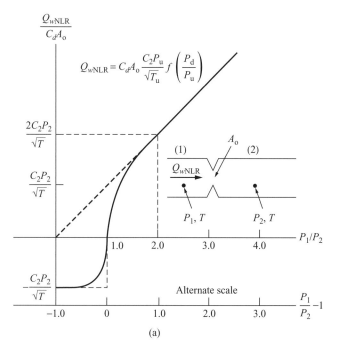

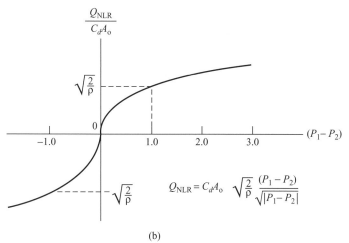

Figure 9.17. Weight flow rate vs. pressure graphs for flow through a sharp-edged orifice of area A_o (a) for air and (b) for a liquid fluid such as water or light oil.

EXAMPLE 9.4

The pneumatic amplifier shown schematically in Fig. 9.18 has been widely used in pneumatic instruments for measurement and automatic control in the process industries, for heating and ventilating controls, and in certain military and aerospace systems. It operates from a pressure source of clean air much as an electric circuit would operate from a battery. Usually the working fluid is air drawn continuously from the atmosphere by a pressure-controlled air compressor. After losing pressure as it flows through a fixed

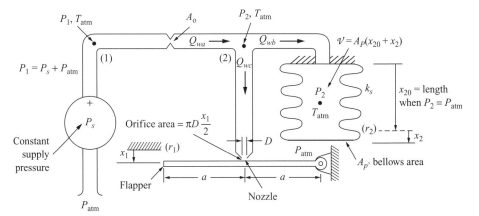

Figure 9.18. Schematic diagram of a pneumatic amplifier system.

orifice of area A_o, the flow rate Q_{wa} approaches the branch point or node (2) from which part may flow as Q_{wb} to the bellows chamber and part may flow as Q_{wc} through the flapper-nozzle orifice back to the atmosphere.

The atmospheric pressure P_{atm} can serve as a secondary reference, but the primary reference is a perfect vacuum so that all pressures (except P_s) are expressed as absolute pressures.

This system is referred to as an amplifier because the energy or power required for actuating the input x_1 is usually only a tiny fraction of the energy or power available to produce the output x_2. As an amplifier it is then useful, when combined with other elements, for executing control functions in complete systems.

During operation at a steady-state normal operating point, the pressure P_2 and the output x_2 are constant, and the flow rate Q_{wb} is zero.

(a) Find the normal operating-point values $\bar{x}_1$ and $\bar{x}_2$ when $\bar{P}_2 = 0.4P_1$, where $P_1 = 50.0 \, \text{lb/in}^2$ absolute.

(b) Write the necessary and sufficient set of describing equations for this system and develop the state-variable equation for this first-order system using P_2 as the state variable and using symbols wherever possible. Assume that changes in P_2 occur slowly (isothermal chamber).

(c) Linearize the describing equations for small perturbations and combine to form a linearized state-variable equation, using $\hat{P}_2$ as the state variable. Write the necessary output equation for $\hat{x}_2$ as the output variable.

(d) Develop the linearized system input–output differential equation relating $\hat{x}_2$ to $\hat{x}_1$ and find and sketch the response of $\hat{x}_2$ to a small step change in x_1, $\hat{x}_1 = D/50$.

SOLUTION

(a) At the normal operating point, $\bar{Q}_{wa} = \bar{Q}_{wc}$, and, because $(\bar{P}_2/\bar{P}_1) < 0.5$, the flow through A_o is choked, so that, from Eq. (9.49),

$$\bar{Q}_{wa} = \frac{C_{d1} A_o C_2 \bar{P}_1}{(T_{atm})^{0.5}}.$$

Because $P_{atm}/P_2 = 14.7/20.0 = 0.735$, the flow through the flapper-nozzle orifice is not choked, so that, from Eq. (9.49),

$$\overline{Q}_{wc} = C_{d2}\pi D \left(\frac{\overline{x}_1}{2}\right) C_2(0.4\overline{P}_1)(2.0)\left[\frac{(0.735)(0.265)}{T_{atm}}\right]^{0.5}.$$

Equating these flows then yields

$$\overline{x}_1 = \frac{A_o}{(0.1764)(3.1416)(0.15)} = 12.03\,A_o. \tag{9.50}$$

For $\overline{x}_2$,

$$\overline{x}_2 = \frac{(\overline{P}_1 - P_{atm})A_p}{k_s} = \frac{45.3\,A_p}{k_s}. \tag{9.51}$$

(b) The describing equations are as follows. For orifice A_o, by use of Eq. (9.49),

$$Q_{wa} = \frac{C_{d1}A_o C_2 P_1}{(T_{atm})^{0.5}}. \tag{9.52}$$

For the flapper-nozzle orifice, by use of Eq. (9.49),

$$Q_{wc} = C_{d2}\pi D x_1 C_2 P_2 \left\{\frac{\left[\left(\frac{P_{atm}}{P_2}\right)\left(1 - \frac{P_{atm}}{P_2}\right)\right]}{T_{atm}}\right\}^{0.5}. \tag{9.53}$$

For the bellows capacitor, by use of Eq. (9.39),

$$Q_{wb} = C_{fw}\frac{dP_2}{dt}, \tag{9.54}$$

where, by use of Eq. (9.45) (isothermal case),

$$C_{fw} = \frac{g A_p}{R T_{atm}}\left(x_{20} + x_2 + \frac{P_2 A_p}{k_s}\right). \tag{9.55}$$

For the bellows spring,

$$x_2 = A_p \frac{P_2 - P_{atm}}{k_s}, \tag{9.56}$$

and continuity at node (2) in Fig. 9.18,

$$Q_{wa} = Q_{wb} + Q_{wc}. \tag{9.57}$$

Combining Eqs. (9.52)–(9.57) yields the nonlinear state-variable equation,

$$\frac{dP_2}{dt} = \left[\left(\frac{-C_{d2}C_2\pi D x_1}{C_{fw}\sqrt{T_{atm}}}\right)\sqrt{\left(\frac{P_{atm}}{P_2}\right)\left(1 - \frac{P_{atm}}{P_2}\right)}\right]P_2 + \left(\frac{C_{d1}C_2}{C_{fw}\sqrt{T_{atm}}}\right)P_1 A_o. \tag{9.58}$$

(c) Linearizing Eq. (9.52) for small perturbations yields

$$\hat{Q}_{wa} = 0. \tag{9.59}$$

For the flapper-nozzle orifice, by use of Eq. (9.53),

$$\hat{Q}_{wc} = k_1\hat{x}_1 + k_2\hat{P}_2, \tag{9.60}$$

where

$$k_1 = C_{d2}\pi DC_2 \left(P_{\text{atm}} \frac{\overline{P}_2 - P_{\text{atm}}}{T_{\text{atm}}} \right), \tag{9.61}$$

$$k_2 = C_{d2}\pi D \left(\frac{\overline{x}_1}{2} \right) \frac{C_2 P_{\text{atm}}}{[T_{\text{atm}} P_{\text{atm}}(\overline{P}_2 - P_{\text{atm}})]^{0.5}}. \tag{9.62}$$

For the bellows capacitor, from Eqs. (9.54) and (9.55),

$$\hat{Q}_{wb} = \frac{g A_p (x_{20} + \overline{x}_2 + \overline{P}_2 A_p/k_s)}{R T_{\text{atm}}} \frac{d\hat{P}_2}{dt}. \tag{9.63}$$

For the bellows spring, from Eq. (9.56),

$$\overline{x}_2 = \frac{A_p \left(\overline{P}_2 - P_{\text{atm}} \right)}{k_s}, \tag{9.64}$$

$$\hat{x}_2 = \left(\frac{A_p}{k_s} \right) \hat{P}_2. \tag{9.65}$$

Combining Eqs. (9.63), (9.64), and (9.65) yields

$$\hat{Q}_{wb} = k_3 \frac{d\hat{P}_2}{dt}, \tag{9.66}$$

where

$$k_3 = \frac{g A_p \left(x_{20} - \frac{A_p P_{\text{atm}}}{k_s} + \frac{2 A_p \overline{P}_2}{k_s} \right)}{R T_{\text{atm}}}. \tag{9.67}$$

Continuity at node (2) implies that

$$\hat{Q}_{wa} = \hat{Q}_{wb} + \hat{Q}_{wc}. \tag{9.68}$$

Combining Eqs. (9.59), (9.60), (9.66), and (9.68) yields

$$\frac{d\hat{P}_2}{dt} = \left(\frac{-k_2}{k_3} \right) \hat{P}_2 + \left(\frac{-k_1}{k_3} \right) \hat{x}_1. \tag{9.69}$$

Rearranging Eq. (9.69), we obtain

$$\frac{d\hat{P}_2}{dt} + \left(\frac{k_2}{k_3} \right) \hat{P}_2 = \left(\frac{-k_1}{k_3} \right) \hat{x}_1. \tag{9.70}$$

Solving Eq. (9.65) for $\hat{P}_2$ in terms of $\hat{x}_2$ yields

$$\hat{P}_2 = \left(\frac{k_s}{A_p} \right) \hat{x}_2. \tag{9.71}$$

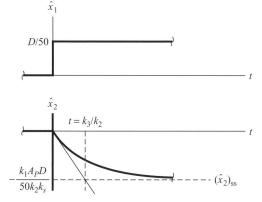

Figure 9.19. Linearized response of bellows in Example 9.4.

(d) Combining Eqs. (9.70) and (9.71) yields

$$\frac{d\hat{x}_2}{dt} + \left(\frac{k_2}{k_3}\right)\hat{x}_2 = \left(\frac{-A_p k_1}{k_s k_3}\right)\hat{x}_1. \tag{9.72}$$

The input is

$$\hat{x}_1(t) = \begin{cases} 0, & \text{for } t < 0 \\ \dfrac{D}{50}, & \text{for } t > 0 \end{cases}.$$

The initial condition for the output is $\hat{x}_2(0) = 0$.

The system time constant $\tau = k_3/k_2$,

$$(\hat{x}_2)_{ss} = \frac{-k_1 A_p D}{50 k_2 k_s}, \tag{9.73}$$

$$\hat{x}_2(t) = \left(\frac{-k_1 A_p D}{50 k_2 k_s}\right)(1 - e^{-t/\tau}). \tag{9.74}$$

The step change in $\hat{x}_1$ and the system's response are plotted in Fig. 9.19.

9.6 SYNOPSIS

The basic fluid system elements were described and shown to be analogs of their corresponding mechanical and electrical A-, T-, and D-type elements: fluid capacitors, fluid inertors, and fluid resistors, respectively. Here, the A variable is pressure and the T variable is volume flow rate. The interconnecting laws of continuity and compatibility needed for dealing with systems of fluid elements correspond to Kirchhoff's current and voltage laws used in electrical system analysis.

When the working fluid is a liquid (or sometimes a slightly compressed gas), the system usually is referred to as a hydraulic system. When the working fluid is a gas, such as air, that undergoes large pressure changes and/or flows with velocities having Mach numbers greater than about 0.2, the system usually is referred to as a pneumatic system. Analysis of pneumatic systems requires modification of the describing equations so that the T variable is mass rate of flow instead of volume rate of flow.

Both linear and nonlinear fluid resistors were introduced, with emphasis on orifice characteristics. For pneumatic orifice flow, the flow rate is a function of the ratio of upstream and downstream absolute pressures, whereas the hydraulic orifice flow relation expresses the flow rate as a function of the difference between the upstream and downstream pressures.

Examples were included to demonstrate the techniques of modeling and analysis for both hydraulic and pneumatic systems including an electrohydraulic servoactuator. Because many systems that use working fluids also incorporate mechanical, electrical, or both, devices, other examples of fluid system analysis will appear in later chapters.

Fluid system components often offer significant advantages over other types of system components, such as speed of response, survivability in difficult environments, safety in hazardous environments, ease of use and/or maintenance, etc.

PROBLEMS

9.1 (a) Develop the system differential equation relating P_{2r} to P_s for the first-order low-pass hydraulic filter shown in Fig. P9.1(a).

(b) Write the expression for the system time constant and sketch the response, $P_{2r}(t)$ vs. t, for a step change ΔP_s from an initial value $P_s(0)$, which has been constant for a very long time [Fig. P9.1(b)].

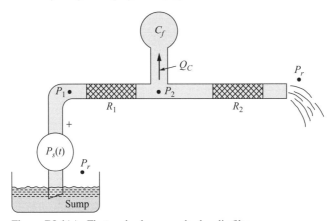

Figure P9.1(a). First-order low-pass hydraulic filter.

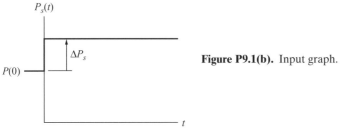

Figure P9.1(b). Input graph.

9.2 (a) Develop the system differential equation relating P_{3r} to P_s for the second-order low-pass hydraulic filter shown in Fig. P9.2(a).

(b) Find expressions for the natural frequency ω_n and damping ratio ζ, and sketch the response versus time for a step change ΔP_s from an initial value $P_s(0)$, which has been constant for a very long period of time, assuming that $\zeta = 0.3$. Show clearly the period of the oscillation, the per-cycle DR, and the final steady-state value of P_{3r} in Fig. P9.2(b).

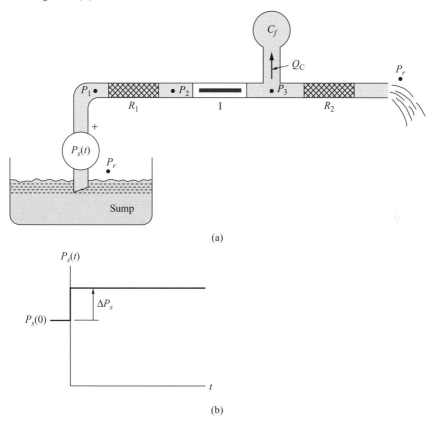

(a)

(b)

Figure P9.2. (a) Second-order low-pass hydraulic filter, (b) input graph.

9.3 The fluid system modeled in Fig. P9.3 represents the process of filling a remote tank or reservoir in a batch process at a chemical plant.

Fluid provided by an ideal pressure source P_s is suddenly turned on or off by a rotary valve that offers very little resistance to flow when it is open. The long line connecting the tank to the shutoff valve has an internal diameter of 2.0 cm and a length of 50 m. The reservoir has an inside diameter of 25 cm.

In analyzing the response of this system to sudden opening of the valve, it has been proposed that the line be modeled as a lumped resistance in series with a lumped inertance, as shown.

(a) Find the estimated lumped resistance R_f, the lumped inertance I, and the lumped capacitance C_f for the line and the capacitance C_t of the reservoir. Assume laminar flow in the line. The fluid has viscosity $\mu = 1.03 \times 10^{-2}$ N s/m^2, weight density $\gamma = 8.74 \times 10^{-3}$ N/m^3, and bulk modulus of elasticity $\beta = 1.38 \times 10^9$ N/m^2.

(b) Write the necessary and sufficient set of describing equations for this system based on the $R_f - I$ model of the line for the time starting with $t = 0$ when the valve is suddenly opened.

(c) Develop the system differential equation for the case in which P_{3r} is the output.

(d) Calculate the natural frequency and damping ratio for this model, using pertinent values found in part (a).

(e) How do you feel about the decision to neglect the lumped capacitance, evaluated in part (a), in modeling this system?

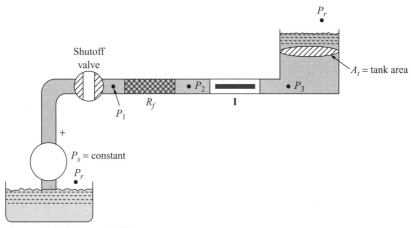

Figure P9.3. Reservoir filling system.

9.4 A variable flow source $Q_s(t)$ is being used to replenish a reservoir that in turn supplies an orifice, as shown in Fig. P9.4(a).

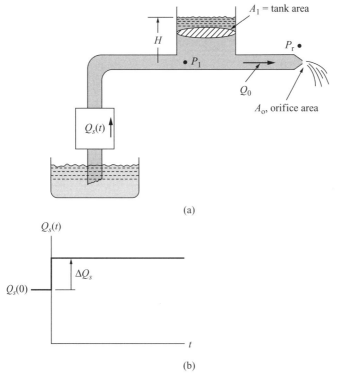

Figure P9.4. (a) Reservoir and discharge orifice replenished by flow source, (b) input graph.

(a) Write the necessary and sufficient set of describing equations for this system.

(b) Linearize and combine to develop the system differential equation relating $\hat{H}$ to $\hat{Q}_s(t)$.

(c) Using only symbols, solve to find and plot the response of $\hat{H}$ to a small step change ΔQ_s from an initial value $Q_s(0)$, which has been constant for a very long time.

9.5 Two tanks are connected by a fluid line and a shutoff valve, as shown in Fig. P9.5. The valve resistance is negligible when it is fully open.

(a) Write the necessary and sufficient set of describing equations for this system.

(b) Combine to form the system differential equation needed to solve for finding P_{2r} in response to suddenly opening the valve at $t = 0$ when $H_1(0)$ is greater than $H_2(0)$. Note that the input here is a sudden change in the system and that both initial conditions are needed in combining the describing equations.

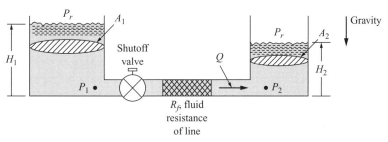

Figure P9.5. Two interconnected tanks.

9.6 The pneumatic system shown schematically in Fig. P9.6 is a model of the air supply system for a large factory. The air compressor with its own on–off controller delivers air on a cyclic basis to the large receiving tank, where the compressed air is stored at a somewhat time-varying pressure P_1. From this receiving tank, the air flows through an orifice having area A_{o1} to a ballast tank that acts, together with A_{o1}, as a pressure filter to provide an output pressure P_2 having much smaller cyclic variations than the fluctuations in P_1 caused by the on–off flow Q_{ws} from the compressor. The object of this problem is to develop the mathematical model needed to compare the predicted variations of P_2 with the variations in P_1.

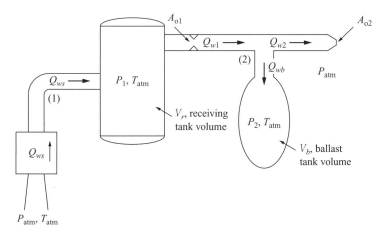

Figure P9.6. Schematic diagram of factory air supply system model.

The second orifice, with area A_{o2} and exhausting to the atmosphere, is provided to simulate the load effects of all the air-consuming devices and processes in the factory, and the value of A_{o2} is chosen to simulate the normal operating (that is, average) factory consumption rate of flow of air, $\overline{Q}_{w2} = 0.05$ N/s (0.0051 kg/s) when the normal operating pressure $\overline{P}_2$ is 5.0×10^5 N/m² absolute.

(a) Compute the value of A_{o2} needed for the given desired normal operating-point conditions.

(b) Write the necessary and sufficient set of describing equations for this system, considering the compressor flow rate $Q_{ws}(t)$ as the system input.

(c) Linearize the equations in part (b) and form the set of state-variable equations having $\hat{P}_1$ and $\hat{P}_2$ as the system state variables.

(d) Develop the system input–output differential equation relating $\hat{P}_2(t)$ to $\hat{Q}_{ws}(t)$ and do the same for relating $\hat{P}_1(t)$ to $\hat{Q}_{ws}(t)$.

9.7 A common problem in household plumbing is the waterhammer in which transient pressure waves cause noisy vibrations in the pipes when a valve is suddenly turned off. The effects of a waterhammer can be greatly mitigated by the appropriate placement of a fluid capacitance near the valve. Figure P9.7 shows a conceptual model that can be used for understanding this phenomenon.

(a) Develop the second-order state-space model that describes this system. Compute the values of the fluid inertance assuming the pipe is 3 m long and 0.03 mm in diameter and the fluid is water at room temperature. Assume that the valve behaves like a sharp-edged orifice in turbulent flow.

(b) Linearize the equations of part (a) about a nominal pressure of P_s and a valve orifice opening of 0.005-m diameter. Use these linearized equations to predict the pressure response for a rapid closing of the valve (assume the area changes in a step-wise fashion).

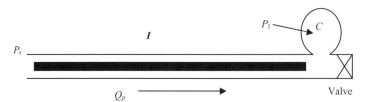

Figure P9.7.

9.8 As stated in the previous problem, a waterhammer is a potentially destructive phenomenon that arises from the sudden closing of a valve. In Problem 9.7, a linearized approach was used to predict the response. Now a computer simulation is used. Use the nonlinear equations derived in Problem 9.7(a) and develop a computer simulation of the system. Use a step-input block to model the sudden change in orifice area. Compare your results with the answer in Problem 9.7(b).

9.9 Further modify the simulation of the waterhammer (see Problems 9.7 and 9.8) with a better model of the long pipe. Break the pipe up into six segments, with each having its own inertance and capacitance. Compare the results you find for this simulation with the answers to Problems 9.7(b) and 9.8.

10

Mixed Systems

LEARNING OBJECTIVES FOR THIS CHAPTER

10–1 To integrate knowledge of mechanical, electrical, thermal, and fluid systems to model physical systems made up of more than one type of element.

10–2 To model coupling transducers as ideal and nonideal devices.

10–3 To use state or input–output models to analyze performance of translational–rotational mechanical, electromechanical, and fluid mechanical systems.

10.1 INTRODUCTION

In previous chapters, various types of systems were discussed, each within its own discipline: mechanical, electrical, thermal, and fluid. However, many engineering systems consist of combinations of these elementary single-discipline system elements: electromechanical, fluid mechanical, and so on. To combine single-discipline systems, it is necessary to use coupling devices that convert one kind of energy or signal to another: mechanical to electrical, fluid to mechanical, and so on.

The general term "transducer" will be used here for ideal coupling devices. In cases in which significant amounts of energy or power are involved, these coupling devices will be referred to as energy-converting transducers; when the amount of energy being transferred is minimal, they will be referred to as signal-converting transducers.

Selected nonideal energy convertors, which are modeled graphically, are also discussed in terms of typical characteristic curves that have been derived from performance tests.

10.2 ENERGY-CONVERTING TRANSDUCERS AND DEVICES

The energy-converting transducers introduced here are ideal in that they are lossless models that contain no energy-storage or energy-dissipation elements. When energy storage or energy dissipation is present, these effects are modeled with lumped ideal elements connected at the terminals of the ideal transducer.

10.2.1 Translational–Mechanical to Rotational–Mechanical Transducers

The symbolic diagram for mechanisms that convert translational motion to rotational motion or vice versa is shown in Fig. 10.1, where n is the coupling coefficient relating output motion to input motion.

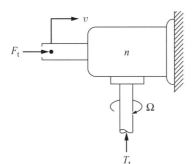

Figure 10.1. Symbolic diagram for an ideal translational-to-rotational transducer.

The elemental equations for the transducer are

$$\Omega = nv, \tag{10.1}$$

$$nT_t = F_t. \tag{10.2}$$

For the device shown, the flow of power is from left to right when the product of vF_t (and the product of ΩT_t) is positive, that is, whenever the through and across variables are of the same sign.

Pulley-and-cable systems, lever-and-shaft mechanisms, and rack-and-gear mechanisms are examples of translational-to-rotational transducers.

10.2.2 Electromechanical Energy Converters

In electromechanical systems, the coupling between mechanical and electrical elements of the system is provided through a magnetic field. Two processes, studied in introductory physics courses, are essential in establishing the coupling. The first process involves a current-carrying wire placed within a magnetic field. Force $\mathbf{F}_e$ is exerted on the wire and the differential of the force is

$$d\mathbf{F}_e = i(d\mathbf{l} \times \mathcal{B}), \tag{10.3}$$

where i is the current in the wire, $d\mathbf{l}$ is the differential length of the wire, and $\mathcal{B}$ is the flux density of the magnetic field. It should be noted that $\mathbf{F}_e, \mathbf{l}$, and $\mathcal{B}$ are vectors and the symbol $\times$ denotes the vector cross product. In many applications, the wire-length vector is perpendicular to the magnetic-field flux vector, and in such cases Eq. (10.3) can be replaced with the following scalar equation:

$$F_e = il\mathcal{B}, \tag{10.4}$$

where the direction of force F_e is determined by the right-hand rule.

The other process responsible for magnetic coupling involves a wire moving within a magnetic field. Voltage e_m is induced in the wire, and the voltage differential is

$$de_m = (\mathbf{v} \times \mathcal{B}) \cdot d\mathbf{l}, \tag{10.5}$$

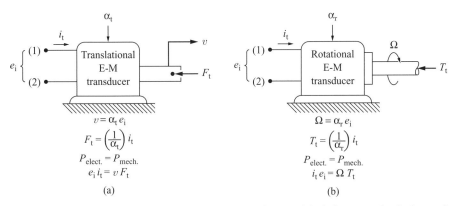

Figure 10.2. Symbolic diagrams with elemental equations for ideal electromechanical transducers: (a) translational–mechanical and (b) rotational–mechanical.

where $\mathbf{v}$ is the velocity vector of the wire and the dot denotes the scalar product. Again, in most applications, the three vectors $\mathbf{v}$, $\mathcal{B}$, and $\mathbf{l}$ are mutually perpendicular, and Eq. (10.5) simplifies to

$$e_m = v\mathcal{B}l. \tag{10.6}$$

Equations (10.4) and (10.6) will be used in Example 10.2 to derive the mathematical model of a dc motor.

The symbolic diagrams used here for ideal energy-converting transducers are shown in Fig. 10.2 for the case of translational–mechanical motion [Fig. 10.2(a)] and for the case of rotational–mechanical motion [Fig. 10.2(b)]. The elemental equations are also given in Fig. 10.2. Because the coupling coefficient α is in many cases controllable, which is a feature that makes electromechanical transducers especially attractive for control systems use, it is shown as an input signal.

The translational version is an ideal solenoid. The direction of power flow for the device as shown [Fig. 10.2(a)] is from left to right when $e_{12}i_t$ is positive.

The rotational version is an ideal electric motor or an ideal electric generator, depending on the usual direction of power flow. The device as shown [Fig. 10.2(b)] operates as a motor when $e_{12}i_t$ is positive and as a generator when $e_{12}i_t$ is negative. A device that usually operates as a motor may temporarily operate as a generator when system transients occur, and vice versa.

Although these models are intended primarily for use with dc devices, they apply reasonably well to the same devices operating on ac as long as rms values of voltage and current are used and the dynamic response of the rest of the system is slow compared with the periodic variation of the ac.

AC induction motors of the squirrel-cage type operate well only at speeds close to their no-load speed, which is synchronous with the ac frequency and is usually constant. A typical torque-versus-speed characteristic of such a motor, operating with constant supply voltage, is shown in Fig. 10.3. This characteristic includes the effects of bearing friction and windage loss in the fluid surrounding the rotating members, as well as resistance and inductance in the windings, so it is not an ideal

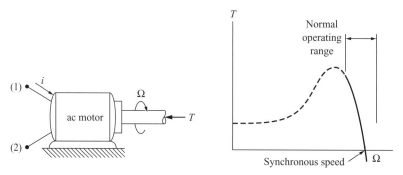

Figure 10.3. Steady-state torque-versus-speed characteristic for a squirrel-cage ac motor operating at constant supply frequency and voltage.

lossless model. Thus it is necessary to know how the efficiency varies with load in order to determine the input current. When the efficiency is known, the input current, when the unit is operating as a motor, is given by

$$i_{\text{rms}} = \frac{1}{\eta_m} \cdot \frac{T\Omega}{e_{i_{\text{rms}}}}, \qquad (10.7)$$

where η_m is the efficiency of the unit operating as a motor.

10.2.3 Fluid Mechanical Energy Converters

The symbolic diagrams to be used here for ideal energy-converting transducers are shown in Fig. 10.4 for the case of translational–mechanical motion [Fig. 10.4(a)] and the case of rotational–mechanical motion [Fig. 10.4(b)]. The elemental equations are also given in Fig. 10.4. The coupling coefficient D is shown as an input signal for both cases, although it is not controllable for the case of translational motion.

The translational version is a fluid cylinder with reciprocating piston, which may operate in pump or motor fashion, depending on the direction of power flow. The power flow is from left to right when $P_{12}Q_t$, is positive for the device shown [Fig. 10.4(a)].

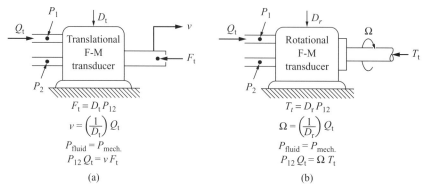

Figure 10.4. Symbolic diagrams for ideal fluid mechanical energy-converting transducers: (a) translational and (b) rotational.

The rotational version is an ideal positive-displacement fluid motor or an ideal fluid pump, depending on the usual direction of power flow. The device as shown [Fig. 10.4(b)] operates as a fluid motor when $P_{12}Q_t$ is positive and as a fluid pump when $P_{12}Q_t$ is negative. A device that usually operates as a pump may temporarily operate as a motor when system transients occur, and vice versa.

The coupling coefficient D is the ideal volume displaced per unit motion of the output shaft (i.e., no leakage occurring) for both the rotational and translational cases.

These models are valid only when the compressibility of the fluid used is negligible, so that the volume rate of flow of fluid entering the transducer is the same as the volume rate of flow leaving it. Because the action of the fluid on the moving members is by means of static fluid pressure alone (i.e., momentum transfer is negligible), these devices are sometimes referred to as hydrostatic energy converters.

Hydrokinetic energy converters, such as centrifugal pumps and turbines, do involve momentum interchange between the moving fluid and moving blades and the fixed walls. These devices are modeled graphically through the use of experimentally derived characteristic curves rather than by means of ideal transducers (similar to the modeling shown for squirrel-cage ac motors). Thus these models are not lossless, and it is necessary to know how their efficiency varies with operating conditions.

The symbolic diagram and typical characteristic curves for a centrifugal pump operating at a series of constant speeds are shown in Fig. 10.5. The pump torque is given by

$$T_p = \frac{1}{\eta_p} \cdot \frac{P_{34} Q_p}{\Omega}, \tag{10.8}$$

where η_p is the pump efficiency.

The symbolic diagram and characteristic curves for operation of a hydraulic turbine at a series of constant pressure drops are shown in Fig. 10.6. The turbine flow rate is given by

$$Q_t = \frac{1}{\eta_t} \cdot \frac{\Omega T_t}{P_{12}}, \tag{10.9}$$

where η_t is the turbine efficiency.

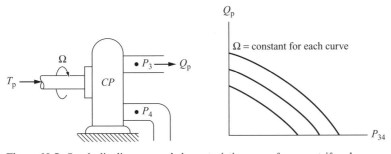

Figure 10.5. Symbolic diagram and characteristic curves for a centrifugal pump.

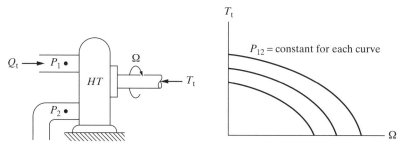

Figure 10.6. Symbolic diagram and typical characteristic curves for a hydraulic turbine.

Linearization of these characteristics for small perturbations about a set of normal operating conditions is readily accomplished by use of the techniques discussed in Chap. 2.

10.3 SIGNAL-CONVERTING TRANSDUCERS

Signal-converting transducers in some cases are simply energy-converting transducers that have negligible load—for instance, a tachometer generator operating into a very high-resistance load. In other cases they are specially designed devices or systems that convert one type of signal to another—for instance, a flyweight speed sensor or a Bourdon gauge pressure sensor. Still another example is the use of a resistance potentiometer, supplied with a constant voltage, to deliver an output wiper-arm signal that is a function of its wiper-arm position.

A partial list of signal-converting transducers is given in the small table.

- Tachometer generator
- Centrifugal pump speed sensor
- Linear variable-differential transformer position sensor
- Variable-reluctance pressure sensor
- Flyweight speed sensor
- Linear velocity sensor
- Thermistor temperature sensor
- Variable-capacitor proximity sensor
- Piezoelectric force sensor

The potential user is referred to a wide range of technical bulletins and specifications prepared by the manufacturers of such equipment. Several texts are currently in print on this topic.[1]

The ideal signal-converting transducers are controlled sources, shown symbolically in Fig. 10.7 together with their describing equations. Figure 10.7(a) shows a controlled A-variable-type transducer, and Fig. 10.7(b) shows a controlled T-variable-type transducer.

The input variable x denotes the variable being sensed, and e_s (or P_s or v_s or T_s) denotes the output of an A-variable-type transducer. For T-variable-type transducers, i_s (or Q_s or F_s or Q_{hs}) denotes the output.

---

[1] As a starting point, consider A. J. Wheeler and A. R. Ganji, *Introduction to Engineering Experimentation*, 2nd ed. (Prentice-Hall, Englewood Cliffs, NJ, 2004), or E. O. Doeblin, *Measurement Systems*, 5th ed. (McGraw-Hill, New York, 2004).

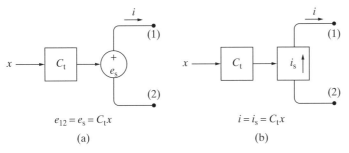

$$e_{12} = e_s = C_t x \qquad\qquad\qquad i = i_s = C_t x$$

(a)   (b)

Figure 10.7. Symbols for signal-converting transducers.

APPLICATION EXAMPLES

EXAMPLE 10.1

The rack-and-pinion system shown in Fig. 10.8 is to be modeled as part of a large system, which is to be simulated on a digital computer. The object here is to set up the state-variable equation(s) for this subsystem, considering T_s and F_s as the system inputs and v as the system output.

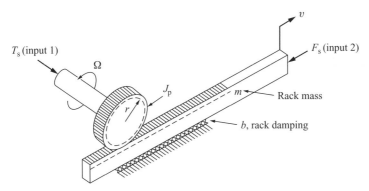

Figure 10.8. Rack-and-pinion system, including pinion inertia, rack mass, and friction.

SOLUTION

The detailed symbolic free-body diagram for this system is shown in Fig. 10.9.
    The elemental equations are as follows: For the pinion inertia,

$$T_s - T_t = J_p \frac{d\Omega}{dt}. \tag{10.10}$$

For the transducer,

$$\Omega = nv, \tag{10.11}$$

$$T_t = (1/n)F_t, \tag{10.12}$$

where $n = 1/r$. For the rack mass,

$$F_t - F_b - F_s = m \frac{dv}{dt}. \tag{10.13}$$

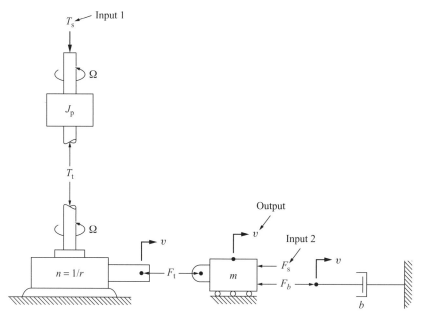

Figure 10.9. Detailed symbolic free-body diagram of rack-and-pinion system.

For the rack friction,

$$F_b = bv. \tag{10.14}$$

Combine Eqs. (10.10), (10.11), and (10.12) to eliminate T_t and Ω:

$$T_s - \frac{1}{n}F_t = nJ_p\frac{dv}{dt}. \tag{10.15}$$

Combine Eqs. (10.13) and (10.14) to eliminate F_b:

$$F_t - bv - F_s = m\frac{dv}{dt}. \tag{10.16}$$

Combine Eqs. (10.15) and (10.16) to eliminate F_t:

$$T_s - \frac{1}{n}\left(m\frac{dv}{dt} + bv + F_s\right) = nJ_p\frac{dv}{dt}. \tag{10.17}$$

Rearranging yields a single state-variable equation:

$$\frac{dv}{dt} = \frac{1}{n^2 J_p + m}\left(-bv + nT_s - F_s\right). \tag{10.18}$$

Note that the inertia from the rotational part of the system becomes an equivalent mass equal to $n^2 J_p$; thus, although this system has two energy-storage elements, they are coupled by the transducer so that they behave together as a single energy-storage element. Therefore only one state-variable equation is required for describing this system.

The multiplication of a system parameter of one domain by the square of the transducer constant to produce an equivalent parameter in the other domain will be seen to occur consistently in all mixed-system analyses.

EXAMPLE 10.2

A permanent magnet dc motor is being used to drive a mechanical load consisting of a load inertia J_1 and load damping B_1 as shown in Fig. 10.10. Develop the system differential input–output equation relating the output speed Ω to the input voltage e_i.

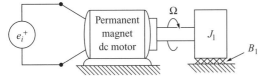

Figure 10.10. Schematic diagram of dc-motor-driven mechanical system.

SOLUTION

A simplified schematic of the dc motor is shown in Fig. 10.11. The armature winding (only one turn of the winding is shown in Fig. 10.11) is placed within a uniform magnetic field of flux density $\mathcal{B}$. The resistance of the winding is R and its inductance is L. When the input voltage e_i is applied, current i flows in the armature winding. Under these conditions (a current-carrying conductor within a magnetic field), force F_e is exerted on both the top and the bottom sections of the armature winding as illustrated in Fig. 10.11. The force F_e is given by Eq. (10.4),

$$F_e = il\mathcal{B},$$

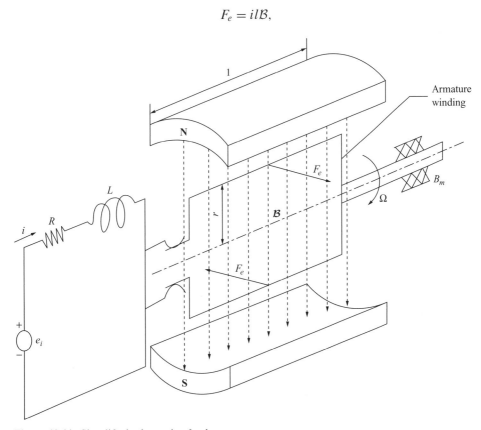

Figure 10.11. Simplified schematic of a dc motor.

where l is the length of the armature coil. Because the winding is free to rotate around its longitudinal axis, force F_e produces a torque:

$$T_e = F_e r. \tag{10.19}$$

This torque acts on each (top and bottom) section of each turn in the armature winding. Assuming that there are N turns in the winding and they are all within the uniform magnetic field, the total induced torque exerted on the armature winding is

$$T_e = 2NF_e r. \tag{10.20}$$

When Eq. (10.20) is compared with the equation for a rotational–electromechanical transducer given in Fig. 10.2, the coupling coefficient for the dc motor can be identified as

$$\alpha_r = \frac{1}{2NlBr}, \tag{10.21}$$

and hence the electrically induced torque can be expressed as

$$T_e = \frac{1}{\alpha_r}i. \tag{10.22}$$

When the armature winding starts to rotate within a uniform magnetic field, voltage e_m is induced in the winding, given by Eq. (10.6):

$$e_m = vBl.$$

Replacing translational velocity v with the rotational velocity Ω times radius r and accounting for N turns and two sections of each turn in the armature winding, one finds that the total voltage induced is

$$e_m = 2N\Omega r Bl \tag{10.23}$$

or

$$e_m = \frac{1}{\alpha_r}\Omega, \tag{10.24}$$

where α_r is the coupling coefficient defined by Eq. (10.21). Equations (10.22) and (10.24) describe mathematically the coupling between mechanical and electrical parts of the system and are often called the coupling equations.

To derive the system equations of motion, use the detailed symbolic circuit and free-body diagram shown in Fig. 10.12.

The equation for the electrical part of the system is

$$e_i = Ri_t + L\frac{di_t}{dt} + e_{12}, \tag{10.25}$$

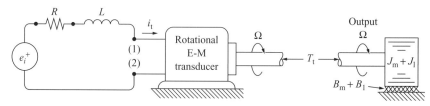

Figure 10.12. Detailed symbolic diagram of a motor-driven system.

where e_{12} is the induced voltage given by Eq. (10.24). Hence the equation for the electrical part can be written as

$$e_i = Ri_t + L\frac{di_t}{dt} + \frac{1}{\alpha_r}\Omega. \tag{10.26}$$

For the mechanical part of the system,

$$(J_m + J_1)\frac{d\Omega}{dt} + (B_m + B_1)\Omega = T_t, \tag{10.27}$$

or, when Eq. (10.22) is used,

$$(J_m + J_1)\frac{d\Omega}{dt} + (B_m + B_1)\Omega = \frac{1}{\alpha_r}i_t. \tag{10.28}$$

Equations (10.26) and (10.28) constitute the set of basic equations of motion for the system and can be easily transformed into the state-variable equations by use of i_t and Ω as the state variables. The objective in this example is to derive the input–output equation relating Ω to e_i. One can accomplish this by combining Eqs. (10.26) and (10.28) to eliminate i_t. From Eq. (10.28), current i_t is

$$i_t = \alpha_r(J_m + J_1)\frac{d\Omega}{dt} + \alpha_r(B_m + B_1)\Omega. \tag{10.29}$$

Differentiating both sides of this equation gives

$$\frac{di_t}{dt} = \alpha_r(J_m + J_1)\frac{d^2\Omega}{dt^2} + \alpha_r(B_m + B_1)\frac{d\Omega}{dt}. \tag{10.30}$$

Substituting the expressions for i_t and di/dt into Eq. (10.26) yields

$$e_i = R\alpha_r(J_m + J_1)\frac{d\Omega}{dt} + R\alpha_r(B_m + B_1)\Omega + L\alpha_r(J_m + J_1)\frac{d^2\Omega}{dt^2}$$
$$+ L\alpha_r(B_m + B_1)\frac{d\Omega}{dt} + \frac{1}{\alpha_r}\Omega. \tag{10.31}$$

Multiplying both sides by α_r and collecting terms gives the system input–output equation:

$$(J_m + J_1)\alpha_r^2 L\frac{d^2\Omega}{dt^2} + [(B_m + B_1)\alpha_r^2 L + (J_m + J_1)\alpha_r^2 R]\frac{d\Omega}{dt}$$
$$+ [(B_m + B_1)\alpha_r^2 R + 1]\Omega = \alpha_r e_i. \tag{10.32}$$

In Example 10.2, the inductor L is transformed into an equivalent spring $1/\alpha_r^2 L$, and the resistor R is transformed into an equivalent damper $1/\alpha_r^2 R$ on the mechanical side of the system. Because the two inertias are connected to each other by a rigid shaft, they become lumped together as a single energy-storage element, and the system is only a second-order system.

EXAMPLE 10.3

A variable-displacement hydraulic motor supplied with a constant flow source is used to vary the output speed of an inertia-damper load, as shown in Fig. 10.13. The displacement of the motor D_m is proportional to the stroke lever angle ψ, and the motor has leakage resistance R_f, rotor inertia J_m, and bearing and windage friction B_m. The load torque T_1 is a second input to the system, in addition to the motor stroke ψ. Develop the state-variable equation(s) for this system.

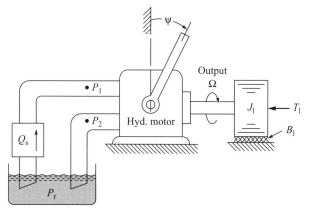

Figure 10.13. Schematic diagram of variable-displacement motor speed control system.

SOLUTION

The detailed symbolic diagram for this system is shown in Fig. 10.14. The elemental equations are as follows: For the leakage resistor,

$$Q_R = Q_s - Q_t = \frac{1}{R_f} P_{12}.$$ (10.33)

For the transducer,

$$T_t = C_1 \psi \, P_{12},$$ (10.34)

$$\Omega = \frac{1}{C_1 \psi} \cdot Q_t.$$ (10.35)

For the inertias and dampers,

$$T_t - T_l = (J_m + J_l) \frac{d\Omega}{dt} + (B_m + B_l)\Omega.$$ (10.36)

Combining Eqs. (10.33), (10.34), and (10.36) and multiplying all terms by $R_f C_1$ yields

$$T_t = C_1 \psi \, R_f Q_s - (C_1 \psi)^2 R_f \Omega.$$ (10.37)

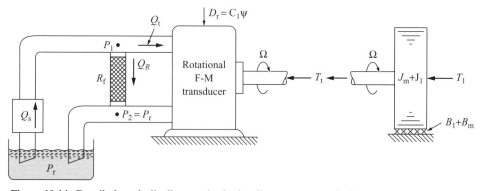

Figure 10.14. Detailed symbolic diagram for hydraulic-motor-controlled system.

Combine Eqs. (10.36) and (10.37) to eliminate T_t:

$$C_1\psi\, R_f Q_s - (C_1\psi)^2 R_f \Omega - T_1 = (J_m + J_1)\frac{d\Omega}{dt} + (B_m + B_l)\Omega. \qquad (10.38)$$

Divide all terms by $(J_m + J_1)$ and rearrange into state-variable format:

$$\frac{d\Omega}{dt} = -\frac{B_m + B_l + (C_1\psi)^2 R_f}{J_m + J_1}\Omega + \frac{C_1\psi\, R_f}{J_m + J_1}Q_s - \frac{1}{J_m + J_1}T_1. \qquad (10.39)$$

In this example, the fluid resistor R_f is transformed into an equivalent displacement-referenced damper $(C_1\psi)^2 R_f$ on the mechanical side of the system.

10.5 SYNOPSIS

In this chapter, several types of ideal transducers were used to model the coupling devices that interconnect one type of system with another type of system, resulting in what is called a mixed system. In this modeling of coupling devices, nonideal characteristics were described by the careful addition of A-, T-, or D-type elements to the ideal transducer. For the cases in which the use of an ideal transducer is not feasible (for instance, ac machinery or hydrokinetic machinery), graphical performance data were used to describe the characteristics of the coupling device as a function of two variables by use of families of curves.

Detailed examples were provided to illustrate the techniques of modeling and analysis associated with the design and development of mixed systems. The transformation of the characteristics of an element on one side of an ideal transducer to an equivalent coupling-coefficient-referenced element on the other side of the transducer was illustrated. The use of state-variable equations in system modeling was reiterated, and in each case input–output differential equations ready for use in analytical solution or computer simulation studies were also provided.

PROBLEMS

10.1 A rack-and-pinion mechanism has been proposed as a means of using a dc motor to control the motion of the moving carriage of a machine tool, as shown schematically in Fig. P10.1(a).

The ideal current source $i_s(t)$ is capable of delivering the current i_s to the motor regardless of the voltage e_{21} required. The nonlinear friction in the rack-and-pinion mechanism has been lumped into the single nonlinear friction force F_{NLD} versus velocity v_2 characteristic shown in Fig. P10.1(b).

(a) Draw a complete free-body diagram of the system showing the ideal inertialess motor, the motor inertia J_m, the motor damper B_m, the shaft spring K, the shaft torque T_s, the gear inertia J_g, the ideal rotational-to-translational transducer, the lumped bar and carriage mass $(m_b + m_c)$, and the NLD. The speed at the motor end of the shaft is Ω_1.

(b) Using only symbols, write the necessary and sufficient set of describing equations for this system.

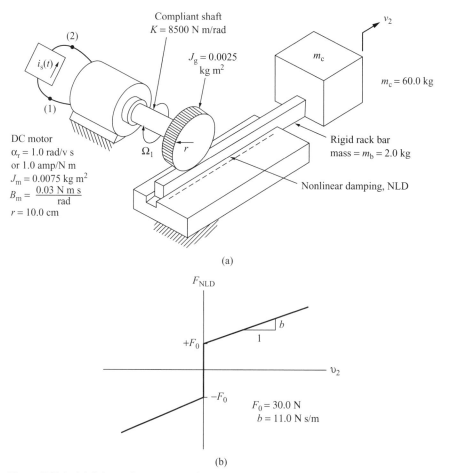

Figure P10.1. (a) Schematic representation of motor-driven rack-and-pinion, (b) nonlinear friction characteristic between rack and ground.

(c) Using only symbols, rearrange the equations developed in part (b) to form the set of nonlinear state-variable equations. Use Ω_1, v_2 and T_s as the state variables.

(d) Find the steady operating-point values $\overline{\Omega}_1$ and $\overline{v}_2$ when $\overline{i}_s = 5$ amp.

(e) Using only symbols, linearize the state-variable equations and combine them to develop the system differential equation relating $\hat{v}_2$ to $\hat{i}_s(t)$.

(f) Find the roots of the characteristic equation and calculate the natural frequency and damping ratio for the linearized model, using the data given in Fig. P10.1 and calculated in part (d).

(g) Write the conditional seatements needed to describe the NLD for all possible values of F_{NLD}.

10.2 The schematic diagram shown in Fig. P10.2(a) represents a model of a small hydro-electric power plant used to convert water power, diverted from a nearby stream, into dc electricity for operating a small factory.

   You have been given the task of determining the dynamic response characteristics of this system. The input is the current source $i_s(t)$, supplied to the field winding of the generator, which establishes the field strength and therefore the value of the coupling

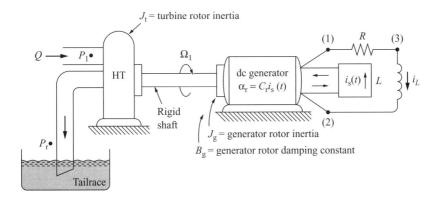

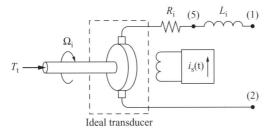

Figure P10.2(a). Schematic diagram of a turbine-driven generator.

coefficient α_r of the generator. The output is the voltage e_{12} with which the generator delivers power to its electrical load modeled here as a series $R\text{-}L$ circuit. (This information about the system dynamic response will be needed later for designing a voltage controller that will keep the ouput voltage constant under varying electrical load conditions.)

Previously, one of your colleagues found an appropriate approximate analytical model for the hydraulic turbine that fitted reasonably well the torque-speed characteristics obtained by a lab technician. The technician's results are shown graphically in Fig. P10.2(b); your colleague's equation is

$$T_t = T_0 - \frac{T_0}{6}\left(1 - \frac{\Omega_1}{\Omega_0}\right)^2 - T_0\left(1 - \frac{P_r}{P_0}\right).$$

Note that in Fig. P10.2(b) the numerical value of Ω_0 is negative.

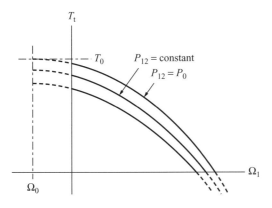

Figure P10.2(b). Steady-state turbine torque versus speed characteristics at various constant speeds.

(a) Draw a free-body diagram of the mechanical part of the system showing the inertialess turbine, the lumped turbine-plus-generator inertia $(J_t + J_g)$, the generator damper B_g, and the ideal inertialess generator. Also draw the complete electrical circuit including the ideal generator, the internal resistance R_i, the internal inductance L_i, and the load resistance and inductance R and L.

(b) Write the necessary and sufficient set of describing equations for this system.

(c) Rearrange to form the set of nonlinear state-variable equations using Ω_1 and i_L as the state variables.

(d) Linearize the state-variable equations and provide an output equation for the system output $\hat{e}_{12}$.

(e) Combine the state-variable equations to form the system differential equation relating $\hat{e}_{12}$ to the input $\hat{i}_s(t)$.

10.3 In the system shown in Fig. P10.3, a positive-displacement pump is driven by an armature-controlled dc motor. The coupling equations for the motor and the pump are included in the figure.

(a) Select state variables and derive a complete set of state-variable equations for the system.

(b) Draw a simulation block diagram for the system.

(c) Find the calibration formula for the system relating output flow rate, Q_o, to the input voltage, e_i, at steady state.

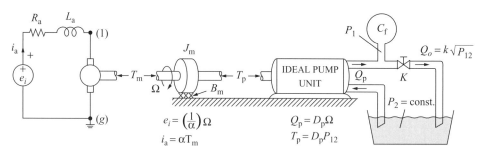

Figure P10.3. Schematic diagram of the system considered in Problem 10.3.

10.4 A system is being designed that will use a field-controlled dc motor to drive the spindle of a variable-speed drilling machine. The drive system is shown schematically in Fig. P10.4. The armature circuit of the motor is driven by a constant voltage source e_s, and the circuit is modeled to include a resistor R_a and an inductor L_a. The field circuit is driven with a variable input voltage source $e_f(t)$, and this circuit is modeled to include a field resistance R_f and a field inductance L_f. The coupling equations for the induced voltage (back emf) e_{3g} and the shaft torque T_m are given in Fig. P10.4. The mechanical load that the motor is driving includes constant torque T_L, the inertia of the spindle lumped with the self-inertia of the motor armature J, and linear viscous friction B, which is due to bearings.

(a) Formulate the set of state-variable equation for the system, using i_a, i_f, and Ω_1 as the state variables.

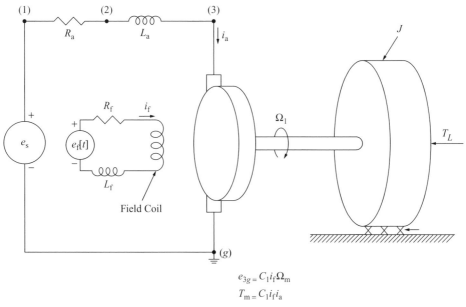

$$e_{3g} = C_1 i_f \Omega_m$$
$$T_m = C_1 i_f i_a$$

Figure P10.4. Field-controlled dc motor.

(b) Determine the normal operating-point values of the state variables for the input voltage $e_f(t) = \bar{e}_f = 90$ V and the power supply voltage $e_s = 120$ V. The values of the system parameters are as follows:

$R_a = 0.5\ \Omega$	$L_a = 4.0$ H	$R_f = 80\ \Omega$
$L_f = 60$ H	$J = 5.0$ N m s^2/rad	$B = 0.04$ N m s/rad
$C_1 = 1.2$ V s/A rad	$T_L = 30$ N m	

(c) Develop the linearized state-variable equations for the system.

10.5 Shown in Fig. P10.5 is a simplified schematic of a mass (m)–spring (k)–damper (b) system with adjustable damping coefficient. The input to the system is voltage e_i, and the output is position of the mass x. The elements of the electrical circuit are resistor R and a coil placed within a magnetic field and attached to the mass. The inductance of the coil can be ignored in the analysis. When current i flows through the coil, force F_e acts on the mass, and, when the mass moves, voltage e_m is induced in the coil.

(a) Derive the state-variable equations for the system.

(b) Combine the state-variable equations into an input–output equation using voltage e_i as the input and displacement x as the output variable.

(c) Show that the system damping ratio ζ can be changed by changing resistance R. In particular, given the system parameters, $m = 0.001$ kg, $b = 0.1$ N s/m, $k = 10$ N/m, and $C_m = 2.0$ V s/m, find the value of the resistance for which the damping ratio is 0.7.

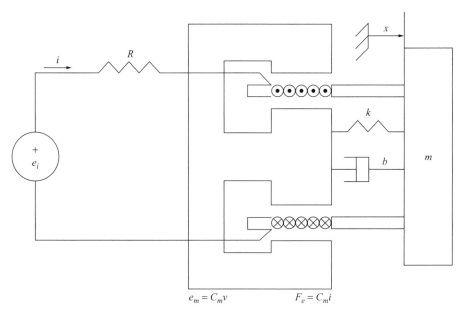

Figure P10.5. Electromechanical system with adjustable damping.

10.6 A valve-controlled motor is used with a constant pressure source P_s to control the speed of an inertia J_1 having a load torque T_1, and linear damper B_1 acting on it, as shown in Fig. P10.6.

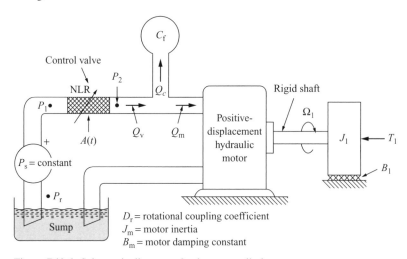

Figure P10.6. Schematic diagram of valve-controlled motor system.

The lumped fluid capacitance C_f arises from the compressibility of the fluid under pressure between the control valve and the working parts (pistons, vanes, or gears) of the hydraulic motor. The valve is basically a variable nonlinear fluid resistor in the form of a variable-area orifice and is described by the equation

$$Q_v = A(t)C_d \left(\frac{2}{\rho}\right)^{0.5} \frac{P_{12}}{|P_{12}|^{0.5}}.$$

The shaft is very stiff, so that the motor inertia J_m may be lumped together with the load inertia J_1, and the motor damper B_m may be lumped together with the load damper B_1; leakage in the motor is negligible.

(a) Draw a free-body diagram of the mechanical part of the system showing the inertialess ideal motor shaft, the lumped inertias, the lumped dampers, and the load torque.

(b) Write the necessary and sufficient set of describing equations for this system.

(c) Rearrange the equations developed in part (b) to form the set of state-variable equations using P_{2r} and Ω_1 as the state variables.

(d) Linearize and combine the state-variable equations to produce the system differential equation relating $\hat{\Omega}_1$ to the input $\hat{A}(t)$.

10.7 A variable-displacement hydraulic pump, driven at constant speed Ω_1, is being used in a hydraulic power supply system for a variable-resistance load R_1, as shown in Fig. P10.7.

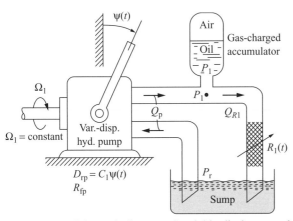

Figure P10.7. Schematic diagram of variable-displacement hydraulic power supply.

The displacement of the pump $D_{rp} = C_1 \psi(t)$, where $\psi(t)$ is the pump stroke input to the system. The internal leakage in the pump is $Q_{lp} = (1/R_{fp})P_{lr}$. The air in the gas-charged accumulator occupies volume $\mathcal{V}$ when the system is at its normal operating point, and the absolute temperature of the gas is approximately the same as that of the atmosphere.

(a) Develop the system differential equation relating the pressure P_{lr} to the two inputs $\psi(t)$ and $R_1(t)$, using only symbols.

(b) Determine the normal operating-point values $\bar{\psi}$ and $\bar{Q}_p$, given the following data:

$$\overline{\Omega}_1 = 180 \, \text{rad/s},$$
$$C_1 = 4.4 \, \text{in}^3/\text{rad per stroke rad},$$
$$\overline{R}_1 = 40 \, \text{lb s/in}^5,$$
$$\overline{P}_{lr} = 1000 \, \text{lb/in}^2,$$
$$R_{fp} = 1200 \, \text{lb s/in}^5.$$

(c) Linearize the system differential equation for small perturbations of all variables about the normal operating point and find the system time constant τ given the data that follow (assume "fast" changes of pressure in the accumulator):

$$\bar{V} = 300\,\text{in}^3,$$
$$T_{\text{atm}} = 530^\circ\,\text{R}.$$

(d) Sketch the response to a small step change:
$\hat{R}_1(t) = -2\,\text{lb s/in}^5$ at $t = 0$, followed by a small step change,
$\hat{\psi}(t) = \bar{\psi}/20$ at $t = \tau/2$.

10.8 A variable-speed hydraulic transmission used to drive the spindle of a large turret lathe is shown schematically, together with constant-speed drive motor and associated output gearing, in Fig. P10.8(a).

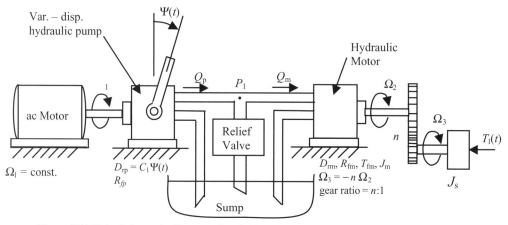

Figure P10.8(a). Schematic diagram of variable-speed drive system.

The variable-displacement pump has displacement $D_{\text{rp}} = C_1\psi(t)$, and the internal leakage of the pump is proportional to the pressure P_{1r} modeled by a leakage resistance R_{fp}. The fixed-displacement motor has displacement D_{rm}, internal leakage resistance R_{fp}, and rotational inertia J_{m}. The friction torque of the motor versus shaft speed Ω_{2g} is shown by the graph in Fig. P10.8(b).

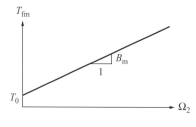

Figure P10.8(b). Graph of motor friction torque versus shaft speed.

The relief valve (RV), which acts only if something happens to cause P_{1r} to exceed a safe operating value, is of no concern in this problem. The combined friction of the gear and spindle bearings and the friction in the gear train are negligible compared with the friction in the motor.

The gear inertias are negligible, and the combined inertia of the lathe spindle, chuck, and workpiece is J_{s}. The torque $T_1(t)$ represents the load torque exerted by the cutting tool on the workpiece being held and driven by the spindle chuck.

(a) As a starting point, it is assumed that the pressurized line between the pump and motor can be modeled as a simple fluid capacitor with negligible inertance having volume $V^2 = AL + V_{int}$, where A is the flow area of the line, L is the length of the line, and V_{int} is the sum of the internal volumes of the pump and the motor. In other words, V is the total volume between the pistons in the pump and the pistons in the motor. From this simplification, prepare a complete system diagram showing all variables and all the essential ideal elements in schematic/free-body diagram form (as in the solution for Example 10.3).

(b) For a desired normal operating pressure $\bar{P}_{1r}$, when the normal operating load torque is $\bar{T}_1$ and the normal spindle speed is $\bar{\Omega}_3$, derive an expression for the motor displacement D_m in terms of $\bar{P}_{1r}$, $\bar{T}_1$, and $\bar{\Omega}_3$. Then develop an expression for the normal flow rate $\bar{Q}_m = \bar{Q}_p$ in terms of $\bar{\Omega}_3$, D_{rm}, and $\bar{P}_{1r}$, and also develop an expression for the normal operating value of pump displacement $\bar{D}_{rp} = C_1\bar{\psi}$ in terms of $\bar{Q}_m$, $\bar{P}_{1r}$, and $\bar{\Omega}_1$.

(c) Write the necessary and sufficient set of describing equations for this system, combine to form the set of state-variable equations having P_{1r} and Ω_2 as state variables, and write the output equation for Ω_2.

(d) Linearize the state-variable equations and the output equation for small perturbations of all variables.

(e) Using the data that follow, find the damping ratio ζ and natural frequency ω_n (or time constants) for this system using the input–output differential equation relating $\hat{\Omega}_3$ to $\hat{\psi}(t)$:

$$\bar{\Omega}_1 = 190\,\text{rad/s},$$
$$\bar{P}_{1r} = 1000\ \text{lb/in}^2,$$
$$R_{fp} = 480\ \text{lb s/in}^5,$$
$$R_{fm} = 520\ \text{lb s/in}^5,$$
$$\bar{V} = 10\,\text{in}^3,$$
$$T_0 = 30\ \text{lb in.},$$
$$B_m = 0.65\ \text{lb in. s},$$
$$J_m = 0.11\ \text{lb in. s}^2,$$
$$n = 3,$$
$$J_s = 6.0\ \text{lb in. s}^2,$$
$$\bar{T}_1 = -1500\,\text{lb in.},$$
$$\bar{\Omega}_3 = -260\,\text{rad/s}.$$

Fluid properties:

$$\gamma = 0.032\,\text{lb/in}^3$$
$$\beta = 200{,}000\,\text{lb/in}^2.$$

(f) Compute the fluid inertance of the hydraulic line using an internal line diameter of 0.5 in. and a line length of 40 in., and compare its motor-displacement-reflected effective inertia $D_{rm}^2 I$ with the motor inertia J_m and with the gear-ratio-reflected inertia $n^2 J_s$. How do you feel about the tentative decision made in part (a) to neglect the fluid inertance of the line?

10.9 The generator–motor arrangement shown in Fig. P10.9(a) is one version of a Ward–Leonard variable-speed drive used to drive loads at variable speeds and loads, using power from a constant-speed source Ω_1. This system is similar in many respects to the hydraulic variable-speed drive studied in Problem 10.5.

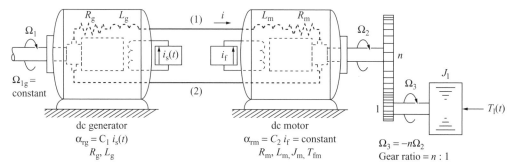

Figure P10.9(a). Schematic diagram of variable-speed drive system.

The coupling coefficient for the generator is $\alpha_{rg} = C_1 i_s(t)$, where $i_s(t)$ is a system input and the generator armature winding has resistance R_g and inductance L_g.

The coupling coefficient for the motor is $\alpha_{rm} = C_2 i_f$, which is constant. The motor armature winding has resistance R_m and inductance L_m. The motor rotor has inertia J_m, and the motor friction torque as a function of speed is shown in the graph of friction torque versus speed in Fig. P10.9(b). The combined friction of gear and load inertia bearings and the friction in the gear train are negligible. The gear inertias are also negligible. The load inertia is J_1. The torque $T_1(t)$ represents the external load, which is a second input to the system.

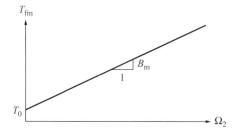

Figure P10.9(b). Graph of motor friction torque versus shaft speed.

(a) It will be assumed that the conductors between the generator and the motor have negligible resistance, inductance, and capacitance. Prepare a complete system diagram showing all variables and all the essential ideal elements in schematic circuit/free-body diagram form (as in Example 10.2).

(b) For a desired normal operating current $\bar{i}$ when the normal operating load is $\overline{T}_1$ and the normal operating speed is $\overline{\Omega}_3$, derive an expression for the motor coupling coefficient $\bar{\alpha}_{rm}$ in terms of $\bar{i}$, $\overline{T}_1$, and $\overline{\Omega}_3$. Then develop an expression for $\bar{e}_{12}$ in terms of $\overline{\Omega}_3$, $\bar{\alpha}_{rm}$, and $\bar{i}$. Also develop an expression for the normal operating value $\bar{\alpha}_{rg} = C_1 \bar{i}_s$ in terms of $\bar{e}_{12}$, $\bar{i}$, and $\overline{\Omega}_1$.

(c) Write the necessary and sufficient set of describing equations for this system and combine to form the state-variable equations having i and Ω_2 as the state variables. Also write the output equation for Ω_3.

(d) Linearize the state-variable equations and the output equation for small perturbations about the normal operating point.

(e) Using the data that follow, find the damping ratio ζ and the natural frequency ω_n (or time constants) for this system, using the system input–output differential equation relating $\hat{\Omega}_3$ to $\hat{i}_s(t)$:

$$\overline{\Omega}_1 = 190\,\text{rad/s},$$
$$i = 30\,\text{amp},$$
$$R_g = 0.5\,\Omega,$$
$$R_m = 0.5\,\Omega,$$
$$L_g = 5\,\text{h},$$
$$L_m = 5\,\text{h},$$
$$T_0 = 10.0\,\text{lb in.},$$
$$B_m = 0.4\,\text{lb in. s},$$
$$J_m = 30\,\text{lb in. s}^2,$$
$$J_l = 6\,\text{lb in. s}^2,$$
$$T_l = -1500\,\text{lb in.},$$
$$\overline{\Omega}_3 = -260\,\text{rad/s},$$
$$n = 3.$$

10.10 Find expressions for the steady–state response of x_2 to a small step change in the input $x_1(t)$ and for the damping ratio for the valve-controlled pneumatic system shown in Fig. P10.10, assuming k_s is negligible and F_1 is constant. Assume that $\overline{P}_2 < 0.45\overline{P}_1$, and use linearized equations for your analysis.

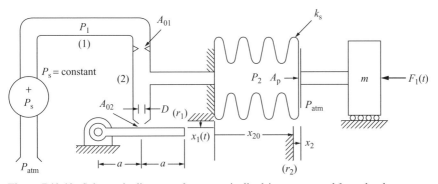

Figure P10.10. Schematic diagram of pneumatically driven mass and force load.

Using parameters supplied by you or your instructor, calculate the system natural frequency ω_n (or time constants) of this system. Note that $A_{02} = \pi D x_1/2$ should be less than $\pi D^2/8$—in other words, x_1 should be less than $D/4$—for the flapper-nozzle valve to function effectively. Furthermore, A_{02}/A_{01} must satisfy the desired $\overline{P}_2/\overline{P}_1$ ratio in order to proceed with the calculation.

10.11 Find expressions for the steady-state response of Ω_1 to a small step change in e_s and for the damping ratio for the amplifier-driven dc motor and load system shown in Fig. P10.11

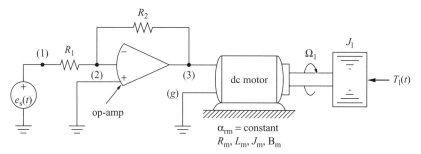

Figure P10.11. Schematic diagram of amplifier-driven motor and load system.

10.12 Consider the system shown in Fig. P10.12(a). The electric motor is connected to the inertia by means of a rigid shaft. The load torque T_1 is an arbitrary input just as is the voltage e_m. The rotational damping coefficient is B. The motor characteristics are given by Fig. P10.12(b). Derive the differential equation that describes the relationship between the two inputs and the speed of the motor.

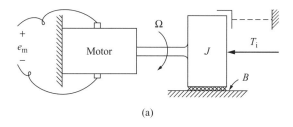

(a)

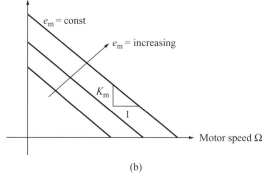

(b)

Figure P10.12. Motor schematic and characteristic curves.

11

System Transfer Functions

LEARNING OBJECTIVES FOR THIS CHAPTER

11-1 To drive system transfer functions as a compact representation of relationships between inputs and outputs in linear systems.

11-2 To predict time-domain system behavior by using the system transfer function.

11-3 To apply the concept of the transfer function to systems with multiple inputs, multiple outputs, or both.

11-4 To develop, manipulate, and interpret system transfer function block diagrams.

11.1 INTRODUCTION

A significant body of theory has been developed for analyzing linear dynamic systems without having to go through the classical methods of solving the input–output differential equations for the system. This body of theory involves the use of the complex variable $s = \sigma + j\omega$, sometimes known as a complex frequency variable. This variable is essentially the same as the Laplace transformation variable s in many respects, but its use does not need to involve the complex (in both senses of the word) transformation problems of ensuring convergence of integrals having limits approaching infinity, nor does it involve the need to carry out the tedious process of inverse transformation by means of partial fraction expansion and the use of a table of transformation pairs. Here the variable s is simply considered to be the coefficient in the exponential input function e^{st}, and the transfer function emerges from solving for the particular or forced part of the response to this input. The notion of transfer function is then combined with the use of simple input–output block diagrams to symbolically express the input–output characteristics of the dynamic behavior of the system.

Because many readers may be familiar with the Laplace transformation (see Appendix 2), an equivalent approach to the derivation of the system transfer function based on the Laplace transformation is also presented in this chapter as an alternative to the approach based on the exponential inputs. It is hoped that, by presenting these different paths to the same end point, the concept of the transfer function, which is often a difficult one for engineering students, will be more accessible.

The systems described and analyzed in previous chapters have nearly all incorporated naturally occurring feedback effects. The presence of these feedback effects

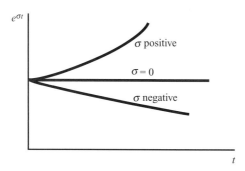

Figure 11.1. Variation of $e^{\sigma t}$ with time for different values of σ.

is readily observed when a simulation block diagram has been prepared for the system. In addition to the natural feedback effects, we will need to deal with intentional, engineered feedback of the kind used in feedback amplifiers and automatic control systems. The use of transfer functions and transfer function block diagrams described in this chapter will prove to be of great value in the analysis of all types of feedback, or closed-loop, systems, beginning with the next chapter on frequency-response analysis.

11.2 APPROACH BASED ON SYSTEM RESPONSE TO EXPONENTIAL INPUTS

Consider a family of time-varying functions of the form e^{st}, called the exponential input functions. This form is particularly versatile, exhibiting different characteristics, depending on whether $s = \sigma + j\omega$ is real ($\omega = 0$), imaginary ($\sigma = 0$), or complex ($s, \omega \neq 0$), as shown in Figs. 11.1 and 11.2.

(a) When both σ and ω are zero, $e^{\sigma t} = 1$, representing a constant input with time—as, for instance, the value of a unit step input for $t > 0$.
(b) When only ω is zero, the input is a growing or decaying real exponential, depending on whether σ is positive or negative, as shown in Fig. 11.1.
(c) When only σ is zero, the input is complex, having sinusoidal real and imaginary parts, and it is represented by a unit vector rotating at speed ωt, as shown in Fig. 11.2.

Examination of Fig. 11.2 readily reveals the basis for the Euler identity

$$e^{j\omega t} = \cos \omega t + j \sin \omega t. \tag{11.1}$$

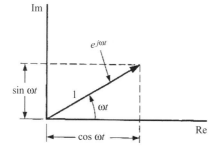

Figure 11.2. Complex plane representation of $e^{j\omega t}$.

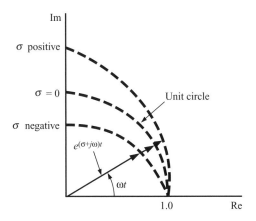

Figure 11.3. Complex plane representation of $e^{(\sigma + j\omega)t}$.

Here it is seen that the input $e^{j\omega t}$ can be used to represent either a sine wave or a cosine wave, or even both, as the occasion demands. As such, $e^{j\omega t}$ in Eq. (11.1) is used to represent sinusoidal inputs in general, and $e^{j\omega t}$ forms the basis for the frequency-response transfer function that will be developed in Chap. 12.

When neither σ nor ω is zero, the input is represented on the complex plane by a rotating vector, which is initially unity at $t = 0$ and then grows or decays exponentially with time, as shown in Fig. 11.3. This input may be used to represent growing or decaying sinusoids, the latter often occurring naturally as the output of an underdamped system responding to its own step input.

For the purposes of this discussion, the homogeneous part of the response of a stable system is considered of minor interest (it usually dies away soon and, in a linear system, in no way affects the particular, or forced, part). The forced part of the response is used here as the basis for developing the notion of a transfer function that relates the output of a linear system to its input.

Recalling that the particular or forced part of the response is of the same form as that of the input and/or its derivatives,[1] we see that the response to an exponential input must also be an exponential of the form $\mathbf{C}e^{st}$, where $\mathbf{C}$ is an undetermined complex coefficient. As an illustration, consider an nth-order system having the following input–output differential equation:

$$a_n \frac{d^n y}{dt^n} + a_{n-1} \frac{d^{n-1} y}{dt^{n-1}} + \cdots + a_1 \frac{dy}{dt} + a_0 y$$
$$= b_m \frac{d^m u}{dt^m} + b_{m-1} \frac{d^{m-1} u}{dt^{m-1}} + \cdots + b_1 \frac{du}{dt} + b_0 u, \qquad (11.2)$$

where the input u and the output y are functions of time and $m \leq n$ for physically realizable systems. Now, with $u = \mathbf{U}e^{st}$, the forced part of the output is taken to be $y = \mathbf{Y}e^{st}$, which is then substituted into Eq. (11.2) to yield[2]

$$(a_n s^n + \cdots + a_1 s + a_0) \mathbf{Y} e^{st} = (b_m s^m + \cdots + b_1 s + b_0) \mathbf{U} e^{st}. \qquad (11.3)$$

---

[1] The additional exponential term Ate^{st} is part of the particular integral for the case when s is precisely equal to one of the system poles. However, this additional term, which represents the mechanism by which the particular solution may grow with time, is of no direct interest here.

[2] Note that the time derivative of the exponential function $\mathbf{C}e^{st}$ is $s\mathbf{C}e^{st}$.

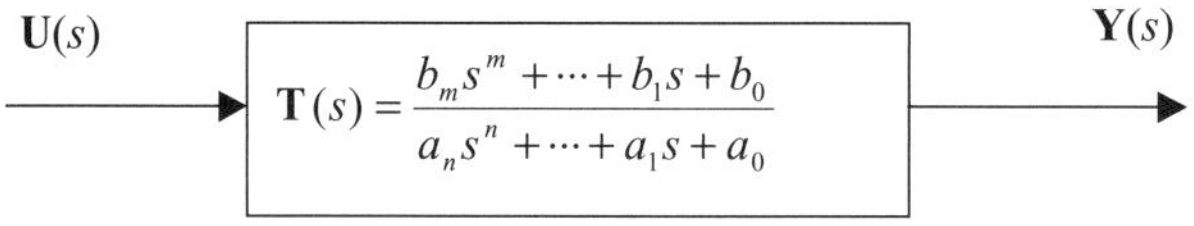

U(s)

$$T(s) = \frac{b_m s^m + \cdots + b_1 s + b_0}{a_n s^n + \cdots + a_1 s + a_0}$$

Y(s)

Figure 11.4. Block-diagram representation of the transfer function relating **U** and **Y**.

Solving for **Y** yields

$$\mathbf{Y} = \frac{b_m s^m + \cdots + b_1 s + b_0}{a_n s^n + \cdots + a_1 s + a_0} \mathbf{U} \qquad (11.4)$$

or

$$\mathbf{Y} = \mathbf{T}(s)\mathbf{U}, \qquad (11.5)$$

where

$$\mathbf{T}(s) = \frac{b_m s^m + \cdots + b_1 s + b_0}{a_n s^n + \cdots + a_1 s + a_0}. \qquad (11.6)$$

The (until now) undetermined complex output-amplitude coefficient **Y** is seen to be a complex coefficient that one readily obtains by multiplying the input-amplitude coefficient **U** by the transfer function **T**(s). With Eq. (11.6) in mind, the system block diagram shown in Fig. 11.4 can be used to express symbolically the relationship between the output **Y** and input **U**. The similarity to the Laplace transform representation for the system with all initial conditions equal to zero, used in the next section, is obvious. However, the Laplace transformation is not necessary here.

It is important to note that the use of transfer functions to model nonlinear systems is not valid unless the system has been linearized and the limitations imposed by the linearization are thoroughly understood.

11.3   APPROACH BASED ON LAPLACE TRANSFORMATION

The transfer function of a linear system can also be derived with the Laplace transformation. For readers who are not familiar with the Laplace transformation or who need to refresh their knowledge, the material presented in Appendix 2 provides a brief but sufficient, for the purposes of this chapter, review of the relevant information.

Consider again a linear nth-order system with an input signal $u(t)$ and an output signal $y(t)$, described by Eq. (11.2). Assume that all initial conditions in the system for time $t = 0^-$ are zero, i.e., $y(0^-) = 0$, $\dot{y}(0^-) = 0$, ... , $y^{(n-1)}(0^-) = 0$. Taking the Laplace transform of both sides of the system input–output equation yields

$$a_n s^n \mathbf{Y}(s) + \cdots + a_1 s \mathbf{Y}(s) + a_0 \mathbf{Y}(s) = b_m s^m \mathbf{U}(s) + \cdots + b_1 s \mathbf{U}(s) + b_0 \mathbf{U}(s), \quad (11.7)$$

where $m \le n$ and $\mathbf{U}(s)$ and $\mathbf{Y}(s)$ are the Laplace transforms of $u(t)$ and $y(t)$, respectively. The system transfer function is defined as the ratio of the Laplace transform of the output signal $\mathbf{Y}(s)$ over the Laplace transform of the input signal $\mathbf{U}(s)$. From Eq. (11.7), the system transfer function $\mathbf{T}(s)$ is

$$\mathbf{T}(s) = \frac{\mathbf{Y}(s)}{\mathbf{U}(s)} = \frac{b_m s^m + \cdots + b_1 s + b_0}{a_n s^n + \cdots + a_1 s + a_0}. \tag{11.8}$$

The expression obtained in Eq. (11.8) is the same as the expression derived based on the system response to exponential inputs and given by Eq. (11.6). *Again, it should be noted here that, regardless of the approach taken in deriving the transfer function, it can be applied only to linear or linearized systems.*

11.4 PROPERTIES OF SYSTEM TRANSFER FUNCTION

It was shown in the two preceding sections that the transfer function of an nth-order linear system described by Eq. (11.2) takes the form of a ratio of polynomials in s as in Eq. (11.8), which can be rewritten as

$$\mathbf{T}(s) = \frac{\mathbf{B}(s)}{\mathbf{A}(s)}, \tag{11.9}$$

where the two polynomials, $\mathbf{A}(s)$ and $\mathbf{B}(s)$, are

$$\mathbf{A}(s) = a_n s^n + \cdots + a_1 s + a_0, \tag{11.10}$$

$$\mathbf{B}(s) = b_m s^m + \cdots + b_1 s + b_0. \tag{11.11}$$

In all existing and realizable engineering systems, the order of the polynomial in the numerator of the transfer function is not higher than the order of the polynomial in the denominator, $m \leq n$.

The roots of the polynomial $\mathbf{B}(s)$, which can be real or complex numbers, are called zeros of the system transfer function. Another form that $\mathbf{B}(s)$ can be presented in is

$$\mathbf{B}(s) = b_m(s - z_1)(s - z_2) \cdots (s - z_m), \tag{11.12}$$

where $z_1, z_2, \ldots, z_m$ are the zeros of the transfer function. Note that, for each of the zeros, the transfer function becomes zero:

$$\mathbf{T}(s)\big|_{s=z_i} = 0, \quad i = 1, 2, \ldots, m. \tag{11.13}$$

The roots of the polynomial in the denominator of the transfer function, $\mathbf{A}(s)$, are called the poles of the system transfer function. The transfer function of an nth-order system has n poles, which can be real or complex numbers, and the polynomial $\mathbf{A}(s)$ can be written as

$$\mathbf{A}(s) = a_n(s - p_1)(s - p_2) \ldots (s - p_n), \tag{11.14}$$

where the poles are solutions of the equation

$$\mathbf{A}(s) = 0, \tag{11.15}$$

or, equally,

$$a_n s^n + \cdots + a_1 s + a_0 = 0. \tag{11.16}$$

Note that the form of this equation is the same as the form of the characteristic equation of an nth-order linear system, Eq. (4.7). Therefore the system poles are the same as the roots of the system characteristic equation. *It can thus be implied that the poles of the system transfer function provide complete information about the inherent dynamic characteristics of the system.* Specifically, each real pole corresponds to a time constant and each pair of complex-conjugate poles corresponds to a natural frequency and a damping ratio of the system.

Finally, when Eqs. (11.2) and (11.8) are compared, it can be observed that the term containing the ith power of s in the denominator of the transfer function corresponds to the term in the system differential equation involving the ith derivative of the output, whereas the term containing the jth power of s in the numerator of the transfer function corresponds to the term in the system differential equation involving the jth derivative of the input. Thus the system differential equation is readily recovered, by simple observation, by inspection the numerator and denominator of the system transfer function.

The practical applications of the properties of the system transfer function presented in this section are demonstrated in Examples 11.1–11.4.

EXAMPLE 11.1

Derive the transfer function for the electric circuit shown in Fig. 11.5. Use voltage across the capacitor, $e_C(t)$, as the output and the source voltage $e_s(t)$ as the input signal. Find poles and zeros of the transfer function.

SOLUTION

The Kirchhoff's voltage law equation for the circuit is

$$e_s = e_R + e_C.$$

The equation for the resistor is

$$e_R = Ri.$$

For the capacitor,

$$i = C\frac{de_C}{dt}.$$

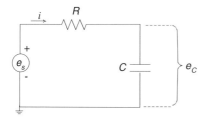

Figure 11.5. Linear RC circuit.

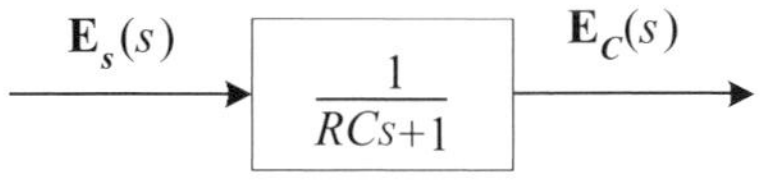

$E_s(s)$ → [$\dfrac{1}{RCs+1}$] → $E_c(s)$

Figure 11.6. Transfer function block diagram of the circuit considered in Example 11.1 relating voltage across the capacitor to input voltage.

Combining the three equations gives the differential input–output equation:

$$RC\frac{de_C}{dt} + e_C = e_s.$$

Assuming zero initial condition, $e_C(0^-) = 0$, and taking the Laplace transform of both sides of the preceding equation yields

$$RCs\mathbf{E}_C(s) + \mathbf{E}_C(s) = \mathbf{E}_s(s),$$

where $\mathbf{E}_C(s)$ and $\mathbf{E}_s(s)$ are Laplace transforms of $e_C(t)$ and $e_s(t)$, respectively. Hence, the circuit transfer function is

$$\mathbf{T}_C(s) = \frac{E_C(s)}{E_s(s)} = \frac{1}{RCs+1}.$$

Note that this first-order system has no zeros $[\mathbf{B}(s) = 1]$ and one pole, $s_1 = -1/RC$, where RC is the circuit time constant, $\tau = RC$, and thus $s_1 = -1/\tau$. The transfer function block diagram for the circuit is shown in Fig. 11.6.

EXAMPLE 11.2

Derive the transfer function for the mass–spring–damper system being acted on by an input force applied to the mass, as shown in Fig. 11.7. Use force $F(t)$ as the input and the mass displacement $x(t)$ as the output variable.

SOLUTION

The input–output equation for a linear mass–spring–dashpot system was derived in Chap. 2, Eq. (2.34):

$$m\frac{d^2x}{dt^2} + b\frac{dx}{dt} + kx = F(t).$$

Assuming zero initial conditions, $x(0^-) = 0$ and $\dot{x}(0^-) = 0$, and taking the Laplace transform of both sides of the input–output equation yields

$$ms^2\mathbf{X}(s) + bs\mathbf{X}(s) + k\mathbf{X}(s) = \mathbf{F}(s);$$

hence, the system transfer function is

$$\mathbf{T}(s) = \frac{\mathbf{X}(s)}{\mathbf{F}(s)} = \frac{1}{ms^2 + bs + k}.$$

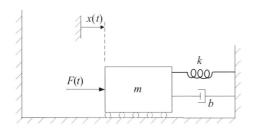

Figure 11.7. Schematic diagram of mass–spring–damper system.

It can be seen that this second-order system has no zeros and two poles, s_1 and s_2, which can be found when the system characteristic equation is solved:

$$ms^2 + bs + k = 0.$$

The poles are easily found by the solution of this quadratic equation:

$$s_1, s_2 = \left(-b \pm \sqrt{b^2 - 4mk}\right)/2m.$$

By use of the damping ratio (see Section 4.4), $\zeta = b/(2\sqrt{mk})$, when $\zeta \geq 1$, the poles are equal to negative inverse time constants,

$$s_1, s_2 = -\frac{1}{\tau_1}, -\frac{1}{\tau_2},$$

and, when $0 < \zeta < 1.0$, the poles are complex-conjugate numbers related to the system damping ratio and natural frequency:

$$s_1, s_2 = -\zeta \omega_n \pm j\omega_n \sqrt{1 - \zeta^2},$$

where

$$\omega_n = \sqrt{k/m}.$$

The availability of the system transfer function offers an alternative way for solving for the response of linear systems (an alternative to analytical solutions of input–output equations or computer solutions of state-variable equations). From the definition of the transfer function, the Laplace transform of the output signal can be calculated as

$$\mathbf{Y}(s) = \mathbf{U}(s)\mathbf{T}(s). \tag{11.17}$$

One can then find the system response in the time domain by taking the inverse transform of $\mathbf{Y}(s)$ by using partial fraction expansion described in Appendix 2 or by using MATLAB, which is the topic of Section 11.7:

$$y(t) = \mathcal{L}^{-1}\{\mathbf{Y}(s)\}. \tag{11.18}$$

In many situations, it is of practical value to engineers to be able to quickly evaluate a general character of the system step response rather than to calculate a detailed and complete response. In such situations, the system transfer function can be used to evaluate the step response immediately after the input signal is applied, at time $t = 0^+$, and at steady state for time approaching infinity. Based on the initial-value theorem (see Appendix 2), the system response at time $t = 0^+$ is

$$y(0^+) = \lim_{s \to \infty} s\mathbf{Y}(s) = \lim_{s \to \infty} s\mathbf{U}(s)\mathbf{T}(s). \tag{11.19}$$

For a step input of magnitude A, we have

$$y(0^+) = \lim_{s \to \infty} s\frac{A}{s}\mathbf{T}(s) = A \lim_{s \to \infty} \mathbf{T}(s). \tag{11.20}$$

Thus the value of the step response at $t = 0^+$ is equal to the limiting value of the transfer function as $s \to \infty$ multiplied by the magnitude of the step input.

The value of the final value of the steady-state step response can be found with the final-value theorem (Appendix 2):

$$y_{ss} = \lim_{t \to \infty} y(t) = \lim_{s \to 0} s\mathbf{Y}(s) = \lim_{s \to 0} s\mathbf{U}(s)\mathbf{T}(s). \qquad (11.21)$$

For a step input of magnitude A, $\mathbf{U}(s) = A/s$,

$$y_{ss} = A \lim_{s \to 0} \mathbf{T}(s). \qquad (11.22)$$

According to this equation, the final value of the steady-state step response is the same as the limit of the value of the system transfer function when $s \to 0$ multiplied by the magnitude of the step input.

EXAMPLE 11.3

Evaluate the response of the RC circuit considered in Example 11.1 to a step change in input voltage from 0 to ΔE_s that occurs at time $t = 0$. Use both $e_C(t)$ and $e_R(t)$ as the output variables.

First, consider the voltage drop across the capacitor. The transfer function relating that voltage to input voltage, derived in Example 11.1, was

$$\mathbf{T}_C(s) = \frac{E_C(s)}{E_s(s)} = \frac{1}{RCs + 1}.$$

By use of Eq. (11.20), the value of voltage across the capacitor immediately after the step change in the source voltage occurs is

$$e_C(0^+) = \Delta E_s \lim_{s \to \infty} \frac{1}{RCs + 1} = 0 \,\text{V}.$$

Recall that in Example 11.1 it was assumed that the initial value of $e_C(t)$ was zero, $e_C(0^-) = 0$. It has just been found that the voltage remains equal to zero at $t = 0^+$, which could be expected if one considers that the amount of energy stored by the capacitor is related to e_C, and if e_C changes suddenly between $t = 0^-$ and $t = 0^+$, so would the amount of energy stored in the system, which is physically impossible.

The final-value theorem finds the value of e_C after the transients die out:

$$\lim_{t \to \infty} e_C(t) = \Delta E_s \lim_{s \to 0} \frac{1}{RCs + 1} = \Delta E_s.$$

Now, consider the voltage drop across resistor R as the output variable. From Kirchhoff's voltage law equation we have

$$e_R(t) = e_s(t) - e_C(t).$$

Assuming that all voltages in this equation are zero at time $t = 0^-$ and taking Laplace transform of both sides yields

$$\mathbf{E}_R(s) = \mathbf{E}_s(s) - \mathbf{E}_C(s).$$

Substituting the expression for $\mathbf{E}_C(s)$ in terms of $\mathbf{E}_s(s)$ derived in Example 11.1 gives

$$\mathbf{E}_R(s) = \mathbf{E}_s(s) - \frac{1}{RCs + 1}\mathbf{E}_s(s).$$

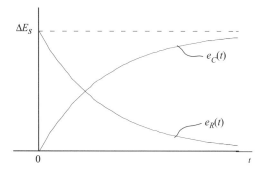

Figure 11.8. Step responses plots for the RC circuit shown in Fig. 11.5.

Hence the transfer function relating voltage across the resistor to the input voltage is

$$\mathbf{T}_R(s) = \frac{E_R(s)}{E_s(s)} = \frac{RCs}{RCs + 1}.$$

Now that the transfer function has been found, the values of e_R at the time immediately after the change in the input voltage occurred and at steady state can be determined. At time $t = 0^+$,

$$e_R(0^+) = \Delta E_s \lim_{s \to \infty} \mathbf{T}_R(s) = \Delta E_s \lim_{s \to \infty} \frac{RCs}{RCs + 1} = \Delta E_s.$$

Thus it can be seen that there is a sudden change in the voltage drop across the resistor from 0 to ΔE_s, but this sudden change is not associated with a sudden change of energy storage and it may indeed take place in the circuit.

The final-value of the steady-state step response is

$$\lim_{t \to \infty} e_R(t) = \Delta E_s \lim_{s \to 0} \frac{RCs}{RCs + 1} = 0.$$

By knowing the values of the voltages in the circuit immediately after the input voltage changes and at steady state, and recognizing that the circuit is a first-order system with a time constant $\tau = RC$, an engineer can make fairly accurate sketches of the circuit responses, as shown in Fig. 11.8.

EXAMPLE 11.4

The transfer function of the RLC bandpass filter shown in Fig. 11.9 is

$$\mathbf{T}(s) = \frac{E_o(s)}{E_i(s)} = \frac{RCs}{LCs^2 + RCs + 1}.$$

Find the differential equation relating output voltage $e_o(t)$ to input voltage $e_i(t)$.

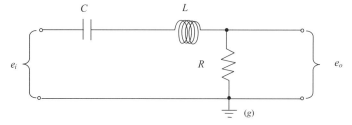

Figure 11.9. Schematic of RLC bandpass filter.

SOLUTION

From the expression for the system transfer function we have

$$\mathbf{E}_o(s)(LCs^2 + RCs + 1) = \mathbf{E}_i(s)RCs$$

or

$$LCs^2\mathbf{E}_o(s) + RCs\,\mathbf{E}_o(s) + \mathbf{E}_o(s) = RCs\,\mathbf{E}_i(s).$$

The inverse Laplace transformation of the preceding equation yields the filter's input–output differential equation:

$$LC\frac{d^2e_o}{dt^2} + RC\frac{de_o}{dt} + e_o = RCe_i.$$

11.5 TRANSFER FUNCTIONS OF MULTI-INPUT, MULTI-OUTPUT SYSTEMS

In Example 11.3, the RC circuit was considered as a system with one input, $e_s(t)$, and two outputs, $e_C(t)$ and $e_R(t)$. Two separate transfer functions were derived, each relating one of the output variables to the common input variable.

The transfer function block diagram of the circuit is shown in Fig. 11.10. In general, in a multi-input, multi-output system, a transfer function is identified between each output and each input signal. In a system with l inputs and p outputs, a total of $l \times p$ transfer functions can be identified. This is illustrated in Fig. 11.11 for a system with two inputs and two outputs. The output signals are

$$\mathbf{Y}_1(s) = \mathbf{U}_1(s)\mathbf{T}_{11}(s) + \mathbf{U}_2(s)\mathbf{T}_{12}(s), \tag{11.23}$$

$$\mathbf{Y}_2(s) = \mathbf{U}_1(s)\mathbf{T}_{21}(s) + \mathbf{U}_2(s)\mathbf{T}_{22}(s). \tag{11.24}$$

Hence, the four system transfer functions can be defined as follows:

$$\mathbf{T}_{11}(s) = \left.\frac{\mathbf{Y}_1(s)}{\mathbf{U}_1(s)}\right|_{\mathbf{U}_2(s)=0}, \tag{11.25}$$

$$\mathbf{T}_{12}(s) = \left.\frac{\mathbf{Y}_1(s)}{\mathbf{U}_2(s)}\right|_{\mathbf{U}_1(s)=0}, \tag{11.26}$$

$$\mathbf{T}_{21}(s) = \left.\frac{\mathbf{Y}_2(s)}{\mathbf{U}_1(s)}\right|_{\mathbf{U}_2(s)=0}, \tag{11.27}$$

$$\mathbf{T}_{22}(s) = \left.\frac{\mathbf{Y}_2(s)}{\mathbf{U}_2(s)}\right|_{\mathbf{U}_1(s)=0}. \tag{11.28}$$

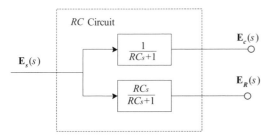

Figure 11.10. Transfer function block diagram of the RC circuit with one input and two outputs.

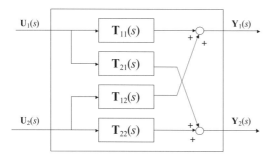

Figure 11.11. Block diagram of a two-input, two-output system.

The general expression for each of the $l \times p$ transfer functions of a multi-input, multi-output system is

$$\mathbf{T}_{ij}(s) = \frac{\mathbf{Y}_i(s)}{\mathbf{U}_j(s)}\bigg|_{\mathbf{U}_k(s)=0}, \qquad (11.29)$$

where $i = 1, 2, \ldots, l$, $j = 1, 2, \ldots, p$, and $k = 1, 2, \ldots, j-1, j+1, \ldots, l$. Expressed verbally, the transfer function for a given pair of output and input signals is equal to the ratio of the Laplace transform of the output signal over the Laplace transform of the input signal in the pair, with all other inputs equal to zero. Note that this assumption, setting all inputs except one to zero, is valid because we are dealing with linear systems for which the principle of superposition holds.

The following example illustrates the concept of transfer function of a multi-input, multi-output system. It also demonstrates a simple method for calculating the component transfer functions from the basic equations of motion derived for the system.

EXAMPLE 11.5

Derive the transfer functions relating displacements $x_1(t)$ and $x_2(t)$ to the external forces $F_1(t)$ and $F_2(t)$ in the system shown in Fig. 11.12.

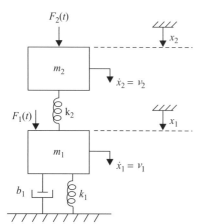

Figure 11.12. Two-input, two-output system considered in Example 11.5.

The following equations of motion were derived for this system in Example 3.4:

$$m_1 \frac{d^2 x_1}{dt^2} + b_1 \frac{dx_1}{dt} + (k_1 + k_2)x_1 - k_2 x_2 = F_1,$$

$$m_2 \frac{d^2 x_2}{dt^2} + k_2 x_2 - k_2 x_1 = F_2.$$

Setting initial conditions to zero and taking the Laplace transform of the basic equations of motion yields

$$m_1 s^2 \mathbf{X}_1(s) + b_1 s \mathbf{X}_1(s) + (k_1 + k_2)\mathbf{X}_1(s) - k_2 \mathbf{X}_2(s) = \mathbf{F}_1(s),$$

$$m_2 s^2 \mathbf{X}_2(s) + k_2 \mathbf{X}_2(s) - k_2 \mathbf{X}_1(s) = \mathbf{F}_2(s).$$

The two preceding equations can be put in a matrix form:

$$\begin{bmatrix} m_1 s^2 + b_1 s + (k_1 + k_2) & -k_2 \\ -k_2 & m_2 s^2 + k_2 \end{bmatrix} \begin{bmatrix} \mathbf{X}_1(s) \\ \mathbf{X}_2(s) \end{bmatrix} = \begin{bmatrix} \mathbf{F}_1(s) \\ \mathbf{F}_2(s) \end{bmatrix}.$$

This matrix equation can be solved for $\mathbf{X}_1(s)$ and $\mathbf{X}_2(s)$ by use of Cramer's rule. According to Cramer's rule, the two unknowns can be calculated as the ratios of determinants:

$$\mathbf{X}_1(s) = \frac{\begin{vmatrix} F_1(s) & -k_2 \\ F_2(s) & m_2 s^2 + k_2 \end{vmatrix}}{\begin{vmatrix} m_1 s^2 + b_1 s + k_1 + k_2 & -k_2 \\ -k_2 & m_2 s^2 + k_2 \end{vmatrix}},$$

$$\mathbf{X}_2(s) = \frac{\begin{vmatrix} m_1 s^2 + b_1 s + k_1 + k_2 & F_1(s) \\ -k_2 & F_2(s) \end{vmatrix}}{\begin{vmatrix} m_1 s^2 + b_1 s + k_1 + k_2 & -k_2 \\ -k_2 & m_2 s^2 + k_2 \end{vmatrix}}.$$

Calculating the values of the determinants gives

$$\mathbf{X}_1(s) = \frac{F_1(s)(m_2 s^2 + k_2) + k_2 F_2(s)}{(m_1 s^2 + b_1 s + k_1 + k_2)(m_2 s^2 + k_2) - k_2^2},$$

$$\mathbf{X}_2(s) = \frac{(m_1 s^2 + b_1 s + k_1 + k_2)F_2(s) + k_2 F_1(s)}{(m_1 s^2 + b_1 s + k_1 + k_2)(m_2 s^2 + k_2) - k_2^2}.$$

From these two equations, all four transfer functions of the two-input, two-output system can be obtained:

$$\mathbf{T}_{11}(s) = \left. \frac{\mathbf{X}_1(s)}{\mathbf{F}_1(s)} \right|_{F_2(s)=0} = \frac{m_2 s^2 + k_2}{m_1 m_2 s^4 + m_2 b_1 s^3 + (m_1 k_2 + m_2 k_1 + m_2 k_2)s^2 + b_1 k_2 s + k_1 k_2},$$

$$\mathbf{T}_{12}(s) = \left. \frac{\mathbf{X}_1(s)}{\mathbf{F}_2(s)} \right|_{F_1(s)=0} = \frac{k_2}{m_1 m_2 s^4 + m_2 b_1 s^3 + (m_1 k_2 + m_2 k_1 + m_2 k_2)s^2 + b_1 k_2 s + k_1 k_2},$$

$$\mathbf{T}_{21}(s) = \left. \frac{\mathbf{X}_2(s)}{\mathbf{F}_1(s)} \right|_{F_2(s)=0} = \frac{k_2}{m_1 m_2 s^4 + m_2 b_1 s^3 + (m_1 k_2 + m_2 k_1 + m_2 k_2)s^2 + b_1 k_2 s + k_1 k_2},$$

$$\mathbf{T}_{22}(s) = \left. \frac{\mathbf{X}_2(s)}{\mathbf{F}_2(s)} \right|_{F_1(s)=0} = \frac{m_1 s^2 + b_1 s + k_1 + k_2}{m_1 m_2 s^4 + m_2 b_1 s^3 + (m_1 k_2 + m_2 k_1 + m_2 k_2)s^2 + b_1 k_2 s + k_1 k_2}.$$

Recall from Section 11.4 that the denominator of the transfer function is the same as the characteristic polynomial (the left-hand side of the system characteristic equation). The four transfer functions derived for this system represent the relations between different pairs of output and input variables within the same system, and it can be seen that the denominator in each transfer function is the same because it represents the inherent dynamic characteristics of the same system. The characteristic equation is

$$m_1 m_2 s^4 + m_2 b_1 s^3 + (m_1 k_2 + m_2 k_1 + m_2 k_2)s^2 + b_1 k_2 s + k_1 k_2 = 0.$$

Furthermore, from the equations previously derived for $X_1(s)$ and $X_2(s)$, one can obtain the system input–output differential equations by multiplying both sides of those equations by the denominator of the expression on the right-hand side and taking the inverse Laplace transform to obtain

$$m_1 m_2 \frac{d^4 x_1}{dt^4} + m_2 b_1 \frac{d^3 x_1}{dt^3} + (m_1 k_2 + m_2 k_1 + m_2 k_2)\frac{d^2 x_1}{dt^2} + b_1 k_2 \frac{dx_1}{dt}$$
$$+ k_1 k_2 x_1 = m_2 \frac{d^2 F_1}{dt} + k_2 F_1 + k_2 F_2$$

for $x_1(t)$ as the output variable and

$$m_1 m_2 \frac{d^4 x_2}{dt^4} + m_2 b_1 \frac{d^3 x_2}{dt^3} + (m_1 k_2 + m_2 k_1 + m_2 k_2)\frac{d^2 x_2}{dt^2} + b_1 k_2 \frac{dx_2}{dt} + k_1 k_2 x_2$$
$$= k_2 F_1 + m_1 \frac{d^2 F_2}{dt^2} + b_1 \frac{dF_2}{dt} + (k_1 + k_2)F_2$$

for $x_2(t)$ as the output variable. It can be verified that the input–output equations derived here with the system transfer functions are the same as the input–output equations for the same system obtained in Example 3.4 by combining the system basic equations of motion in the time domain.

11.6 TRANSFER FUNCTION BLOCK-DIAGRAM ALGEBRA

The development of the concept of a system transfer function operating within a single block to represent an input–output relationship for a given linear system makes it possible to use transfer function block diagrams as "building blocks." The building blocks may then be used to assemble a complete system from a number of individual elementary parts in a manner somewhat similar to that used for the simulation block diagrams in Chap. 2. However, there are several important differences between the simulation block diagrams and the transfer function block diagrams. First, the two kinds of block diagrams are set in different domains, one in the time domain and the other in the domain of complex variable s. Second, the simulation block diagrams can be used to model linear and nonlinear systems, whereas the transfer function block diagrams can be used only with systems that are assumed to be linear. Third, the types of elementary blocks used in the two kinds of block diagrams are different. In addition to the summation block that appears in both block diagrams, the other type of block used in the transfer function block diagrams is a block representing a transfer function $T(s)$, as shown in Fig. 11.4.

The process of assembling elementary blocks to develop a complete transfer function model of a system will be illustrated in an example of an electromechanical system involving a dc motor to control an angular position of an antenna dish. The

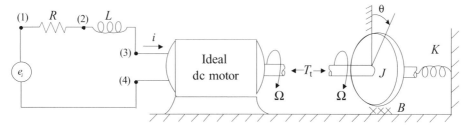

Figure 11.13. Schematic diagram of a motor-driven antenna dish system.

system is shown schematically in Fig. 11.13. The motor is driven by the input voltage source e_i applied to the armature winding having resistance R and inductance L. The coupling coefficient for the motor is α_r, and the motor torque is T_t. The combined mechanical inertia of the rotor and the antenna dish is J, and the combined rotational damping in the system is B. The rotational motion of the antenna is constrained by the torsional spring K. The output variable is the angular displacement of the antenna dish θ.

The elemental equations for the system are as follows: For the armature winding of the motor,

$$e_i - e_{34} = Ri + L\frac{di}{dt}. \tag{11.30}$$

For the ideal dc motor acting as an electromechanical transducer,

$$e_{34} = \frac{1}{\alpha_r}\Omega, \tag{11.31}$$

$$T_t = \frac{1}{\alpha_r}i. \tag{11.32}$$

For the mechanical part of the system,

$$T_t = J\frac{d^2\theta}{dt^2} + B\frac{d\theta}{dt} + K\theta. \tag{11.33}$$

Assuming zero initial conditions and converting Eqs. (11.30)–(11.33) to the s domain yields

$$\mathbf{E}_i(s) - \mathbf{E}_{34}(s) = R\mathbf{I}(s) + Ls\mathbf{I}(s), \tag{11.34}$$

$$\mathbf{T}_t(s) = \frac{1}{\alpha_r}\mathbf{I}(s), \tag{11.35}$$

$$\mathbf{E}_{34}(s) = \frac{1}{\alpha_r}\mathbf{\Omega}(s), \tag{11.36}$$

$$\mathbf{T}_t(s) = Js^2\mathbf{\theta}(s) + Bs\mathbf{\theta}(s) + K\mathbf{\theta}(s). \tag{11.37}$$

Figure 11.14 shows the process of constructing the transfer function block diagram for the system, step by step.

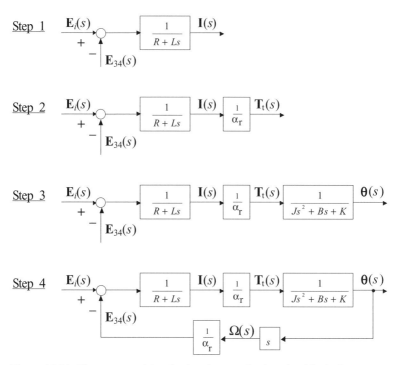

Figure 11.14. The process of developing the transfer function block diagram.

In Step 1, the part of the block diagram described by Eq. (11.34) is drawn with input $E_i(s)$ and the outgoing signal from this part of the system $I(s)$. In Step 2, the block converting current $I(s)$ to torque $T_t(s)$ per Eq. (11.35) is added. In Step 3, the transfer function relating displacement $\theta(s)$ to torque $T_t(s)$ is attached. One completes the process in Step 4 by closing the loop from the output displacement to the voltage induced in the armature winding $E_{34}(s)$ by first using a differentiator to convert the displacement to angular velocity $\Omega(s)$ and then inserting a gain block representing Eq. (11.36).

The detailed transfer function block diagram developed for the antenna dish system is an interconnection of elementary blocks described by the basic equations of motion in the s domain, Eqs. (11.34)–(11.37). Although such a detailed illustration of the system components and the internal interactions among the system variables are usually of value in system analysis, it is also desirable to know the overall equivalent system transfer function relating the output variable to the input variable. The detailed block diagram, like the one shown in Fig. 11.14, can be reduced to a single-block equivalent transfer function by use of the rules of block-diagram algebra. Seven basic rules of block-diagram algebra are subsquently presented. Rules 1, 2, and 3 are used to replace different combinations of blocks with an equivalent single-block transfer function. Rules 4–7 are used in rearranging block diagrams by moving summing points and branch (pickoff) points around blocks to simplify the interconnections among the blocks so that Rule 1, 2, or 3 can then be applied. The overarching principle in making changes within a system is that the system output signal(s) must not be affected by the changes.

(a)

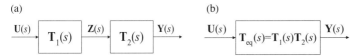

(b)

Figure 11.15. (a) Two blocks in series and (b) an equivalent single-block representation.

Rule 1: Combining blocks in series (cascaded). For two blocks of transfer functions $\mathbf{T}_1(s)$ and $\mathbf{T}_2(s)$ in series, as shown in Fig. 11.15(a), the equivalent transfer function is

$$\mathbf{T}_{eq}(s) = \frac{\mathbf{Y}(s)}{\mathbf{U}(s)} = \frac{\mathbf{Y}(s)}{\mathbf{Z}(s)}\frac{\mathbf{Z}(s)}{\mathbf{U}(s)} = \mathbf{T}_2(s)\mathbf{T}_1(s). \qquad (11.38)$$

In general, the equivalent transfer function for n blocks in series is

$$\mathbf{T}_{ser}(s) = \prod_{i=1}^{n} \mathbf{T}_i(s). \qquad (11.39)$$

Rule 2: Combining blocks in parallel. For two blocks connected in parallel, as shown in Fig. 11.16(a), the equivalent transfer function is

$$\mathbf{T}_{eq}(s) = \frac{\mathbf{Y}(s)}{\mathbf{U}(s)} = \frac{\mathbf{Y}_1(s) + \mathbf{Y}_2(s)}{\mathbf{U}(s)} = \frac{\mathbf{U}(s)\mathbf{T}_1(s) + \mathbf{U}(s)\mathbf{T}_2(s)}{\mathbf{U}(s)} = \mathbf{T}_1(s) + \mathbf{T}_2(s). \qquad (11.40)$$

In general, the equivalent transfer function for n blocks in parallel is

$$\mathbf{T}_{par}(s) = \sum_{i=1}^{n} \mathbf{T}_i(s). \qquad (11.41)$$

Rule 3: Combining blocks in a feedback system. In a feedback system, the system output variable $\mathbf{Y}(s)$ is transferred through a feedback transfer function $\mathbf{H}(s)$ back to the input of the system where it is subtracted from or added to the input signal, thus closing a negative- or positive-feedback loop, respectively. Figure 11.17(a) shows a block diagram of a closed-loop feedback system. The output variable can be expressed as

$$\mathbf{Y}(s) = \mathbf{E}(s)\mathbf{G}(s) = [\mathbf{U}(s) \pm \mathbf{H}(s)\mathbf{Y}(s)]\mathbf{G}(s) = \mathbf{U}(s)\mathbf{G}(s) \pm \mathbf{H}(s)\mathbf{Y}(s)\mathbf{G}(s). \qquad (11.42)$$

(a)   (b)

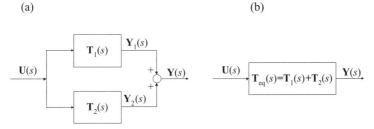

Figure 11.16. (a) Two blocks in parallel and (b) an equivalent single-block representation.

(a)   (b)

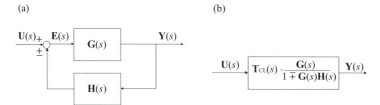

Figure 11.17. (a) Feedback system and (b) an equivalent single-block representation.

Hence the equivalent transfer function of a closed-loop feedback system is

$$\mathbf{T}_{CL}(s) = \frac{\mathbf{Y}(s)}{\mathbf{U}(s)} = \frac{\mathbf{G}(s)}{1 + \mathbf{G}(s)\mathbf{H}(s)} \tag{11.43}$$

for a negative-feedback system and

$$\mathbf{T}_{CL}(s) = \frac{\mathbf{Y}(s)}{\mathbf{U}(s)} = \frac{\mathbf{G}(s)}{1 - \mathbf{G}(s)\mathbf{H}(s)} \tag{11.44}$$

for a positive-feedback system.

Rule 4: Moving a summing point forward around a block. When a summing point is being moved forward around a block, it is necessary to make sure that the output $\mathbf{Y}(s)$ remains the same after the change is made. The output signal in the original system shown in Fig. 11.18(a) is

$$\mathbf{Y}(s) = [\mathbf{U}(s) - \mathbf{Z}(s)]\mathbf{T}(s). \tag{11.45}$$

The output signal in the modified system, Fig. 11.18(b), is

$$\mathbf{Y}(s) = \mathbf{U}(s)\mathbf{T}(s) - \mathbf{Z}(s)\mathbf{T}(s) = [\mathbf{U}(s) - \mathbf{Z}(s)]\mathbf{T}(s). \tag{11.46}$$

This result verifies that, in spite of moving the summing point, the system output remains the same.

Rule 5: Moving a summing point backward around a block. This rule is illustrated in Fig. 11.19. The output signal in the original system [Fig. 11.19(a)] is

$$\mathbf{Y}(s) = \mathbf{U}(s)\mathbf{T}(s) - \mathbf{Z}(s). \tag{11.47}$$

The output remains the same in the modified system [Fig. 11.19(b)]:

$$\mathbf{Y}(s) = \left[\mathbf{U}(s) - \mathbf{Z}(s)\frac{1}{\mathbf{T}(s)}\right]\mathbf{T}(s) = \mathbf{U}(s)\mathbf{T}(s) - \mathbf{Z}(s). \tag{11.48}$$

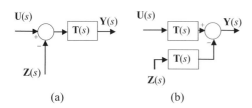

(a)   (b)

Figure 11.18. Moving a summing point forward: (a) original system and (b) modified system.

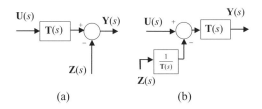

Figure 11.19. Moving a summing point backward: (a) original system and (b) modified system.

Rule 6: Moving a branch point forward around a block. The original and modified systems are shown in Fig. 11.20. The outputs in the original system [Fig. 11.20(a)] are

$$\mathbf{Y}_1(s) = \mathbf{U}(s)\mathbf{T}(s), \qquad \mathbf{Y}_2(s) = \mathbf{U}(s). \tag{11.49}$$

The output signals in the modified system [Fig. 11.20(b)] are

$$\mathbf{Y}_1(s) = \mathbf{U}(s)\mathbf{T}(s), \qquad \mathbf{Y}_2(s) = \mathbf{U}(s)\mathbf{T}(s)\frac{1}{\mathbf{T}(s)} = \mathbf{U}(s), \tag{11.50}$$

which proves that neither output has been changed as a result of moving the branch point.

Rule 7: Moving a branch point backward around a block. This rule is illustrated in Fig. 11.21. The two outputs in the original system [Fig. 11.21(a)] are obviously identical:

$$\mathbf{Y}_1(s) = \mathbf{Y}_2(s) = \mathbf{U}(s)\mathbf{T}(s), \tag{11.51}$$

and they both remain unchanged in the modified system [Fig. 11.21(b)]:

$$\mathbf{Y}_1(s) = \mathbf{U}(s)\mathbf{T}(s), \qquad \mathbf{Y}_2(s) = \mathbf{U}(s)\mathbf{T}(s). \tag{11.52}$$

EXAMPLE 11.6

Find the overall transfer function of the antenna dish system represented by the block diagram developed in Fig. 11.14 and repeated in Fig. 11.22 for convenience.

The objective now is to reduce the detailed block diagram shown in Fig. 11.22 to a single-block representation with the overall system transfer function $\mathbf{T}(s)$ relating the angular position of the antenna dish to the input voltage:

$$\mathbf{T}(s) = \frac{\Theta(s)}{\mathbf{E}_i(s)}.$$

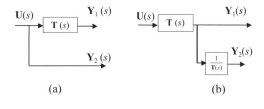

Figure 11.20. Moving a branch point forward: (a) original system and (b) modified system.

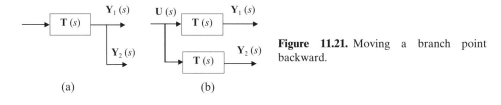

Figure 11.21. Moving a branch point backward.

(a)   (b)

First, it can be seen that there are three blocks connected in series in the forward path, which can be reduced to a single block by use of Eq. (11.39):

$$\mathbf{T}_1(s) = \frac{1}{R + Ls} \frac{1}{\alpha_r} \frac{1}{Js^2 + Bs + K}$$

$$= \frac{1}{J L \alpha_r s^3 + \alpha_r (JR + BL)s^2 + \alpha_r (BR + KL)s + K R \alpha_r}.$$

There are also two blocks in series in the feedback path, and their equivalent transfer function is

$$\mathbf{T}_2(s) = \left(\frac{1}{\alpha_r}\right) s = \frac{s}{\alpha_r}.$$

The system transfer function block diagram is now reduced to the form shown in Fig. 11.23. This is a negative-feedback configuration that one can reduce to a single block by setting $\mathbf{G}(s) = \mathbf{T}_1(s)$ and $\mathbf{H}(s) = \mathbf{T}_2(s)$ and using Eq. (11.43) to obtain the overall system transfer function:

$$\mathbf{T}(s) = \frac{\Theta(s)}{\mathbf{E}_i(s)} = \frac{\mathbf{T}_1(s)}{1 + \mathbf{T}_1(s)\mathbf{T}_2(s)}$$

$$= \frac{\alpha_r}{J L \alpha_r^2 s^3 + (JR + BL)\alpha_r^2 s^2 + [(BR + KL)\alpha_r^2 + 1]s + K R \alpha_r^2}.$$

The denominator of the transfer function gives the system characteristic equation,

$$J L \alpha_r^2 s^3 + (JR + BL)\alpha_r^2 s^2 + [(BR + KL)\alpha_r^2 + 1]s + K R \alpha_r^2 = 0.$$

Furthermore, based on the definition of the system transfer function, the system input–output equation in the time domain is found to be

$$J L \alpha_r^2 \frac{d^3\theta}{dt^3} + (JR + BL)\alpha_r^2 \frac{d^2\theta}{dt^2} + [(BR + KL)\alpha_r^2 + 1]\frac{d\theta}{dt} + K R \alpha_r^2 \theta = \alpha_r e_i.$$

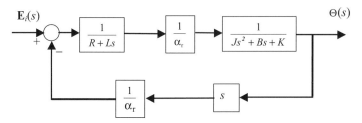

Figure 11.22. Block diagram of the antenna dish system.

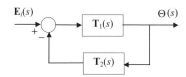

Figure 11.23. The dish antenna system as a simple negative-feedback system.

Note that, as expected, this is a third-order model, because of three independent energy-storing elements included in the system – inductance L, mechanical inertia J, and rotational spring K.

11.7 MATLAB REPRESENTATION OF TRANSFER FUNCTION

As will be discussed in the following chapters, the transfer function representation is a powerful tool in system modeling and control. Several computer-aided analysis packages such as MATLAB have incorporated techniques and algorithms to aid in this analysis. In this section, the functionality of MATLAB and the Control Systems Toolbox, a common extension of MATLAB, is presented.

In the previous sections of this chapter, it was established that transfer functions are ratios of polynomials in s. These polynomials have coefficients that are constant and have no imaginary parts. MATLAB has several built-in functions that operate on polynomials that are useful when one is dealing with transfer function representations.

First, consider the following polynomial in s:

$$A(s) = a_n s^n + a_{n-1} s^{n-1} + \cdots + a_1 s + a_0.$$

The essential information contained in this polynomial is completely contained in the coefficients of the polynomial. Consider the following vector:

$$\mathbf{A} = \begin{bmatrix} a_n & a_{n-1} & \ldots & a_1 & a_0 \end{bmatrix}.$$

The vector is an adequate representation of the polynomial as long as a consistent interpretation of the vector is maintained. In other words, as long as a convention is established such that the coefficient of the highest order of the polynomial is listed first and that all coefficients (even those whose value is zero) are included, the vector is a complete representation of the polynomial.

In the previous sections, the importance of the poles of the transfer function was established. Finding the poles of a transfer function involves finding the roots (often complex) of a polynomial in the denominator of the transfer function. MATLAB has a very robust polynomial root-solving algorithm that is easy to use. The MATLAB command is

```
>> roots(a)
```

where a is a vector containing the coefficients of a polynomial.

For example, take the following cubic polynomial:

$$3s^3 + 5s^2 + s + 10.$$

The roots of this polynomial can easily be found with the following MATLAB command:

```
>> roots([3 5 1 10])
```

The results are

```
ans =
-2.2025
0.2679 + 1.2007i
0.2679 - 1.2007i
```

The importance of listing zero coefficients can best be illustrated by an example as well. Consider this fourth-order polynomial:

$$A(s) = s^4 + 10s^2 + 5s + 30.$$

The incautious reader would be tempted to use the following command to find the roots:

```
>> roots([1 10 5 30])
```

which gives the following results:

```
ans =
-9.8021
-0.0989 + 1.7466i
-0.0989 - 1.7466i
```

Clearly, however, four roots are to be expected, not three. By ignoring the coefficient of the s^3 term (in this case, zero), MATLAB interprets the vector as representing a third-order polynomial and gives the wrong answers. The correct command is

```
>> roots([1 0 10 5 30])
```

which gives the correct answers:

```
ans =
0.6388 + 2.7138i
0.6388 - 2.7138i
-0.6388 + 1.8578i
-0.6388 - 1.8578i
```

Although the complete information of a transfer function is contained in two vectors, each of which represents a polynomial, MATLAB offers a data structure that can more easily be manipulated and used for complicated analysis and design operations. The details of this data structure are not important to the beginning user and are best left to individual study. The command to create such an object is simply

```
>> mytf=tf(b,a)
```

where b is a vector representing the numerator polynomial and a represents the denominator polynomial.

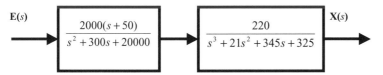

Figure 11.24. Transfer function block diagram of amplifier–motor pair.

In subsequent chapters (and in Appendix 3), the power of this approach is illustrated fully; the following two examples demonstrate the use of these functions for simplifying series and feedback transfer functions.

EXAMPLE 11.7

In electromechanical control applications, a common task is to model the dynamics of actuators (e.g., a dc motor). For high-performance applications, it is sometimes necessary to model the amplifier as well. Figure 11.24 shows two cascaded transfer functions, one that represents the amplifier, one the motor. Using both the rules outlined in Section 11.6 and the MATLAB data structure for manipulating system transfer functions, find the equivalent transfer function for the amplifier–motor pair.

SOLUTION

As discussed in Section 11.6, two transfer functions in series (cascaded) are combined by multiplication of the numerator and denominator polynomials. The algebra involved is straightforward, but somewhat tedious. It is left as an exercise to the reader to verify that the process of multiplying the polynomials leads to the following overall transfer function:

$$T(s) = \frac{X(s)}{E(s)} = \frac{440000(s + 50)}{s^5 + 321s^4 + 26645s^3 + 523825s^2 + 6997500s + 6500000}.$$

With MATLAB, the process is straightforward. First, enter the four polynomials as vectors:

```
>> n1=2000*[1 50]
>> d1=[1 300 20000]
>> n2=220
>> d2=[1 21 345 325]
```

Then use the tf( ) function to form two transfer function objects in MATLAB:

```
>>tf1=tf(n1,d1)
Transfer function:
2000 s + 100000
-------------------------
s^2 + 300 s + 20000
```

Note that MATLAB echoes the transfer function in a format that is easy to read:

```
>>tf2=tf(n2, d2)
Transfer function:
          220
--------------------------------
s^3 + 21 s^2 + 345 s + 325
```

Combining the two transfer functions is a simple case of multiplying the two objects together:

```
tf3=tf1*tf2
Transfer function:
                      440000 s + 2.2e007
       -------------------------------------------------------------
       s^5 + 321 s^4 + 26645 s^3 + 523825 s^2 + 6.998e006 s + 6.5e006
```

The results are the same as those obtained by hand.

EXAMPLE 11.8

Now consider the same two transfer functions, but with the amplifier in a feedback loop, as shown in Fig. 11.25.

According to the rules of Section 11.6, the overall transfer function would be given by

$$\mathbf{T}(s) = \frac{\mathbf{X}(s)}{\mathbf{E}(s)} = \frac{\mathbf{G}(s)}{1 + \mathbf{G}(s)\mathbf{H}(s)},$$

where $\mathbf{G}(s)$ is the transfer function in the forward (top) block and $\mathbf{H}(s)$ is the transfer function in the feedback (bottom) block. Again, you carry out the process by manipulating the polynomials and multiplying them together, meticulously following the algebraic rules. Note that the $\mathbf{G}(s)\mathbf{H}(s)$ term in the denominator is the same as the results in the previous example. Even starting from that point, however, the process would be a lengthy operation of multiplying, grouping, and carefully checking the math. In MATLAB, however, it's quite simple. Recall that, as a result of the previous example, three transfer function objects are already in the MATLAB workspace, tf1 [our $\mathbf{G}(s)$], tf2 [our $\mathbf{H}(s)$], and tf3 [$\mathbf{G}(s)\,\mathbf{H}(s)$].

The following command solves the problem in a single stroke:

```
>> tf4=tf1/(1+tf3)
Transfer function:

 2000 s^6 + 742000 s^5 + 8.539e007 s^4 + 3.712e009 s^3 + 6.638e010 s^2 + 7.128e011s + 6.5e011
 - - - - - - - - - - - - - - - - - - - - - - - - - - - - - - - - - - - - - - - - - - - - - -
 s^7 + 621 s^6 + 142945 s^5 + 1.494e007 s^4 + 6.975e008 s^3 + 1.274e010 s^2 + 1.573e011s + 5.7e011
```

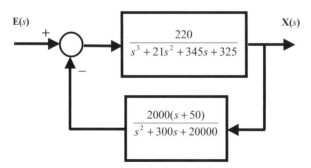

Figure 11.25. Two transfer functions in a feedback configuration.

The results of the MATLAB operation demonstrate an important pitfall of numerical computations. If you follow through the process by hand, you should end up with a fourth-order numerator polynomial and a fifth-order denominator polynomial. So what happened in MATLAB? The answer lies in the fact that all computers have a finite representation of numbers and there is often some round-off error in computations, particularly those that have numbers that cover a wide dynamic range, such as the coefficients do in this problem. The issue becomes clearer when you look at the poles and zeros of the new transfer function.

First, use `tfdata()` to extract the numerator and denominator polynomials back into vectors:

```
>> [num den]=tfdata(tf4,'v')
num =
      1.0e+011 *
      0                 0.0000 0.0000 0.0009 0.0371 0.6638 7.1275 6.5000
den =
      1.0e+011 *
      0.0000        0.0000 0.0000 0.0001 0.0070 0.1274 1.5730 5.7000
```

Now use the `roots` command to find the poles (roots of denominator) and zeros (roots of numerator). Because we are investigating issues related to round-off error, we would also like to view more significant digits in the results, so we set the format to `long`:

```
>> format long
>> poles4=roots(den)
poles4 =
1.0e+002 *
-2.00000000000376
-1.99908539617988
-1.00266811228711
-1.00000000000020
-0.07752455677075 + 0.14393720759809i
-0.07752455677075 - 0.14393720759809i
-0.05319737798755
>> zeros4=roots(num)
zeros4 =
1.0e+002 *
-2.00000000000000
-1.00000000000000
-0.50000000000000
-0.10000000000000 + 0.15000000000000i
-0.10000000000000 - 0.15000000000000i
-0.01000000000000
>>
```

Now examine the first and fourth poles and compare their values with the first and second zeros. You will note that they are exceptionally close to one another in value, differing only in the 12th decimal place and beyond. For all intents and purposes, they are equal (indeed, they are) and should cancel out, leaving a fourth-order numerator and fifth-order denominator as expected.

Fortunately, MATLAB has an easy solution to this problem. The function `minreal ()` offers the user an option to carry out pole-zero cancellation to a specified tolerance. The default tolerance is often sufficient:

```
<< tf5 = minreal(tf4)
Transfer function:
2000s^4 + 1.42e005s^3 +2.79e006s^2 + 3.515e007s + 3.25e007
-----------------------------------------------------------
s^5 + 321s^4 + 2.664e004s^3 + 5.238e005s^2 + 7.437e006s + 2.85e007
```

which is the expected, and correct, result.

Example 11.8 is yet another demonstration of the importance of validation of any result obtained by numerical means. In this case, the results were not accepted solely on face value, but rather were compared with the form of the result that was expected. One can easily compute the expected order of the result by seeing how the various polynomials are combined in the simplification (but not necessarily carrying out the simplification by hand). It is essential that all results be critically evaluated in order that the user may gain confidence in the answer.

11.8 SYNOPSIS

The concept of a linear system transfer function was developed here for the case in which the input to the system is an exponential of the form $u(t) = Ue^{st}$, where U is a constant that may be complex or real, s is the complex variable $\alpha + j\omega$, and the particular solution is $y(t) = Ye^{st}$, where Y is a complex coefficient.

An alternate approach was presented for readers familiar with the Laplace transformation, in which the system transfer function was defined as a ratio of the Laplace transform of the output signal over the Laplace transform of the input signal with all initial conditions set to zero.

For a linear system having an input–output differential equation of the form

$$a_n \frac{d^n y}{dt^n} + a_{n-1} \frac{d^{n-1} y}{dt^{n-1}} + \cdots + a_0 y = b_m \frac{d^m u}{dt^m} + b_{m-1} \frac{d^{m-1} u}{dt^{m-1}} + \cdots b_0 u, \quad (11.53)$$

the system transfer function $T(s)$ is expressed in block diagram form in Fig. 11.26. Note that the term containing the ith power of s in the denominator of the transfer function corresponds to the term in the system input–output differential equation involving the ith derivative of the output, whereas the term containing the jth power of s in the numerator of the transfer function corresponds to the term in the system input–output differential equation involving the jth derivative of the input. Thus

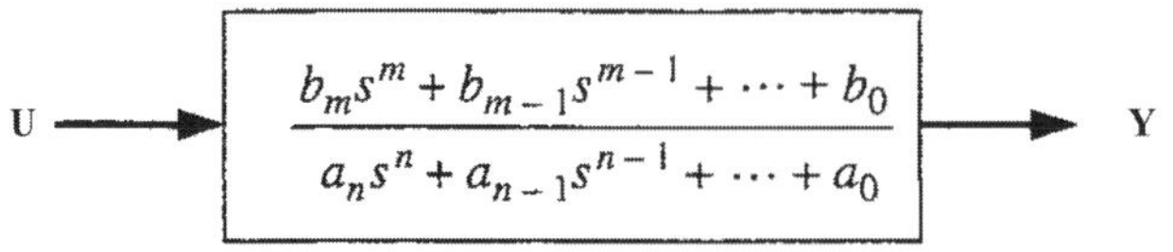

Figure 11.26. General form of the ratio-of-polynomial-type system transfer function, where $m \leq n$.

the system input–output differential equation is readily recovered, by simple obser-vation, from inspection of the numerator and denominator of the system transfer function.

The denominator of $\mathbf{T}(s)$, when set equal to zero, is a form of the system char-acteristic equation

$$a_n s^n + a_{n-1} s^{n-1} + \cdots + a_0 = 0, \tag{11.54}$$

which may also be expressed in terms of its roots (poles) as

$$a_n (s - p_1)(s - p_2) \cdots (s - p_n) = 0. \tag{11.55}$$

The real poles p_i correspond to the inverse system time constants,

$$p_i = \frac{1}{\tau_i}, \tag{11.56}$$

and each pair of complex-conjugate roots p_k, p_{k+1} correspond to exponential coef-ficients of a damped sinusoid:

$$p_k, p_{k+1} = \sigma_k \pm j\omega_{dk} = -\zeta_k \omega_{nk} \pm j\omega_{nk}\sqrt{1 - \zeta_k^2}, \tag{11.57}$$

where

$$\zeta_k = \frac{\sigma_k}{\sqrt{\sigma_k^2 + \omega_{dk}^2}},$$

$$\omega_{nk} = \sqrt{\sigma_k^2 + \omega_{dk}^2}.$$

The concept of a transfer function developed for a system with one input and one output was extended to linear multi-input, multi-output systems.

One of the useful properties of the transfer function presented in this chapter is that it can be conveniently used (by use of the final-value and initial-value theorems) to determine conditions that exist in a system subjected to a step input immediately after the input signal is applied, or at $t = 0^+$, and at steady state. This property is often used in the preliminary stages of system analysis or to check solutions obtained by computer simulation.

Basic rules of the transfer function block diagram algebra were introduced. The rules are useful in reducing detailed transfer function block diagrams to a single-block representation. Finally, representation of the system transfer function in MATLAB was outlined.

In the next chapter, a form of the transfer function, the sinusoidal transfer func-tion $\mathbf{T}(j\omega)$, will be derived with $s = j\omega$, which will be used for dealing with the steady response of a system to steady sinusoidal inputs.

PROBLEMS

11.1 (a) Prepare a detailed transfer function block diagram for the linearized equations of the rotational system discussed in Example 2.7.

(b) Express the overall input–output transfer functions with block diagrams for the lin-earized system relating (1) $\mathbf{T}_K$ to the input torque $\mathbf{T}_e$, with $\mathbf{T}_w = 0$, and (2) $\mathbf{T}_K$ to the other input torque $\mathbf{T}_w$, with $\mathbf{T}_e = 0$.

11.2 (a) Prepare detailed transfer function block diagrams for each of the two subsystems of Example 2.6 [i.e., for the subsystems described by Eqs. (2.42) and (2.43)] after linearizing the nonlinear damper to obtain an incremental damping coefficient:

$$b_{\text{inc}} = \left. \frac{df_{\text{NL}}}{dv_3} \right|_{\bar{v_3}}.$$

(b) Combine the diagrams prepared in part (a) into a single transfer function block diagram with input $\mathbf{X}_1$ and outputs $\mathbf{X}_2$ and $\mathbf{X}_3$.

(c) Apply the rules of the block-diagram algebra to the block diagram prepared in part (b) to obtain transfer functions relating $\mathbf{X}_2$ to $\mathbf{X}_1$ and $\mathbf{X}_3$ to $\mathbf{X}_1$.

11.3 (a) Prepare a detailed transfer function block diagram for the electric circuit discussed in Example 7.1.

(b) Express the overall transfer function with a block diagram relaling $\mathbf{E}_o$ to $\mathbf{E}_s$.

11.4 (a) Prepare a detailed transfer function block diagram for the linearized state-variable equations in Example 7.3.

(b) Express the overall transfer function with a block diagram relating $\mathbf{E}_{2g}$ to $\mathbf{E}_s$.

11.5 (a) For the case in which $T_w = C_2|\Omega_2|\Omega_2$, prepare a detailed transfer function block diagram for the linearized rotational system discussed in Example 2.7.

(b) Combine the linearized equations to develop the input–output system differential equation relating $\hat{\Omega}_2$ to $\hat{T}_e$.

(c) Express the overall transfer function with a block diagram relating Ω_2 to T_e.

11.6 (a) Prepare a detailed transfer function block diagram for the motor-driven inertia system discussed in Example 10.2.

(b) Express the overall transfer function with a block diagram relating Ω_1 to $\mathbf{E}_s$.

11.7 (a) Prepare a detailed transfer function block diagram for the linearized fluid control system discussed in Example 9.2.

(b) Express the overall transfer function with a block diagram relating $\mathbf{P}_{2r}$ to $\mathbf{A}_o$.

11.8 (a) Prepare a detailed transfer function block diagram for the third-order system having the following three state-variable equations:

$$\frac{dx_1}{dt} = a_{11}x_1 + a_{12}x_2 + b_{11}u_1,$$

$$\frac{dx_2}{dt} = a_{21}x_1 + a_{23}x_3,$$

$$\frac{dx_3}{dt} = a_{31}x_1 + a_{32}x_2 + a_{33}x_3.$$

(b) Use Cramer's rule with determinants to find the transfer function relating $\mathbf{X}_3$ to $\mathbf{U}_1$, and express the overall transfer function with a block diagram for this input–output combination.

(c) For an output $y_1 = c_{12}x_2 + c_{13}x_3$, find the transfer function relating $\mathbf{Y}_1$ to $\mathbf{U}_1$, and express the overall transfer function with a block diagram for this input–output combination.

11.9 If two systems have the following transfer functions,

$$\mathbf{T}_1(s) = \frac{13s + 1}{10s^2 + 3s + 1},$$

$$\mathbf{T}_2(s) = \frac{130s + 1}{1000s^2 + 30s + 1},$$

which system responds faster? Verify your answer by using the MATLAB system object and the step() function.

11.10 Consider the radar-tracking control system shown in Fig. P11.10.

(a) Find the closed-loop transfer function relating the output $\mathbf{Y}(s)$ to the input $\mathbf{U}(s)$.

(b) Compare the poles of the closed-loop transfer function with those of the system for which $\mathbf{X}(s)$ is the input and $\mathbf{Y}(s)$ is the output. What do you conclude?

(c) Using MATLAB's system objects and the step() command to compare the closed-loop step response with the open-loop $[\mathbf{Y}(s)/\mathbf{X}(s)]$ step response.

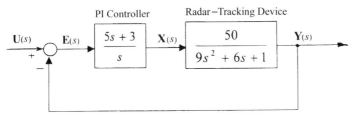

Figure P11.10. Transfer function block diagram of a radar-tracking control system.

11.11 The transfer function block diagram of a system designed to control the liquid level in a chemical process is shown in Fig. P11.11. In this diagram, k_p is an adjustable gain that can be set by the machine operator.

(a) Derive the closed-loop transfer function relating $\mathbf{H}_a(s)$ to $\mathbf{H}_d(s)$.

(b) How does the value of k_p affect the locations of the closed-loop poles relative to the poles of the open-loop system?

(c) Write a script in MATLAB that will compute the values of the closed-loop poles for a range of values of k_p and plot them on the complex plane. From this plot and your knowledge of system response, can you choose a value of k_p that will yield desirable results for a system such as this?

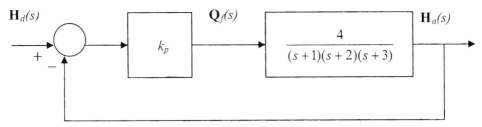

Figure P11.11. Transfer function block diagram of a liquid-level control system.

Frequency Analysis

LEARNING OBJECTIVES FOR THIS CHAPTER

12–1 To apply frequency-response transfer functions to determine response of linear systems to sinusoidal or other periodic inputs.

12–2 To construct (analytically and/or with MATLAB) and to interpret magnitude and phase characteristics (Bode diagrams) of system frequency response.

12–3 To construct (analytically and/or with MATLAB) and to interpret polar plots of frequency response (Nyquist diagrams).

12–4 To develop, manipulate, and simplify transfer function block diagrams of complex systems made up of many subsystems.

12.1 INTRODUCTION

The response of linear systems to sinusoidal inputs forms the basis of an extensive body of theory dealing with the modeling and analysis of dynamic systems. This theory was developed first in the field of communications (telephone and radio) and later became extended and then widely used in the design and development of automatic control systems. The theory is useful not only in determining the sinusoidal response of a system but also in specifying performance requirements (as in sound-system components or in a radar-tracking system), in finding the responses to other periodic inputs (square wave, sawtooth wave, etc.),[1] and for predicting the stability of feedback control systems (amplifiers, regulators, and automatic controllers). A graphical portrayal of the input and output sinusoids of a typical linear system is provided in Fig. 12.1.

12.2 FREQUENCY-RESPONSE TRANSFER FUNCTIONS

Recall from Chap. 4 that the response of a linear system to an input signal $u(t)$ is the sum of the homogeneous (free) and particular (forced) solutions:

$$y(t) = y_h(t) + y_p(t). \tag{12.1}$$

Here the homogeneous part of the response of a stable system is of minor interest (because it usually dies away soon and, in a linear system, in no way affects the

---

[1] See Appendix 1 for a review of the Fourier series and its use in describing a periodic time function in terms of its harmonic components.

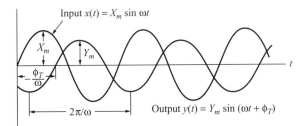

Figure 12.1. Sinusoidal input and output waveforms for a linear dynamic system.

particular solution). Thus, after the transient period, the steady-state response is equal to the particular solution, $y_{ss}(t) = y_p(t)$. Also, recall from Chap. 4 that the particular solution is of the same form as that of the input, its derivatives, or both. In Chap. 11, it was shown that any input signal can be represented by an exponential form:

$$u(t) = \mathbf{U}(s)e^{st}, \tag{12.2}$$

where $\mathbf{U}(s)$ is the complex input-amplitude coefficient. Therefore the steady-state, or particular, solution will take the same exponential form,

$$y_{ss}(t) = y_p(t) = \mathbf{Y}(s)e^{st}, \tag{12.3}$$

where $\mathbf{Y}(s)$ is the complex output-amplitude coefficient. Substituting $\mathbf{Y}(s) = \mathbf{T}(s)\mathbf{U}(s)$ from Eq. (11.5) into Eq. (12.3) yields

$$y_{ss}(t) = \mathbf{Y}(s)e^{st} = \mathbf{T}(s)\mathbf{U}(s)e^{st}. \tag{12.4}$$

Furthermore, it was shown in Chap. 11 that exponential form (12.2) can be used to describe a sinusoid by setting $s = j\omega$. Thus, for sinusoidal inputs, the steady-state response is

$$y_{ss}(t) = \mathbf{Y}(j\omega)e^{j\omega t} = \mathbf{T}(j\omega)\mathbf{U}(j\omega)e^{j\omega t}, \tag{12.5}$$

where $\mathbf{U}(j\omega)$, $\mathbf{Y}(j\omega)$, and $\mathbf{T}(j\omega)$ are sinusoidal input, the output, and the frequency-response transfer function, respectively. Expressing $\mathbf{U}$, $\mathbf{Y}$, and $\mathbf{T}$ in complex exponential form, i.e., $\mathbf{U}(j\omega) = U_m e^{j\phi_u}$, $\mathbf{Y}(j\omega) = Y_m e^{j\phi_y}$, and $\mathbf{T}(j\omega) = T e^{j\phi_T}$ and substituting into Eq. (12.5) gives

$$Y_m e^{j\phi_y} e^{j\omega t} = T e^{j\phi_T} U_m e^{j\phi_u} e^{j\omega t}. \tag{12.6}$$

By convention, the time axis is set such that the phase angle of the input sinusoid $\phi_u = 0$ so that $e^{j\phi_u}$ is equal to unity. Further simplifying by dividing both sides of Eq. (12.6) by $e^{j\omega t}$ yields

$$Y_m e^{j\phi_y} = T U_m e^{j\phi_T}. \tag{12.7}$$

Hence the amplitude of the output sinusoid is

$$Y_m = T U_m, \tag{12.8}$$

and the phase angle of that sinusoid relative to the input is

$$\phi_y = \phi_T. \tag{12.9}$$

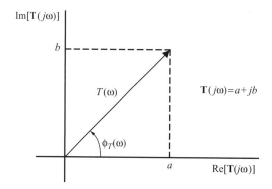

Figure 12.2. Complex plan representation of $\mathbf{T}(j\omega)$.

To better understand this important result, recall that the frequency-response transfer function $\mathbf{T}(j\omega)$ follows directly from the system transfer function $\mathbf{T}(s)$ if the complex variable s is allowed to be limited to a pure imaginary number $j\omega$, i.e.,

$$\mathbf{T}(j\omega) = \mathbf{T}(s)|_{s=j\omega}. \tag{12.10}$$

At a given frequency ω, the transfer function is a complex number having (generally) both real and complex parts. As seen in Fig. 12.2, a complex number can also be expressed in terms of its magnitude (or absolute value) and phase angle:

$$T(\omega) = |\mathbf{T}(j\omega)|, \tag{12.11}$$

$$\phi_T(\omega) = \angle\mathbf{T}(j\omega). \tag{12.12}$$

Thus, when a linear system is subject to a sinusoidal input, its steady-state response is a sinusoid of the same frequency. The amplitude of the output is equal to the amplitude of the input times the magnitude of the transfer function evaluated at that frequency. The phase of the output sinusoid relative to the input is the angle of the transfer function evaluated at that frequency. The frequency-response transfer function $\mathbf{T}(j\omega)$ thus provides the information needed to describe the steady-state response of linear time-invariant systems to sinusoidal inputs.

Note that the absolute value in Eq. (12.11) represents the magnitude of the complex number and the angle in Eq. (12.12) is the angle between the radial line drawn to that complex number and the positive x axis in the complex plane, as shown in Fig. 12.2.

In Cartesian coordinates, the frequency-response transfer function can be expressed in terms of its real and imaginary parts:

$$\mathbf{T}(j\omega) = \mathrm{Re}[\mathbf{T}(j\omega)] + j\,\mathrm{Im}[\mathbf{T}(j\omega)] = T(\omega)\cos\phi_T(\omega) + j\,T(\omega)\sin\phi_T(\omega). \tag{12.13}$$

Figure 12.3 shows a schematic block diagram of a single-input–single-output system represented by a frequency-response transfer function.

Figure 12.3. Block-diagram representation of a frequency-response transfer function.

EXAMPLE 12.1

An electric motor operating at a constant speed of 1000 rpm is generating a vertical vibratory force F_m that is due to an imbalance in its rotating elements. The amplitude of force F_m is 400 N. To absorb the vibration, the motor is supported by rubber shock mounts of stiffness $k = 50{,}000$ N/m and damping $b = 800$ N s/m. The mass of the motor is $m = 25$ kg. A simplified schematic of the system is shown in Fig. 12.4. Find the amplitude and phase of the vertical displacement of the motor, $x(t)$, and determine the ratio of the magnitude of the force transmitted through the shock mounts to the magnitude of the vibratory force.

SOLUTION

The basic equation of motion for the vertical motion of the system is

$$m\ddot{x} + b\dot{x} + kx = F_m, \tag{12.14}$$

where force F_m is a sinusoidal function of time having amplitude 400 N and frequency $\omega = 2\pi 1000/60 = 104.7\,\text{rad/s}$,

$$F_m = 400\sin(104.7t). \tag{12.15}$$

The amplitude and phase of displacement $x(t)$ can be found by use of Eqs. (12.8) and (12.9). The transfer function relating the displacement to force F_m is obtained from Eq. (12.14), assuming zero initial conditions, $x(0) = 0$ and $\dot{x}(0) = 0$:

$$\mathbf{T}(s) = \frac{\mathbf{X}(s)}{\mathbf{F}_m(s)} = \frac{1}{ms^2 + bs + k}. \tag{12.16}$$

One obtains the frequency-response transfer function of the system by putting $s = j\omega$ in Eq. (12.16):

$$\mathbf{T}(j\omega) = \mathbf{T}(s)|_{s=j\omega} = \frac{1}{-m\omega^2 + j\omega b + k}. \tag{12.17}$$

For the given values of the system parameters and the frequency of vibration, $\omega = 104.7$ rad/s, the frequency-response transfer function is

$$\mathbf{T}(j\omega)|_{\omega=104.7} = \frac{1}{-224{,}052 + j83{,}760} = 1.75 \times 10^{-11}(-224{,}052 - j83{,}760). \tag{12.18}$$

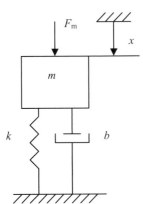

Figure 12.4. A model of vibration of the electric motor considered in Example 12.1.

Note that, in Eq. (12.18), the complex number was rationalized (eliminating the imaginary part from the denominator) by multiplying the numerator and denominator by the complex conjugate of the denominator. Hence the magnitude of the transfer function at the frequency of interest is

$$|\mathbf{T}(j\omega)| = 1.75 \times 10^{-11}\sqrt{(-224{,}052)^2 + (-83{,}760)^2} = 4.18 \times 10^{-6}\ \text{m/N}. \qquad (12.19)$$

From Eq. (12.8), the amplitude of the vertical displacement is

$$|\mathbf{X}(j\omega)| = |\mathbf{F}_m(j\omega)||\mathbf{T}(j\omega)| = 400 \times 4.18 \times 10^{-6} = 1.67 \times 10^{-3}\ \text{m}. \qquad (12.20)$$

From Eq. (12.9), the phase angle of the vertical displacement is equal to the phase angle of the frequency-response transfer function. Note that, from Eq. (12.18), it can be seen that this complex number is in the third quadrant (both imaginary and real parts are negative) and the proper trigonometric equation must be used:

$$\phi_T = -\left[\pi - \tan^{-1}\left(\frac{-83{,}760}{-224{,}052}\right)\right] = -2.78\ \text{rad}. \qquad (12.21)$$

Thus a complete mathematical expression for the vertical displacement of the motor operating at a constant speed of 1000 rpm is

$$x(t) = 0.00167\sin(104.7t - 2.78)\ \text{m}. \qquad (12.22)$$

Now, the force transmitted through the shock mounts is

$$F_t = b\dot{x} + kx = 800\dot{x} + 50{,}000x. \qquad (12.23)$$

The transfer function relating the transmitted force to the input force F_m is

$$\mathbf{T}_F(s) = \frac{\mathbf{F}_T(s)}{\mathbf{F}_m(s)} = \frac{800s + 50{,}000}{25s^2 + 800s + 50{,}000}. \qquad (12.24)$$

The frequency-response transfer function relating the two forces is

$$\mathbf{T}_F(j\omega) = \frac{800\,j\omega + 50{,}000}{-25\omega^2 + 800\,j\omega + 50{,}000}, \qquad (12.25)$$

and for the frequency of vibration,

$$\mathbf{T}_F(j\omega)|_{\omega=104.7} = \frac{50{,}000 + j83{,}760}{-224{,}052 + j83{,}760}. \qquad (12.26)$$

Hence, the magnitude of the force transfer function is

$$|\mathbf{T}_F(j\omega)||_{\omega=104.7} = \frac{\sqrt{50{,}000^2 + 83{,}760^2}}{\sqrt{(-224{,}052)^2 + 83{,}760^2}} = 0.41. \qquad (12.27)$$

Thus the amplitude of the force transmitted through the shock mounts will be equal to just over 40 percent of the amplitude of the force produced by the imbalance of the rotating elements in the motor.

It would be important to engineers considering the installation of the motor to realize that the second-order model of the system is underdamped, with a damping ratio $\zeta = 0.36$ and a natural frequency $\omega_n = 44.7$ rad/s, the frequency corresponding to a motor speed of 427 rpm. One should therefore expect the highest amplitude of vertical vibration to occur during start-up or shutdown of the motor, when its rotational velocity is near the resonant frequency of 427 rpm.

As illustrated in Example 12.1, the frequency-response transfer function is a compact representation that allows one to predict the steady-state response of a system to sinusoidal inputs of any frequency. Although a large number of practical situations may be represented in this manner, the true power of this approach is more apparent when one considers the implications of the Fourier series (i.e., all periodic signals can be expressed as a sum of sinusoids) and the principle of superposition. Superposition, the characteristic that defines linearity, states that the response of a linear system to a sum of multiple inputs is equal to the sum of the individual responses to each input. In the discussion that follows, the salient features of $\mathbf{T}(j\omega)$ as described by its magnitude $T(\omega)$ and its phase $\phi_T(\omega)$ will be discussed in detail with emphasis on how they vary with frequency for various types of systems.

12.3 BODE DIAGRAMS

Because of the pioneering work of H. W. Bode on feedback amplifier design,[2] the use of logarithmic charts to portray the magnitude and phase characteristics has led to the development of Bode diagrams. These diagrams are now widely used in describing the dynamic performance of linear systems.

With the Bode diagrams, the magnitude $T(\omega)$ versus frequency ω and the phase angle $\phi_T(\omega)$ versus frequency ω characteristics are drawn on separate plots that share a logarithmic frequency axis. The amplitude axis is either logarithmic ($\log T$) or quasi-logarithmic (decibels), whereas the phase axis is linear.

The emphasis here is on the use of Bode diagrams for analytical studies. However, these diagrams are also used in portraying the results of frequency-response measurements that one obtains experimentally by applying an input sinusoid through a range of frequencies and measuring the resultant magnitude and phase of the output signal at each of the applied input frequencies. Each measurement is made after a sinusoidal steady state has been achieved.

The procedure for preparing a set of Bode diagram curves analytically is as follows:

Step 1. Determine the system transfer function $\mathbf{T}(s)$.

Step 2. Convert to the sinusoidal transfer function $\mathbf{T}(j\omega)$ by letting $s = j\omega$.

Step 3. Develop expressions for the magnitude, $T(\omega) = |\mathbf{T}(j\omega)|$, and phase angle, $\phi_T(\omega) = \angle\mathbf{T}(j\omega)$.

Step 4. Plot $\log \mathbf{T}(\omega)$ versus $\log \omega$ and ϕ_T versus $\log \omega$ or modify the $\log T(\omega)$ by multiplying it by 20. This modification makes use of the decibel scale, a scale widely used in the study of acoustics and amplifier design. Most

---

[2] H. W. Bode, *Network Analysis and Feedback Amplifier Design* (Van Nostrand, Princeton, NJ, 1945).

commercial plotting and spreadsheet programs make log–log and semi-log plots easy to do.

This procedure is illustrated in Example 12.2.

EXAMPLE 12.2

Prepare Bode diagram curves for the first-order system described by the differential equation

$$\frac{dy}{dt} + \frac{y}{\tau} = \frac{kx}{\tau}, \tag{12.28}$$

where $x(t)$ and $y(t)$ are the input and output, respectively.

SOLUTION

Step 1. Transforming to the s domain yields

$$s\mathbf{Y}(s) + \frac{\mathbf{Y}(s)}{\tau} = \frac{k\mathbf{X}(s)}{\tau}. \tag{12.29}$$

The system transfer function is

$$\mathbf{T}(s) = \frac{\mathbf{Y}(s)}{\mathbf{X}(s)} = \frac{k}{\tau s + 1}. \tag{12.30}$$

Step 2. The sinusoidal transfer function is

$$\mathbf{T}(j\omega) = \mathbf{T}(s)|_{s=j\omega} = \frac{k}{j\omega\tau + 1}. \tag{12.31}$$

Step 3. To simplify the process of obtaining expressions for $T(\omega)$ and $\phi_T(\omega)$, note that $\mathbf{T}(j\omega)$ can be presented as a ratio of complex functions $\mathbf{N}(j\omega)$ and $\mathbf{D}(j\omega)$:

$$\mathbf{T}(j\omega) = \frac{\mathbf{N}(j\omega)}{\mathbf{D}(j\omega)} = \frac{N(\omega)e^{j\phi_N(\omega)}}{D(\omega)e^{j\phi_D(\omega)}} \tag{12.32}$$

or

$$\mathbf{T}(j\omega) = \frac{N(\omega)}{D(\omega)}e^{j[\phi_N(\omega)-\phi_D(\omega)]}, \tag{12.33}$$

so that the magnitude $T(\omega)$ is given by

$$T(\omega) = \frac{N(\omega)}{D(\omega)}. \tag{12.34}$$

In this case, $N(\omega) = k$ and $D(\omega) = |j\omega\tau + 1|$, so that

$$T(\omega) = \frac{k}{|j\omega\tau + 1|} = \frac{k}{\sqrt{1 + \omega^2\tau^2}} \tag{12.35}$$

and the phase angle ϕ_T is given by

$$\phi_T(\omega) = \phi_N(\omega) - \phi_D(\omega), \tag{12.36}$$

where

$$\phi_N(\omega) = \tan^{-1}(0/k) = 0, \tag{12.37}$$

$$\phi_D(\omega) = \tan^{-1}(\omega\tau/1) = \tan^{-1}(\omega\tau). \tag{12.38}$$

Hence

$$\phi_T(\omega) = 0 - \tan^{-1}(\omega\tau) = -\tan^{-1}(\omega\tau). \tag{12.39}$$

Step 4. The Bode diagram curves for amplitude and phase can now be plotted with calculations based on the expressions for $T(\omega)$ and $\phi_T(\omega)$ developed in Step 3. The resulting
magnitude and phase curves for this first-order system are shown in Fig. 12.5.

In a preliminary system analysis, it is often sufficient to approximate the amplitude and phase curves by use of simple straight-line asymptotes, which are very easily
sketched by hand. These straight-line asymptotes are included in Fig. 12.5. The first
asymptote for the magnitude curve corresponds to the case in which ω is very small,
approaching zero. Using Equation (12.35) for the magnitude as $\omega \to 0$ yields

$$\lim_{\omega\to0}[\log\ T(\omega)] = \lim_{\omega\to0}\left(\log\frac{k}{\sqrt{1+\omega^2\tau^2}}\right) = \log\ k. \tag{12.40}$$

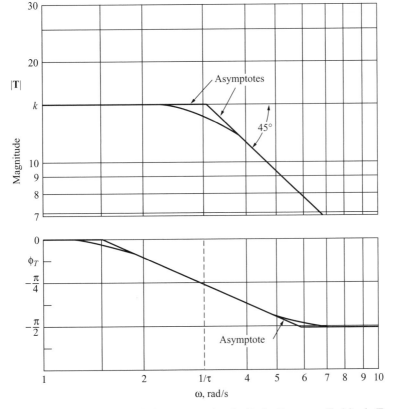

Figure 12.5. Amplitude and phase curves for the Bode diagram called for in Example 12.2.

The second asymptote for the amplitude curve is based on the case in which ω is very large. Using Equation (12.35) with ω approaching infinity yields

$$\lim_{\omega \to \infty} [\log T(\omega)] = \lim_{\omega \to \infty} \left(\log \frac{k}{\sqrt{1 + \omega^2 \tau^2}} \right) = \log k - \log \omega \tau. \tag{12.41}$$

Hence, at high frequencies,

$$\log[T(\omega)] = \log k - \log \tau - \log \omega. \tag{12.42}$$

Note that the slope of the second asymptote on log–log paper is $-45°$, which corresponds to -6 dB per octave on the decibel plot.

The two asymptotes described by Eqs. (12.40) and (12.42) thus constitute the asymptotic magnitude versus frequency characteristic. The maximum error incurred by use of the asymptotic approximation occurs at the "break frequency" where $\omega = 1/\tau$, i.e., where the two asymptotes intersect – sometimes also referred to as the "corner frequency."

The asymptotic approximation for the phase angle curve consists of three straight lines. The first, for very low frequency, is given by

$$\lim_{\omega \to \infty} [- \tan^{-1}(\omega \tau)] = - \tan^{-1}(0) = 0. \tag{12.43}$$

The second is a straight-line tangent to the phase angle curve at its corner frequency, having a slope given by

$$\frac{d\phi_T(\omega)}{d(\log \omega)} = \frac{d\phi_T(\omega)}{d\omega} \frac{d\omega}{d(\log \omega)} = -\frac{\omega \tau}{1 + \omega^2 \tau^2}. \tag{12.44}$$

Thus, at the corner frequency $\omega = 1/\tau$, the phase angle is $-45°$ and the slope is given by

$$\frac{d\phi_T(\omega)}{d(\log \omega)} \bigg|_{\omega=1/\tau} = -1/2. \tag{12.45}$$

The third asymptotic approximation occurs at very large values of ω, where

$$\lim_{\omega \to \infty} [\tan^{-1}(\omega \tau)] = - \tan^{-1}(\infty) = -\pi/2. \tag{12.46}$$

It can be seen in Fig. 12.5 that the system behaves as a low-pass filter, having an output that drops off increasingly as the corner frequency is exceeded. Thus the system is unable to respond at all as the frequency approaches infinity. The range of frequencies over which this system can deliver an effective output is referred to as its bandwidth; the bandwidth of a low-pass filter extends from a very low frequency up to its corner frequency, where the output has dropped to about 0.7 of its low-frequency value. Thus the bandwidth is determined by the value of $1/\tau$; the smaller the time constant τ, the larger the bandwidth.

This means that the bandwidth needs to be large enough (i.e., the time constant small enough) to enable the output to follow the input without excessive attenuation. However, providing extended bandwidth not only is costly but also makes it possible for the system to follow undesirable high-frequency noise in its input. These conflicting requirements necessitate either some compromise in the use of this system or selection of a better (probably higher-order) system that will follow the desired range of input frequencies but filter out the undesired higher-frequency noise.

Example 12.3 is provided to reinforce the use of the required analytical procedures and to demonstrate how a higher-order system might be devised to help meet these conflicting design requirements.

EXAMPLE 12.3

Prepare Bode diagram curves for the second-order system described by

$$\frac{d^2 y}{dt^2} + 2\zeta \omega_n \frac{dy}{dt} + \omega_n^2 y = \omega_n^2 x. \tag{12.47}$$

Step 1. The system transfer function is

$$\mathbf{T}(s) = \frac{\mathbf{Y}(s)}{\mathbf{X}(s)} = \frac{\omega_n^2}{s^2 + 2\zeta \omega_n s + \omega_n^2}. \tag{12.48}$$

Step 2. The sinusoidal transfer function is

$$\mathbf{T}(j\omega) = \frac{\omega_n^2}{(j\omega)^2 + 2\zeta \omega_n j\omega + \omega_n^2} = \frac{1}{1 - (\omega/\omega_n)^2 + j2\zeta(\omega/\omega_n)}. \tag{12.49}$$

Step 3. Develop expressions for $T(\omega)$ and $\phi_T(\omega)$. First, the magnitude is

$$T(\omega) = \frac{1}{\sqrt{[1 - (\omega/\omega_n)^2]^2 + 4\zeta^2(\omega/\omega_n)^2}}. \tag{12.50}$$

Then the phase angle is

$$\phi_T(\omega) = -\tan^{-1} \frac{2\zeta(\omega/\omega_n)}{1 - (\omega/\omega_n)^2}. \tag{12.51}$$

Step 4. Prepare the Bode diagram curves, as shown in Fig. 12.6.

In Example 12.3, a normalized frequency scale based on (ω/ω_n) has been used so that the frequency response of underdamped $(\zeta < 1)$ linear systems can be emphasized. For overdamped second-order systems, the characteristic equation has real roots, and two first-order terms are used. The procedure is given in Example 12.4. Note that the slope of the high-frequency straight-line approximation is now -2 on the log–log plot (-12 dB per octave on the decibel plot), the high-frequency phase angle is $-180°$, and the corner frequency phase angle is $-90°$.

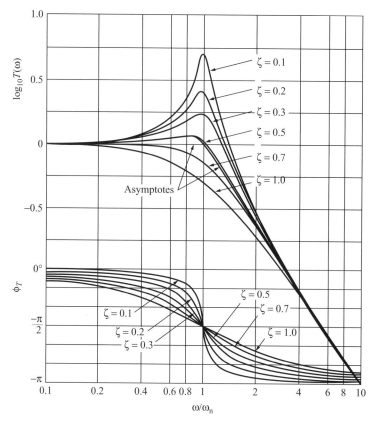

Figure 12.6. Bode diagram magnitude and phase curves for the underdamped second-order system described in Example 12.3.

EXAMPLE 12.4

Prepare asymptotic magnitude and phase characteristics for the system having the following transfer function: .

$$\mathbf{T}(s) = \frac{b_1 s + b_0}{s(a_2 s^2 + a_1 s + a_0)}.$$ (12.52)

SOLUTION

The numerator has a single root $r = -(b_0/b_1)$ so that its time constant is

$$\tau_1 = -(1/r) = (b_1/b_0).$$

The denominator has three roots, or poles, p_1, p_2, and p_3, with $p_1 = 0$. Assuming that the second-order factor in the denominator is overdamped (in contrast to the underdamped case in Example 12.3), the damping ratio $\zeta = a_1/2\sqrt{a_0 a_2}$ is greater than 1.

Thus the poles associated with this second-order factor are real, and the factor may be expressed in terms of its time constants τ_2 and τ_3:

$$a_2 s^2 + a_1 s + a_0 = a_0(\tau_2 s + 1)(\tau_3 s + 1),$$

where

$$\tau_2 = -\frac{1}{p_2} = \frac{a_1}{2a_0}\left(1 + \sqrt{1 - \frac{4a_0a_2}{a_1^2}}\right),$$

$$\tau_3 = -\frac{1}{p_3} = \frac{a_1}{2a_0}\left(1 - \sqrt{1 - \frac{4a_0a_2}{a_1^2}}\right).$$

Therefore the transfer function may be written as

$$\mathbf{T}(s) = \frac{b_0}{a_0}\frac{(\tau_1 s + 1)}{s(\tau_2 s + 1)(\tau_3 s + 1)}, \qquad (12.53)$$

and the sinusoidal transfer function is

$$\mathbf{T}(j\omega) = \frac{b_0}{a_0}\frac{(j\omega\tau_1 + 1)}{j\omega(j\omega\tau_2 + 1)(j\omega\tau_3 + 1)}, \qquad (12.54)$$

which can be expressed in terms of five individual transfer functions:

$$\mathbf{T}(j\omega) = \frac{\mathbf{N}_1(j\omega)\mathbf{N}_2(j\omega)}{\mathbf{D}_1(j\omega)\mathbf{D}_2(j\omega)\mathbf{D}_3(j\omega)}, \qquad (12.55)$$

where $\mathbf{N}_1 = b_0/a_0$, $\mathbf{N}_2 = j\omega\tau_1 + 1$, $\mathbf{D}_1 = j\omega$, $\mathbf{D}_2 = j\omega\tau_2 + 1$, and $\mathbf{D}_3 = j\omega\tau_3 + 1$.

Now, the individual magnitudes and phase angles may be used to obtain

$$\mathbf{T}(j\omega) = \frac{N_1(\omega)e^{j\phi_{N_1}}N_2(\omega)e^{j\phi_{N_2}}}{D_1(\omega)e^{j\phi_{D_1}}D_2(\omega)e^{j\phi_{D_2}}D_3(\omega)e^{j\phi_{D_3}}},$$

or

$$\mathbf{T}(j\omega) = \frac{N_1 N_2}{D_1 D_2 D_3}e^{j(\phi_{N_1} + \phi_{N_2} - \phi_{D_1} - \phi_{D_2} - \phi_{D_3})}. \qquad (12.56)$$

The magnitude of the overall transfer function is

$$T(\omega) = \frac{N_1(\omega)N_2(\omega)}{D_1(\omega)D_2(\omega)D_3(\omega)}, \qquad (12.57)$$

where $N_1 = b_0/a_0$, $N_2 = \sqrt{\omega^2\tau_1^2 + 1}$, $D_1 = \omega$, $D_2 = \sqrt{\omega^2\tau_2^2 + 1}$, and $D_3 = \sqrt{\omega^2\tau_3^2 + 1}$, and the phase angle of the overall transfer function is

$$\phi_T = \phi_{N_1} + \phi_{N_2} - \phi_{D_1} - \phi_{D_2} - \phi_{D_3}, \qquad (12.58)$$

where $\phi_{N_1} = 0$, $\phi_{N_1} = \tan^{-1}(\omega\tau_1)$, $\phi_{D_1} = \pi/2$, $\phi_{D_2} = \tan^{-1}(\omega\tau_2)$, and $\phi_{D_3} = \tan^{-1}(\omega\tau_3)$.

The magnitude and phase curves for the overall transfer function are then prepared by summation of the ordinates of the individual transfer functions, by use of the straight-line asymptotic approximations developed earlier. Usually the small departures at the corner frequencies are not of interest and thus can be ignored. Also, it is very easy to implement these procedures on a digital computer and use an x–y plotter to bypass a lot of tedious calculation and plotting by hand, in which case the results will be accurate at all frequencies.

The magnitude and phase curves obtained with straight-line asymptotic approximations are presented in Fig. 12.7, where $\tau_1 > \tau_2 > \tau_3$.

The preceding examples show how a general picture of the frequency response of even complicated systems can be easily sketched with just a few computations. On the

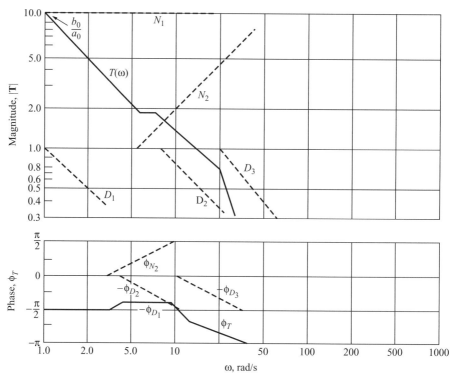

Figure 12.7. Bode diagram characteristics for magnitude and phase of the system in Example 12.4.

other hand, simple spreadsheet applications can be used to develop highly accurate plots with minimal effort. Also, specialized programs such as MATLAB allow the user to develop very accurate plots with single commands, as will be demonstrated in Section 12.6.

12.4 RELATIONSHIP BETWEEN TIME RESPONSE AND FREQUENCY RESPONSE

Chapter 11 introduced the concept of a transfer function based on the particular solution of a linear time-invariant model to an input that can be represented as a complex exponential. In this chapter, the idea is refined to focus on those inputs that are sinusoids. In this section, the relationship between the frequency response and the sinusoidal response is discussed.

Figure 12.6 shows a family of frequency-response curves for a second-order system having a transfer function given by Eq. (12.48). This discussion will be greatly simplified if one specific curve is used. Figure 12.8 shows a Bode diagram for a second-order system with a natural frequency ω_n of 1.0 rad/s and a damping ratio ζ of 0.2. Vertical lines are drawn on the plot at three frequencies, 0.2, 1.0, and 5.0 rad/s.

Table 12.1 summarizes frequency-response data at the highlighted frequencies as extracted from the plot. Recall that the decibel scale is $20 \times \log_{10}$ of the magnitude. Extracting the magnitude from the decibel reading is the inverse (divide by 20, raise 10 to the power of the result.)

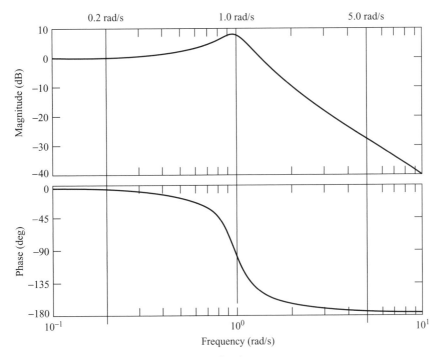

Figure 12.8. Frequency response for second-order system.

In light of the discussion in Section 12.2, these numbers can be interpreted as follows. If the system represented by the transfer function is driven by a sinusoidal input of unity magnitude and zero phase, at a frequency of 0.2 rad/s, the steady-state output of the system (after response to initial conditions has died out) will be a sinusoidal signal of the same frequency (0.2 rad/s), with a magnitude of 1.06 (just slightly larger) and lagging behind the input signal by 5°. Figure 12.9 shows the input and output sinusoids at this frequency.

Similarly, at a driving frequency of 1.0 rad/s (corresponding to the system natural frequency), the steady-state output will be 2.5 times as large as the input and will lag behind the input by 90° (1/4 of a cycle of the sine wave). Figure 12.10 illustrates this case.

Finally, at a driving frequency of 5.0 rad/s, the table indicates that the output will be very small (0.04) and will be nearly 180° out of phase with the input, as seen in Fig. 12.11.

Table 12.1. Frequency-response magnitude and phase angle

Frequency (rad/s)	Magnitude (dB)	Magnitude	Phase (deg)
0.2	0.5	1.06	−5
1.0	8.0	2.5	−90
5.0	−27	0.04	−175

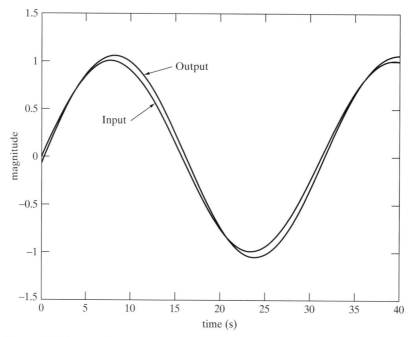

Figure 12.9. Input and output signals when the driving frequency is lower than the natural frequency.

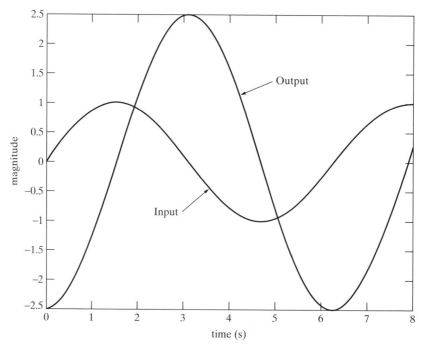

Figure 12.10. Input and output signals when the driving frequency is the same as the natural frequency.

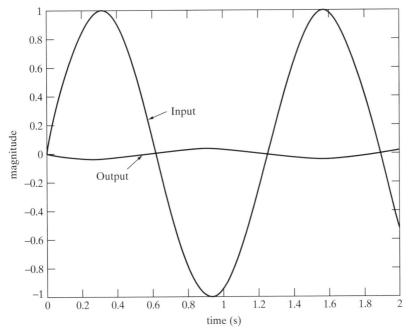

Figure 12.11. Input and output signals for frequencies much higher than the natural frequency.

Thus, given the system frequency response, whether derived analytically by methods presented in this chapter or discovered experimentally through repetitive measurements, the system response to a sinusoidal input or, by use of superposition, any periodic input can be computed by the technique illustrated in this section.

12.5 POLAR PLOT DIAGRAMS

In Section 12.3, the magnitude $T(\omega)$ and phase angle $\phi_T(\omega)$ characteristics of the frequency-response transfer function $\mathbf{T}(j\omega)$ were depicted by separate curves on a Bode diagram. Because each of these characteristics is a unique function of frequency ω, they are also directly related to each other. This relationship was expressed earlier by Eq. (12.13), which showed how the transfer function $\mathbf{T}(j\omega)$ may be represented on the complex plane, in either Cartesian or polar coordinates. Figure 12.2 illustrates how $\mathbf{T}(j\omega)$ at a given frequency ω appears on the complex plane. Note that this method of representing $\mathbf{T}(j\omega)$ does not show the frequency explicitly.

The end points of the vectors $\mathbf{T}(j\omega)$ plotted for successive values of ω from zero to infinity form the basis of a characteristic curve called the polar plot. This form of representing $\mathbf{T}(j\omega)$ has proved to be very useful in the development of system stability theory and in the experimental evaluation and design of closed-loop control systems.

EXAMPLE 12.5

Prepare a polar plot for a first-order system having the system transfer function

$$\mathbf{T}(s) = \frac{10}{2s + 1}. \tag{12.59}$$

The frequency-response transfer function is

$$\mathbf{T}(j\omega) = \frac{10}{2j\omega + 1}. \tag{12.60}$$

On the one hand, the polar plot represents graphically the relationship between the magnitude $T(\omega)$ and the phase angle $\phi_T(\omega)$; on the other hand, it shows the relationship between the real part $T \cos \phi_T$ and the imaginary part $T \sin \phi_T$ of the transfer function $\mathbf{T}(j\omega)$. When the latter is chosen, the expressions for the real and imaginary parts of $\mathbf{T}(j\omega)$ are found to be

$$\mathrm{Re}[\mathbf{T}(j\omega)] = \frac{10}{4\omega^2 + 1}, \tag{12.61}$$

$$\mathrm{Im}[\mathbf{T}(j\omega)] = -\frac{20\omega}{4\omega^2 + 1}. \tag{12.62}$$

The numerical results for computations carried out at several frequencies are given in Table 12.2, which also includes corresponding values of $T(\omega)$ and $\phi_T(\omega)$. The polar plot prepared from Table 12.2 is shown in Fig. 12.12. It can be shown analytically that this curve is a semicircle with its center at the 5.0 point on the real axis.

Note that the sign convention for phase angle is counterclockwise positive. Thus the phase angle $\phi_T(\omega)$ for this system is always negative. Although each computed point for this polar plot corresponds to a certain frequency, the frequency ω does not appear explicitly as an independent variable (unless successive tick marks are labeled for the curve, as was done in Fig. 12.12).

For complicated transfer functions, the process for computing the values of the data points is straightforward but becomes tedious. The next section shows how MATLAB easily generates both Bode diagrams and polar plots with a few simple commands.

Polar plot diagrams display most of the same information as Bode diagrams, especially if the frequency tick marks are included, but in a more compact form. Although they are not as easy to sketch by hand as the straight-line asymptotes of the Bode diagrams, they are especially useful in determining the stability of feedback control systems, which are discussed in Chap 13.

Table 12.2. Numerical data for plot of $T(j\omega) = 10/(2 j\omega + 1)$

ω (rad/s)	0.0	0.1	0.25	0.5	1.0	5.0	10.0	...	∞
Re $[\mathbf{T}(j\omega)]$	10	9.62	8.0	5.0	2.0	0.10	0.025	...	0
Im $[\mathbf{T}(j\omega)]$	0	−1.92	−4.0	−5.0	−4.0	−1.0	−0.5	...	0
$T(\omega)$	10	9.8	8.94	7.07	4.47	1.0	0.5	...	0
$\phi_T(\omega)$, deg	0	−11.3	−26.6	−45.0	−63.4	−84.3	−87.1	...	−90

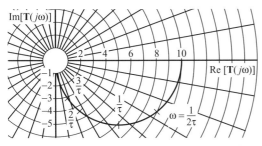

Figure 12.12. Polar plot for $\mathbf{T}(j\omega) = 10/(2j\omega + 1)$.

A few comments and hints are in order here to aid in the preparation and/or verification (especially with computer-generated data) of polar plots. First, the points for zero frequency and infinite frequency are readily determined. The point for zero frequency is obtained by use of $\omega = 0$ in $\mathbf{T}(j\omega)$. The point for infinite frequency is always at the origin because of the inability of real physical systems to have any response to very high frequencies. Stated mathematically,

$$\lim_{\omega \to \infty} \mathbf{T}(j\omega) = 0. \tag{12.63}$$

Second, as the frequency approaches infinity, the terminal phase angle is $(\pi/2)$ $(m - n)$, where m is the order of the numerator and n is the order of the denominator (assuming that the transfer function is a ratio of polynomials). For example, if

$$\mathbf{T}(s) = \frac{b_0 + b_1 s + \cdots + b_m s^m}{a_0 + a_1 s + \cdots + a_n s^n}, \ m \leq n,$$

the limit of the phase angle for frequency approaching infinity is given by

$$\lim_{\omega \to \infty} \phi_T(\omega) = (m - n)\frac{\pi}{2}. \tag{12.64}$$

Suggestion: Use this rule to check the validity of the curve shown in Fig. 12.12.

12.6 FREQUENCY-DOMAIN ANALYSIS WITH MATLAB

In Section 11.7, MATLAB's ability to store and manipulate transfer functions as objects in the workspace was presented. In this section, a set of tools is discussed that greatly simplifies the numerical operations and graphical presentations that are presented in this chapter.

12.6.1 Complex Numbers and MATLAB

A fundamental concept throughout the discussion of frequency-domain tools is that of the complex number. MATLAB can manipulate complex numbers easily and efficiently. Unlike this text, MATLAB's convention for the square root of -1 is i. Therefore, to enter a complex number into the workspace, use i as a symbol:

```
>> mynum=2+5*i
mynum =
2.0000 + 5.0000 i
```

MATLAB'S built-in functions `abs` and `angle` allow both the magnitude and angle of complex numbers to be easily computed:

```
>> abs(mynum)
ans =
5.3852
>> angle(mynum)
ans =
1.1903
```

12.6.2 Frequency Response and Transfer Function Evaluation

Section 12.3 described the process by which the sinusoidal transfer function can be manipulated to find the magnitude of the transfer function, $T(\omega)$, and the phase angle $\phi_T(\omega)$ for various values of driving frequency ω. Because the complete picture requires a large number of evaluations, computer methods are strongly indicated. MATLAB's Control Systems Toolbox provides four routines that operate on transfer function objects and perform the complex number evaluations required for interpreting these objects. Section 11.7 and Appendix 3 discuss the manner in which MATLAB represents transfer functions as a single data structure, the TF object. Table 12.3 summarizes the four routines that can manipulate these TF objects to evaluate and graph frequency-response data.

The following example illustrates the use of these commands.

EXAMPLE 12.6

For this example, use the transfer function introduced in Example 12.5:

$$\mathbf{T}(s) = \frac{10}{2s + 1}.$$

Table 12.3. Summary of MATLAB sinusoidal transfer-function-related commands

Syntax	Description	Returns	Comments
R=Evalfr (sys, w)	Evaluates frequency response of a TF object for a single frequency w	Single complex number that is the value of the TF at frequency w	Appropriate for quickly checking the response at one or a few frequencies
H= Freqresp (sys, w)	Evaluates the frequency response of a TF object over a grid of frequencies (contained in the vector w)	A vector of complex numbers (H) that are the TF evaluated at each frequency contained in w	Generates the response for a range of user-specified frequencies
Bode(sys)	Plots the Bode plot of a TF object.	Can return the Bode data (see help)	Generates the Bode plot
Nyquist(sys)	Plots a polar (Nyquist) frequency-response plot of the TF object	Can return response data (see help)	Generates the polar frequency-response plot

First, enter the transfer function as a TF object:

```
>>extf = tf([10], [2 1])
Transfer function:
  10
 ----
 2s+1
```

Use the `evalfr` command to verify entries in Table 12.2 for a frequency of 0.25 rad/s.

```
>>evalfr(extf, 0.25*i)
ans =
8.0000 - 4.0000i
```

Note that the frequency passed to the function is specified as an imaginary number. This is consistent with the interpretation that the sinusoidal transfer function is found by substituting $s = j\omega$. Note also that the result is the same that can be found in Table 12.2 for a frequency of 0.25 rad/s.

Now, use the `freqresp` function to generate the entire table with a single computation. The first step is to build a vector of frequencies:

```
>>wex = i*[0 .1 .25 .5 1 5 10 1000]
```

Use the function to generate the response:

```
>>h=freqresp(extf,wex)
```

It is left to the reader to verify that the results correspond to the entries previously presented in Table 12.2.

12.6.3 Bode and Nyquist Plots of Frequency Response

Sections 12.3 and 12.5 described in some detail the graphical representation of the frequency response for a transfer functions in form of Bode diagrams or Nyquist plots.

Both the Bode and Nyquist plots are generated by MATLAB with a single command:

```
>>bode(extf)
```

results in the Bode diagram seen in Fig. 12.13.

Similarly,

```
>>nyquist(extf)
```

generates the polar plot seen in Fig 12.14. When comparing Figs. 12.12 and 12.14, it should be noted that the MATLAB-generated plot shows a complete circle, whereas the polar plot generated in Fig 12.12 shows just the lower half. The difference lies in the fact that a Nyquist plot is a polar plot for both positive and negative frequencies. Although a negative frequency has no physical interpretation, the mathematics of complex analysis is such that the inclusion of negative frequencies allows for a more comprehensive analysis. This topic will be treated in greater detail in Chap. 13.

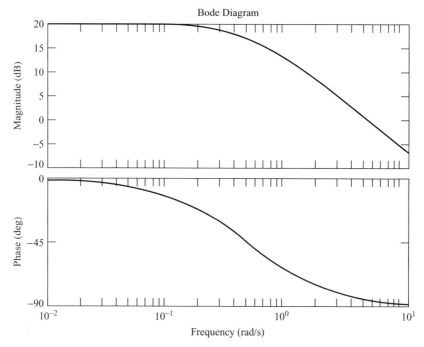

Figure 12.13. Bode plot of the transfer function in Example 12.5.

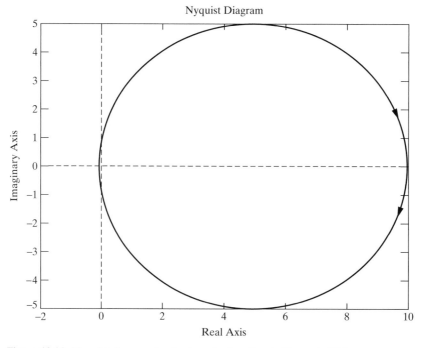

Figure 12.14. Nyquist diagram for the transfer function in Example 12.5.

12.7 SYNOPSIS

The concept of using a transfer function to describe a linear system was applied in this chapter to systems subjected to steady sinusoidal inputs through the use of $s = j\omega$. The steady-state output $\mathbf{Y}e^{j\omega t}$ of a linear system is the product of the input expressed by $\mathbf{U}e^{j\omega t}$ multiplied by the sinusoidal transfer function $\mathbf{T}(j\omega)$. Because the phase of the input $\angle\mathbf{U}$ is usually zero, $\mathbf{U} = U$ and thus the phase angle of the output $\angle\mathbf{Y}$ is equal to the phase angle of the transfer function $\angle\mathbf{T}(j\omega)$. Thus, when the input is a sine wave, $u(t) = \mathrm{Im}\,[Ue^{j\omega t}] = U\sin\omega t$, the output is a phase-shifted sine wave of the same frequency, $y(t) = \mathrm{Im}\,[Ye^{(j\omega t + \phi)}] = Y\sin(\omega t + \phi)$, where $Y = U|\mathbf{T}(j\omega)|$ and $\phi = \angle\mathbf{Y} = \angle\mathbf{T}(j\omega)$.

Two techniques for graphing the variation of the transfer function $\mathbf{T}(j\omega)$ with varying frequency ω were introduced: (a) Bode diagrams showing amplitude and phase versus frequency on separate plots and (b) polar plots showing amplitude and phase in a single plot.

Several rules were provided to assist in the routine preparation of frequency-response characteristics, and examples were given to illustrate the techniques involved. Finally the use of specialized MATLAB functions to generate Bode diagrams and Nyquist plots was demonstrated.

The frequency-response methods presented in this chapter are very useful in studying stability and dynamic response characteristics of automatic control systems. They are also commonly used in marketing and in specifying products for use in industry.

PROBLEMS

12.1 Sketch asymptotic Bode diagrams for the following transfer functions:

(a) $\mathbf{T}(s) = k/s$,
(b) $\mathbf{T}(s) = k/[s(\tau s + 1)]$,
(c) $\mathbf{T}(s) = ks/(\tau s + 1)$,
(d) $\mathbf{T}(s) = k/(\tau s + 1)^2$,
(e) $\mathbf{T}(s) = k(\tau_1 s + 1)/[s(\tau_2 s + 1)]$,
(f) $\mathbf{T}(s) = 10s(0.1s + 1)/(s + 1)^2$,

12.2 Sketch polar plots for the following transfer functions:

(a) $\mathbf{T}(s) = k/s$,
(b) $\mathbf{T}(s) = k/[s(\tau s + 1)]$,
(c) $\mathbf{T}(s) = \omega_n^2/(s^2 + 2\zeta\,\omega_n s + \omega_n^2)$,
(d) $\mathbf{T}(s) = ke^{-s\tau_o}$,
(e) $\mathbf{T}(s) = (s + 1)/([s + 2](s + 5))$,
(f) $\mathbf{T}(s) = 100/[s(s + 5)(s + 10)]$,

12.3 The Bode diagrams and the polar plot for the same system are shown in Fig. P12.3.

(a) Use the Bode diagrams to find the values of the frequencies ω_ϕ and ω_g marked on the polar plot.

(b) Find the value of the static gain k.

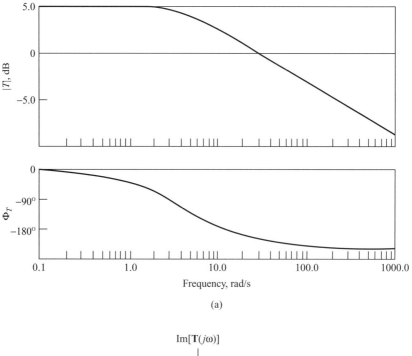

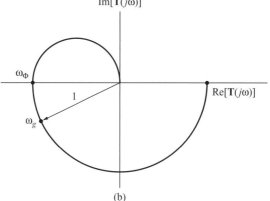

(b)

Figure P12.3. Polar plot and Bode diagrams for Problem 12.3.

12.4 A given system input–output equation is

$$a_2\frac{d^2 y}{dt^2} + a_1\frac{dy}{dt} + a_0 y = b_2\frac{d^2 u}{dt^2} + b_1\frac{du}{dt} + b_0 u.$$

Find the expression for the output $y(t)$ when the input is a sinusoidal function of time, $u(t) = U \sin \omega_f t$ with $\omega_f = (a_0/a_2)^{0.5}$.

12.5 The detailed transfer function block diagram of a certain second-order system is shown in Fig. P12.5. The values of the system parameters are

$$a_{11} = -0.1\,\mathrm{s}^{-1}, \qquad a_{12} = -4.0\,\mathrm{N/m^3},$$
$$a_{21} = 2.0\,\mathrm{m^3/N\,s^2}, \qquad a_{22} = 0,$$
$$b_{11} = 35.0\,\mathrm{N/m^2\,s}.$$

(a) Find the system transfer function $\mathbf{T}(s) = \mathbf{Y}(s)/\mathbf{U}(s)$.

(b) Find the amplitude A_0 and phase ϕ of the output $y(t)$ when the input is $u(t) = 0.001$ sin 1.5 when the input is $u(t) = 0.001 \sin 1.5t$.

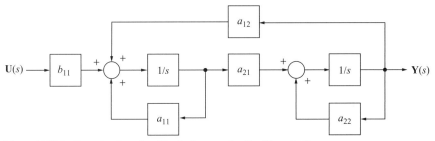

Figure P12.5. Transfer function block diagram for Problem 12.5.

12.6 In a buffer tank, shown in Fig. P12.6, the input flow rate $Q_i(t)$ has a constant component equal to 0.5 m³/s, an incremental sinusoidal component of amplitude 0.1 m³/s, and a frequency of 0.4 rad/s. The density of the liquid is $\rho = 1000$ kg/m³; the output flow conditions are assumed to be linear, with hydraulic resistance $R_L = 39{,}240$ N s/m⁵.

(a) Write the state-variable equation for this system using a liquid height in the tank of $h(t)$ as the state variable.

(b) Find the average liquid height in the tank, $\bar{h}$.

(c) Determine the condition for the cross-sectional area of the tank necessary to limit the amplitude of liquid height oscillations, as a result of sinusoidal oscillations in the input flow rate, to less than 0.2 m.

(d) What will be the time delay between the peaks of the sinusoidal input flow rate and the corresponding peaks of the liquid height if the cross-sectional area of the tank is $A = 1$ m²?

(e) Find the expression for the output flow rate $Q_L(t)$ for $A = 1$ m².

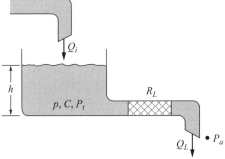

Figure P12.6. Buffer tank considered in Problem 12.6.

12.7 A closed-loop system consisting of a process of transfer function $\mathbf{T}_p(s)$ and a controller of transfer function $\mathbf{T}_c(s)$ has been modeled as shown by the transfer function block diagram in Fig. P12.7, where

$$\mathbf{T}_p(s) = \frac{5}{5s + 1}, \quad \mathbf{T}_c(s) = \frac{2(s + 1)}{(2s + 1)}.$$

(a) Find the system closed-loop transfer funtion $\mathbf{T}_{CL}(s)$.

(b) Find the amplitude and the phase angle of the output signal of the closed-loop system, $y(t)$, when $u(t) = 0.2 \sin 3t$.

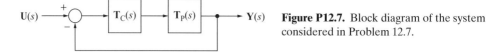

Figure P12.7. Block diagram of the system considered in Problem 12.7.

12.8 A hot-water storage tank has been modeled by the equation

$$8000\frac{dT_w}{dt} = 3T_a - 3T_w,$$

where T_w is the temperature of the water and $\mathbf{T}_a$ is the temperature of the ambient air. For several days, the ambient air temperature has been varying in a sinusoidal fashion from a maximum of $10\,^\circ$C at noon to a minimum of $-10\,^\circ$C at midnight of each day. Determine the maximum and minimum temperatures of the water in the storage tank during those days and find at what times of the day the maximum and minimum temperatures have occurred.

12.9 Figure P12.9 is a schematic of an ecological water system in Alaska. A small river with a steady flow rate of $0.05\ \mathrm{m^3/s}$ flows through two mountain reservoirs. The hydraulic capacitance of the top reservoir is $C_1 = 0.02\ \mathrm{m^5/N}$. The resistances to flow between the reservoirs are $R_1 = R_2 = 400{,}000\ \mathrm{N\ s/m^5}$. The bottom reservoir is cylindrical in shape with cross-sectional area $A_2 = 200\ \mathrm{m^2}$ and depth $h_2 = 2.5$ m.

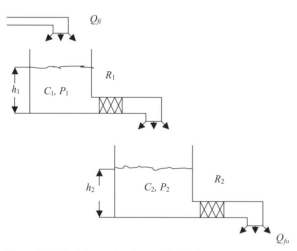

Figure P12.9. Schematic of an ecological water system.

(a) Derive the state-variable equations for the system. Use water heights in the two reservoirs, h_1 and h_2, as the state variables.

(b) Show that, under the current conditions, that is, when $Q_{fi} = 0.05\ \mathrm{m^3/s} = $ const, the size of the bottom reservoir is sufficient to prevent flooding of a nearby town.

(c) It is anticipated that in the future, as a result of global warming, a large iceberg near the top reservoir may start melting, producing an additional water flow rate in the river flowing by. It is estimated that the additional flow rate would by sinusoidal with an average value of $0.02\ \mathrm{m^3/s}$ and an amplitude of $0.01\ \mathrm{m^3/s}$. The period of sinusoidal oscillations would be 24 h. What is the minimum depth of the bottom reservoir necessary to prevent flooding if these predictions are accurate?

12.10 Use MATLAB'S bode and nyquist commands to plot both Bode diagrams and polar plots for the following transfer functions:

(a) $T(s) = \dfrac{2}{s}$,

(b) $T(s) = \dfrac{1}{s(5s + 1)}$,

(c) $T(s) = \dfrac{10s}{(2s + 1)}$,

(d) $T(s) = \dfrac{6}{(5s + 1)^2}$,

(e) $T(s) = \dfrac{2(5s + 1)}{s(0.2s + 1)}$,

(f) $T(s) = \dfrac{10s(0.1s + 1)}{(s + 1)^2}$,

12.11 Use MATLAB to solve for the frequency response of the system in Fig. P12.5 for the following frequencies: 0.5, 1.0, 1.5, 2.0, 3.0, and 5.0 rad/s.

12.12 For the system described in Problem 12.7, consider the following change for the controller transfer functions:

$$T_c(s) \frac{K(s + 1)}{(2s + 1)}$$

where K is a design parameter. Find the value of K that satisfies the following condition: The frequency at which the magnitude of the response falls below unity (1.0, or 0 dB) is as high as possible but less than the frequency at which the phase angle becomes more negative than $-180°$.

12.13 Consider the linearized model of the field-controlled dc motor considered in Problem 10.4 and shown in Fig. P10.4. The input field voltage is subject to low-frequency sinusoidal noise, $\hat{e}_f = 10 \sin(\omega t)$. Determine the sensitivity of the output Ω_1 to the noise as follows:

(a) Derive the transfer function of the linearized model using $\hat{e}_f$ as input and Ω_1 as the output.

(b) Compute the magnitude of the transfer function for the following frequencies: 0.1, 0.3, and 1.5 rad/s. For which noise frequency is the system most sensitive?

(c) Find the poles of the transfer function. How many peaks does the plot of the magnitude of the transfer function have? At what frequencies?

(d) Use MATLAB to obtain plots of the magnitude of the transfer function to verify your responses in part (c).

12.14 Consider the frequency response of the open-loop system shown in Fig. P12.14.

(a) Determine the system response to the following inputs:
   (1) $u(t) = 5.0$,
   (2) $u(t) = \sin(0.1t)$,
   (3) $u(t) = \sin(1.0t)$,
   (4) $u(t) = \sin(10t)$,
   (5) $u(t) = \sin(10000t)$,

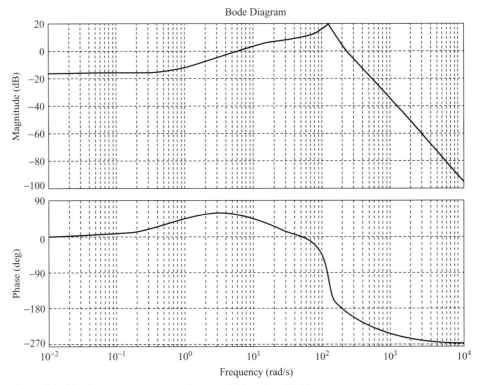

Figure P12.14. Frequency response of system in Problem 12.14.

(b) If the input $u(t)$ is a triangle wave between 0 and 1 with a period of 0.0021 s, sketch the response of the system as best you can.

(c) Write down everything you can deduce about the system by observing the Bode plot.

Closed-Loop Systems and System Stability

LEARNING OBJECTIVES FOR THIS CHAPTER

13–1 To understand the concept of system stability and its implication for dynamic feedback systems.

13–2 To apply algebraic stability criteria for linear dynamic models.

13–3 To apply frequency-domain stability criteria for linear dynamic models.

13–4 To assess relative stability of linear dynamic systems by using phase margin and gain margin.

13–5 To plot and interpret the root locus of linear dynamic systems.

13.1 INTRODUCTION

Up to this point, the modeling and analysis in this text have dealt mainly with systems resulting from the straightforward interconnection of A-type, T-type, and D-type elements together with energy-converting transducers. The graphical representation of these system models by use of simulation block diagrams has revealed the widespread natural occurrence of closed loops containing one or more integrators, each loop of which involves feedback to a summing point. The techniques of analysis used so far with these passive systems have led to descriptions of their dynamic characteristics by means of sets of state-variable equations and/or input–output differential equations and transfer functions. These systems are considered passive because no attempts have been made intentionally to close additional loops with signal-amplifying or signal-modifying devices. They are simply collections of naturally occurring phenomena that, to be naturally occurring, must be inherently stable in order to survive.

A system is considered stable if the following circumstances apply:

(a) The system remains in equilibrium at a steady normal operating point when left undisturbed – in other words, after all transients resulting from previous inputs have died out.

(b) It responds with finite variations of all of its state variables when forced by a finite disturbance.

(c) It regains equilibrium at a steady operating point after the transient response to a step or pulse input has decayed to zero; or its state variables vary cyclically about steady operating-point values when the input varies cyclically about a steady normal operating-point value (as, for instance, a system responding to a sinusoidal input variation about a constant value).

The intentional use of feedback in industrial systems seems to have started in the latter part of the 18th century, when James Watt devised a flyweight governor to sense the speed of a steam engine and use it as a negative feedback to control the flow of steam to the engine, thereby controlling the speed of the engine. It was almost 100 years later that James Clerk Maxwell modeled and analyzed such a system in a celebrated paper presented to the Royal Society in 1868. Since that time, inventors, engineers, and scientists have increasingly used sensors and negative feedback[1] to improve system performance, to improve control of the quality of products, and to improve rates of production in industry. And in many instances, the hardware preceded the modeling and analysis. Thus the use of ingenuity in design and the use of physical reasoning have been very important factors in the development of the field of automatic control. These factors, when combined with the techniques of modeling and analysis, have made it possible to propose new concepts and to evaluate them before trying to build and test them, thereby saving time, effort, and money.

The use of negative feedback to control a passive system results in an active system that has dynamic characteristics for which maintaining stability may be a serious problem. The improved system traits resulting from intentional use of negative-feedback control, such as faster response and decreased sensitivity to loading effects, will be discussed in Chap. 14.

The following example of a very commonly encountered feedback control system, a biomechanical system consisting of an automobile and driver, may be helpful in gaining an understanding of the way in which negative feedback can affect system stability.

Consider the process of steering an automobile along a straight stretch of open road with strong gusts of wind blowing normal to the direction of the roadway. The driver observes the deviation of the heading of the automobile relative to the roadway as the wind gusts deflect the heading, and the driver then manipulates the steering wheel in such a way as to reduce the deviation, keeping the automobile on the road. A simplified block-diagram model of this system is shown in Fig. 13.1.

The desired input signal represents the direction of the roadway, and the output signal represents the heading of the automobile as it proceeds along the highway. Both signals are processed as the human visual system and brain seem to process them – that is, they are compared with the effects of the heading signal acting negatively with respect to the road direction signal. This perceived deviation is acted on by the brain and nervous system so as to generate a motion by the hands on the steering wheel that causes the automobile to change its heading, thereby reducing the deviation of the automobile heading from the road direction.

Barring extreme weather conditions and excessively high speed of the vehicle, this negative feedback by a human controller works well in the hands of moderately skilled but calm drivers. However, if a driver becomes nervous and overreacts to the wind gusts, the deviations of the vehicle heading from the road direction may

---

[1] Positive feedback is seldom used because in most cases its use leads to degradation of system performance.

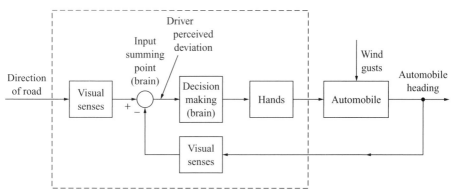

Figure 13.1. Simplified block diagram of the driver–automobile feedback system.

become excessive, possibly causing the vehicle to leave the road or, even worse, to collide with another vehicle.

This example illustrates the common everyday use of negative feedback based on a simplified model of the sensory and control capabilities of a human operator and illustrates how inappropriate feedback can lead to system instability. (A more complex model of the human operator has shown that an experienced driver does more than simply observe the heading of the vehicle relative to the road and that the control capability of such a driver is tempered by experience and the ability to make rapid adjustments to unusual situations – for instance, to sense the incipient effects of wind gusts before the vehicle responds to them.)

There is a persistent problem with stability that is likely to arise when feedback is intentionally used to produce closed-loop or automatic control, a problem that seldom occurs with passive, open-loop systems. Designing a system that will remain stable while achieving desired speed of response and reduction of errors constitutes a problem of major importance in the development of automatic control systems.

This chapter presents basic methods for the analysis of stability of linear dynamic systems. Analysis of the stability of nonlinear systems is beyond the scope of this text. In Section 13.2, basic definitions related to stability analysis are introduced and general conditions for stability of linear systems are formulated. The next two sections are devoted to analytical methods for determining the stability of dynamic systems having zero inputs. In Section 13.3, algebraic stability criteria are presented, including the Hurwitz and Routh methods. The algebraic stability criteria are very simple and easy to use, but their applicability is limited to systems whose mathematical models are known. The Nyquist criterion described in Section 13.4 can be applied to closed-loop control systems whose open-loop frequency characteristics are known, either in an analytical form or in the form of sets of experimentally acquired frequency-response data. In Section 13.5, gain and phase margins for stability are introduced. A brief outline of the root-locus method, a very powerful tool in analysis and design of feedback systems, is given in Section 13.6.

Finally, Section 13.7 presents the MATLAB tools that automate both the computation of relative stability characteristics and the creation of root-locus plots.

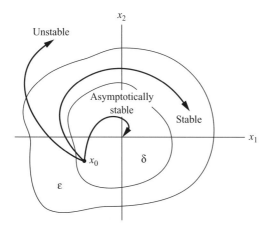

Figure 13.2. Illustration of stability in the sense of Lyapunov.

13.2   BASIC DEFINITIONS AND TERMINOLOGY

A linear system is commonly considered to be stable if its response meets the conditions listed in Section 13.1. Although this definition is intuitively correct, a much more precise definition of stability was proposed by Lyapunov. To explain Lyapunov's definition of stability, consider a dynamic system described by the state equation

$$\dot{\mathbf{q}} = \mathbb{A}\mathbf{q} + \mathbb{B}\mathbf{u}, \tag{13.1}$$

where $\mathbf{q}$ is a state vector, $\mathbb{A}$ is a system matrix, $\mathbf{u}$ is an input vector, and $\mathbb{B}$ is an input matrix. The system described by this vector equation is stable in the sense of Lyapunov under the following conditions: For a given initial state $\mathbf{q}_0$ inside a hypersphere δ, there exists another hypersphere ϵ such that, if the input vector is zero, $\mathbf{u} = 0$, the system will always remain in the hypersphere ϵ. Figure 13.2 illustrates the Lyapunov definition of stability for a two-dimensional system. In this case, a dynamic system is considered stable if, for initial conditions inside the area δ, there exists an area ϵ such that the system will never move outside this area as long as the input signal is zero.

A term stronger than stability is *asymptotic stability*. A dynamic system described by Eq. (13.1) is asymptotically stable in the sense of Lyapunov if, for any initial state $\mathbf{q}_0$ inside δ and with zero input, $\mathbf{u} = 0$, the system state vector will approach zero as time approaches infinity:

$$\lim_{t \to \infty} \|\mathbf{q}\| = 0 \quad \text{for} \quad \mathbf{u} = 0, \tag{13.2}$$

where $\|\mathbf{q}\|$ denotes the norm of the state vector $\mathbf{q}$.[2] If a system is assumed to be linear, its dynamics can be represented by a linear nth-order differential equation:

$$a_n \frac{d^n y}{dt^n} + \cdots + a_1 \frac{dy}{dt} + a_0 y = b_m \frac{d^m u}{dt^m} + \cdots + b_1 \frac{du}{dt} + b_0 u, \tag{13.3}$$

---

[2] The norm of the state vector $\mathbf{q}$ in Euclidean space is defined by

$$\|\mathbf{q}\| = \sqrt{\sum_{i=1}^{n} q_i^2},$$

which is the length of $\mathbf{q}$.

where $m \leq n$. A linear system described by Eq. (13.3) is asymptotically stable if, whenever $u(t) = 0$, the limit of $y(t)$ for time approaching infinity is zero for any initial conditions, that is,

$$\lim_{t \to \infty} y(t) = 0 \quad \text{if} \quad u(t) = 0. \tag{13.4}$$

If $u(t)$ is equal to zero, the system input–output equation becomes homogeneous:

$$a_n \frac{d^n y}{dt^n} + \cdots + a_1 \frac{dy}{dt} + a_0 y = 0. \tag{13.5}$$

Assuming for simplicity that all roots of the characteristic equation, $r_1, r_2, \ldots, r_n$, are distinct, which does not limit the generality of further considerations, the solution of Eq. (13.5) is

$$y(t) = \sum_{i=1}^{n} K_i e^{r_i t}. \tag{13.6}$$

The output signal given by Eq. (13.6) will satisfy the condition given in Eq. (13.4) if and only if all roots of the characteristic equation, $r_1, r_2 \ldots r_n$, are real and negative or are complex and have negative real parts. This observation leads to a theorem defining conditions for stability of linear systems. According to this theorem, a linear system described by Eq. (13.3) is asymptotically stable if and only if all roots of the characteristic equation lie strictly in the left half of the complex plane, that is, if

$$\text{Re}[r_i] < 0 \quad \text{for} \quad i = 1, 2, \ldots, n. \tag{13.7}$$

It can thus be concluded that stability of linear systems depends on only the location of the characteristic roots in the complex plane. Note also that stability of linear systems does not depend on the input signals. In other words, if a linear system is stable, it will remain stable regardless of the type and magnitude of input signals applied to the system. It should be stressed that this is true only for linear systems. Stability of nonlinear systems, which is beyond the scope of this book, is affected not only by the specific system properties but also by the input signals.

13.3 ALGEBRAIC STABILITY CRITERIA

The transfer function of the linear system described by Eq. (13.3) is a ratio of polynomials:

$$\mathbf{T}(s) = \frac{\mathbf{B}(s)}{\mathbf{A}(s)}, \tag{13.8}$$

where the polynomials $\mathbf{A}(s)$ and $\mathbf{B}(s)$ are

$$\mathbf{A}(s) = a_n s^n + \cdots + a_1 s + a_0, \tag{13.9}$$

$$\mathbf{B}(s) = b_m s^m + \cdots + b_1 s + b_0. \tag{13.10}$$

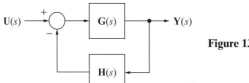

Figure 13.3. A linear feedback system.

The roots of the system characteristic equation are, of course, the same as the roots of the equation $\mathbf{A}(s) = 0$, which are the poles of the system transfer function. The transfer function of the feedback system shown in Fig. 13.3 is

$$\mathbf{T}_{CL}(s) = \frac{\mathbf{G}(s)}{1 + \mathbf{G}(s)\mathbf{H}(s)}, \qquad (13.11)$$

where $\mathbf{G}(s) = \mathbf{N}_g(s)/\mathbf{D}_g(s)$ and $\mathbf{H}(s) = \mathbf{N}_h(s)/\mathbf{D}_h(s)$. The system characteristic equation is thus

$$1 + \mathbf{G}(s)\mathbf{H}(s) = 0 \quad \text{or} \quad \mathbf{N}_g(s)\mathbf{N}_h(s) + \mathbf{D}_g(s)\mathbf{D}_h(s) = 0. \qquad (13.12)$$

All roots of Eq. (13.12) must be located in the left half of the complex plane for asymptotic stability of the system.

An obvious way to determine whether the conditions of asymptotic stability are met would be to find the roots of Eq. (13.12) and check if all have negative real parts. This direct procedure may be cumbersome, and it usually requires a computer if a high-order system model is involved. Several stability criteria have been developed that allow for verification of the stability conditions for a linear system without calculating the roots of the characteristic equation. In this section, two such methods, the Hurwitz and the Routh stability criteria, are introduced. Both methods are classified as algebraic stability criteria because they both require some algebraic operations to be performed on the coefficients of the system characteristic equation. These algebraic methods are widely used because of their simplicity, although their applicability is limited to problems in which an analytical form of the system characteristic equation is known.

In both the Hurwitz and the Routh methods, the same necessary condition for stability is formulated. Consider an nth-order characteristic equation,

$$a_n s^n + \cdots + a_1 s + a_0 = 0. \qquad (13.13)$$

A necessary (but not sufficient) condition for all the roots of Eq. (13.13) to have negative real parts is that all the coefficients a_i, $i = 0, 1, 2, \ldots, n$, have the same sign and that none of the coefficients vanishes. This necessary condition for stability can therefore be verified by inspection of the characteristic equation, and this inspection serves as an initial means of screening for instability (or as a means of detecting that an algebraic sign error has been made in the analysis of an inherently stable passive system).

13.3.1 Hurwitz Criterion

The Hurwitz necessary and sufficient set of conditions for stability is based on the set of determinants, which is formed as follows:

$$D_1 = |a_{n-1}|,$$

$$D_2 = \begin{vmatrix} a_{n-1} & a_{n-3} \\ a_n & a_{n-2} \end{vmatrix},$$

$$D_3 = \begin{vmatrix} a_{n-1} & a_{n-3} & a_{n-5} \\ a_n & a_{n-2} & a_{n-4} \\ 0 & a_{n-1} & a_{n-3} \end{vmatrix},$$

$$D_4 = \begin{vmatrix} a_{n-1} & a_{n-3} & a_{n-5} & a_{n-7} \\ a_n & a_{n-2} & a_{n-4} & a_{n-6} \\ 0 & a_{n-1} & a_{n-3} & a_{n-5} \\ 0 & a_n & a_{n-2} & a_{n-4} \end{vmatrix}.$$

The coefficients not present in the characteristic equation are replaced with zeros in the Hurwitz determinants.[3]

The necessary and sufficient set of conditions for all the roots of Eq. (13.13) to have negative real parts is that all the Hurwitz determinants D_j, $j = 1, 2, \ldots, n$, must be positive.

It can be shown that if the necessary condition for stability is satisfied – that is, if all the coefficients a_i $(i = 0, 1, \ldots, n)$ have the same sign and none of them vanishes – the first and the nth Hurwitz determinants, D_1, D_n, are positive. Thus, if the necessary condition is satisfied, it is sufficient to check if D_2, D_3, $\ldots$, D_{n-1} are positive.

EXAMPLE 13.1

Determine the stability of systems that have the following characteristic equations:

(a) $7s^4 + 5s^3 - 12s^2 + 6s - 1 = 0$. The coefficients in this characteristic equation have different signs and thus the set of necessary conditions for stability is not met. The system is unstable, and thus there is no need to look at the Hurwitz determinants.

(b) $s^3 + 6s^2 + 11s + 6 = 0$. The necessary condition is satisfied. Next, the Hurwitz determinants must be examined (a more stringent test). Because the necessary condition is satisfied, only D_2 needs to be checked in this case. It is

$$D_2 = \begin{vmatrix} 6 & 6 \\ 1 & 11 \end{vmatrix} = 66 - 6 = 60 > 0.$$

Thus all determinants are positive and the system is stable.

(c) $2s^4 + s^3 + 3s^2 + 5s + 10 = 0$. The necessary condition is satisfied. The determinants D_2 and D_3 are

$$D_2 = -7, \quad D_3 = -45.$$

The set of necessary and sufficient conditions is not satisfied because both D_2 and D_3 are negative; the system is unstable.

---

[3] B. C. Kuo, *Automatic Control Systems*, 7th ed. (Prentice-Hall, Englewood Cliffs, NJ, 1995), pp. 339–41.

EXAMPLE 13.2

Determine the range of values for the gain K for which the closed-loop system shown in Fig. 13.4 is stable.

SOLUTION

First, find the system transfer function:

$$\mathbf{T}_{CL}(s) = \frac{\dfrac{K(s+40)}{s(s+10)}}{1 + \dfrac{K(s+40)}{s(s+10)(s+20)}}.$$

Hence the system characteristic equation is

$$s(s+10)(s+20) + K(s+40) = 0,$$

which, after multiplying, becomes

$$s^3 + 30s^2 + (200+K)s + 40K = 0.$$

The necessary conditions for stability are

$$200 + K > 0, \quad 40K > 0.$$

Both inequalities are satisfied for $K > 0$, which satisfies the necessary conditions for system stability. But this is not enough. The Hurwitz necessary and sufficient conditions for stability of this third-order system are $D_1 > 0$ and $D_2 > 0$. The first inequality is satisfied because $D_1 = a_2 = 30$. The second Hurwitz determinant D_2 is

$$D_2 = \begin{vmatrix} a_2 & a_0 \\ a_3 & a_1 \end{vmatrix} = \begin{vmatrix} 30 & 40K \\ 1 & (200+K) \end{vmatrix} = (6000 - 10K).$$

Combining the necessary and sufficient set of conditions yields the range of values of K for which the system is stable:

$$0 < K < 600.$$

13.3.2 Routh Criterion

The Hurwitz criterion becomes very laborious for higher-order systems for which large determinants must be evaluated. The Routh criterion offers an alternative method for checking the sufficient conditions of stability.

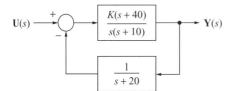

Figure 13.4. Closed-loop system considered in Example 13.2.

The Routh method involves a different set of necessary and sufficient stability conditions based on an array of the form

$$
\begin{array}{c|cccccc}
s^n & a_n & a_{n-2} & a_{n-4} & a_{n-6} & \cdots \\
s^{n-1} & a_{n-1} & a_{n-3} & a_{n-5} & a_{n-7} & \cdots \\
s^{n-2} & b_1 & b_2 & b_3 & b_4 & \cdots \\
s^{n-3} & c_1 & c_2 & c_3 & c_4 & \cdots \\
\vdots & & & & & \\
s^1 & \cdots & & & & \\
s^0 & \cdots & & & &
\end{array}
$$

where the a's are the coefficients in the characteristic equation [Eq. (13.13)] and the other coefficients, $b, c, \ldots$, are calculated as follows:

$$b_1 = (a_{n-1}a_{n-2} - a_n a_{n-3})/a_{n-1}$$
$$b_2 = (a_{n-1}a_{n-4} - a_n a_{n-5})/a_{n-1}$$
$$\vdots$$
$$c_1 = (b_1 a_{n-3} - a_{n-1}b_2)/b_1$$
$$c_2 = (b_1 a_{n-5} - a_{n-1}b_3)/b_1$$
$$\vdots$$

The necessary and sufficient set of conditions for stability is that all elements in the first column of the Routh array, $a_n, a_{n-1}, b_1, c_1, \ldots$, must have the same sign. If this set of conditions is not satisfied, the system is unstable and the number of sign changes in the first column of the Routh array is equal to the number of roots of the characteristic equation located in the right half of the complex plane. If an element in the first column is zero, the system is unstable, but additional coefficients for column 1 may be found by if a small number is substituted for the zero at this location and proceeding. For a more detailed discussion relating to the occurrence of zero coefficients, see Kuo.[4]

EXAMPLE 13.3

Determine the stability of a system that has the characteristic equation

$$s^3 + 2s^2 + 4s + 9 = 0.$$

SOLUTION

The Routh array is

$$
\begin{array}{c|cc}
s^3 & 1 & 4 \\
s^2 & 2 & 9 \\
s^1 & -0.5 & \\
s^0 & 9 &
\end{array}
$$

---

[4] B. C. Kuo, *op cit.*, pp. 340–1.

The system is unstable and there are two roots in the right half of the complex plane, as indicated by two sign changes in the first column of the Routh array, from 2 to -0.5 and from -0.5 to 9.

EXAMPLE 13.4

Determine the range of values of K for which the system having the following characteristic equation is stable:

$$s^3 + s^2 + (5 + K)s + 3K = 0.$$

SOLUTION

The Routh array is

$$
\begin{array}{c|cc}
s^3 & 1 & 5 + K \\
s^2 & 1 & 3K \\
s^1 & 5 - 2K & \\
s^0 & 3K &
\end{array}
$$

There will be no sign changes in the first column of the Routh array if

$$5 - 2K > 0 \quad \text{and} \quad 3K > 0,$$

which give the following range of K for stability:

$$0 < K < 2.5.$$

13.4 NYQUIST STABILITY CRITERION

The method proposed by Nyquist[5,6] many years ago allows for the determination of the stability of a closed-loop system on the basis of the frequency response of the open-loop system. The Nyquist criterion is particularly attractive because it does not require knowledge of the mathematical model of the system. Thus it can be applied also in all those cases in which a system transfer function is not available in an analytical form but the system open-loop frequency response may be obtained experimentally.

To introduce the Nyquist criterion, the polar plots that, so far, have been drawn for positive frequency only, $0 < \omega < \infty$, must be extended to include negative frequencies, $-\infty < \omega < +\infty$. It can be shown that the polar plot of $\mathbf{T}(j\omega)$ for negative frequencies is symmetrical about the real axis with the polar plot of $\mathbf{T}(j\omega)$ for positive frequencies. An example of a polar plot for ω varying from $-\infty$ to $+\infty$ is shown in Fig. 13.5. Note that, to close the contour, a curve is drawn from the point corresponding to $\omega = 0^-$ to the point $\omega = 0^+$ in a clockwise direction, retaining the direction of increasing frequency.

---

[5] H. Nyquist, "Regeneration theory," *Bell Syst. Tech. J.* **11**, 126–47 (1932).
[6] D. M. Auslander, Y. Takahashi, and M. J. Rabins, *Introducing Systems and Control* (McGraw-Hill, New York, 1974), pp. 371–5.

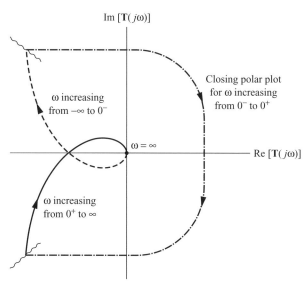

Figure 13.5. Example of polar plot of $\mathbf{T}(j\omega)$ for $-\infty < \omega < \infty$.

Consider the same closed-loop system shown in Fig. 13.3. The Nyquist criterion can be stated as follows: If an open-loop transfer function $\mathbf{G}(s)\mathbf{H}(s)$ has k poles in the right half of the complex plane, then for stability of the closed-loop system the polar plot of the open-loop transfer function must encircle the point $(-1, j0)$ k times in the clockwise direction. An important implication of the Nyquist criterion is that if the open-loop system is stable and thus does not have any poles in the right half of the complex plane, $k = 0$, the closed-loop system is stable if $\mathbf{G}(j\omega)\mathbf{H}(j\omega)$ does not encircle the point $(-1, j0)$.

If an open-loop system is stable, one can measure the system response to a sinusoidal input over a wide enough range of frequency and then plot the experimentally obtained polar plot to determine whether the curve encircles the critical point $(-1, j0)$: This experimental procedure can be applied only if the open-loop system is stable; otherwise, no measurements can be taken from the system.

As one gains experience, it becomes possible for one to use the Nyquist criterion by inspecting the polar plot of the open-loop transfer function for positive frequency only. The general rule is that, for a system to be stable, the critical point $(-1, j0)$ must be to the left of an observer following the polar plot from $\omega = 0^+$ to $\omega = +\infty$.

Figure 13.6 shows polar plots for the system having the open-loop transfer function $\mathbf{T}_{\text{OL}} = \mathbf{G}(s)\mathbf{H}(s) = K/(s^3 + s^2 + s + 1)$ for three different values of gain K. According to the Nyquist criterion, the closed-loop system is asymptotically stable for $K = K_1$; it is unstable for $K = K_3$. For $K = K_2$, the closed-loop system is said to be marginally stable.

EXAMPLE 13.5

Determine the stability of the closed-loop system that has the open-loop transfer function $\mathbf{T}_{\text{OL}}(s) = K/s(s + 1)(2s + 1)$.

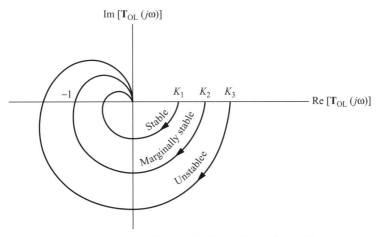

Figure 13.6. Polar plots for stable, marginally stable, and unstable systems.

SOLUTION

The sinusoidal transfer function of the open-loop system, $\mathbf{T}_{OL}(j\omega)$, is

$$\mathbf{T}_{OL}(j\omega) = \frac{K}{j\omega(j\omega + 1)(2j\omega + 1)}. \tag{13.14}$$

The real and imaginary parts of $\mathbf{T}_{OL}(j\omega)$ are

$$\mathrm{Re}[\mathbf{T}_{OL}(j\omega)] = -\frac{3K}{9\omega^2 + (2\omega^2 - 1)^2}, \tag{13.15}$$

$$\mathrm{Im}[\mathbf{T}_{OL}(j\omega)] = -\frac{K(2\omega^2 - 1)}{9\omega^3 + \omega(2\omega^2 - 1)^2}. \tag{13.16}$$

According to the Nyquist criterion, the closed-loop system will be stable if

$$\left|\mathbf{T}_{OL}(j\omega_p)\right| < 1$$

where ω_p is the frequency at the point of intersection of the polar plot with the negative real axis. The imaginary part of $\mathbf{T}_{OL}(j\omega)$ at this point is zero, and so the value of ω_p can be found by the solution of

$$\mathrm{Im}[\mathbf{T}_{OL}(j\omega)] = 0. \tag{13.17}$$

Substituting into Eq. (13.17) the expression for $\mathrm{Im}[\mathbf{T}_{OL}(j\omega)]$, Eq. (13.16), yields

$$2\omega_p^2 - 1 = 0,$$

and hence $\omega_p = 1/\sqrt{2}\,\mathrm{rad/s}$. To ensure stability of the closed-loop system, the real part of $\mathbf{T}_{OL}(j\omega)$ must be less than 1 at $\omega = \omega_p$, i.e.,

$$\left|\mathrm{Re}[\mathbf{T}(j\omega_p)]\right| < 1. \tag{13.18}$$

Now substitute into Eq. (13.13) the expression for the real part, Eq. (13.15), to obtain

$$\frac{3K}{9\omega_p^2 + (2\omega_p^2 - 1)^2} < 1,$$

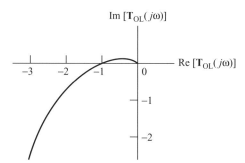

Figure 13.7. Polar plot for the system considered in Example 13.5.

and solve for K:

$$0 < K < 1.5.$$

The closed-loop system will be stable for K less than 1.5. The polar plot of the open-loop system transfer function for $K = 1.5$ is shown in Fig. 13.7.

13.5 QUANTITATIVE MEASURES OF STABILITY

As stated earlier in this chapter, an unstable control system is useless. And asymptotic stability constitutes one of the basic requirements in the design of control systems. However, the mere fact of meeting the stability conditions does not guarantee a satisfactory performance of the system. A system designed to barely meet the stability condition is very likely to become unstable because its actual parameter values may be slightly different from the values used in the design. In addition, very often the actual system parameters change in time as a result of aging, which again may push the system over the stability limit if that limit is too close. It is therefore necessary in the design of control systems to determine not only whether the system is stable but also how far the system is from instability. To answer this question, some quantitative measure of stability is needed. Two such measures, a gain margin and a phase margin, are derived from the Nyquist stability criterion.

If the $\mathbf{G}(s)\mathbf{H}(s)$ curve passes through the point $(-1, j0)$, which represents critical conditions for stability in the Nyquist criterion, the open-loop system gain is equal to 1 and the phase angle is $-180°$, which can be written mathematically as

$$|\mathbf{G}(j\omega)\mathbf{H}(j\omega)| = 1, \tag{13.19}$$

$$\angle\mathbf{G}(j\omega)\mathbf{H}(j\omega) = -180°. \tag{13.20}$$

If both the magnitude condition [Eq. (13.19)] and the phase condition [Eq. (13.20)] are satisfied, the closed-loop system is said to be marginally stable. The use of stability margins provides the means for indicating how far the system is from one of the critical conditions, Eq. (13.19) or (13.20), while the other condition is met.

The gain margin k_g is defined as

$$k_g = \frac{1}{\left|\mathrm{Re}[\mathbf{G}(j\omega_\mathrm{p})\mathbf{H}(j\omega_\mathrm{p})]\right|}, \tag{13.21}$$

where ω_p is the frequency for which the phase angle of the open-loop transfer function is $-180°$. When the gain margin has been expressed in decibels, it is calculated as

$$k_{gdB} = 20 \log \left\{ \frac{1}{|\text{Re}[\mathbf{G}(j\omega_p)\mathbf{H}(j\omega_p)]|} \right\}. \tag{13.22}$$

The gain margin can be thought of as the factor by which the open-loop gain can be increased while stability is maintained in the closed-loop system. For example, if the magnitude of an open-loop transfer function is 0.2 at ω_p, then the gain margin is 5.0 [see Eq. (13.21)]. Because marginal stability of the closed-loop system occurs when the magnitude of the open-loop transfer function at ω_p is 1, then the gain can be increased by as much as 5 (the new open-loop transfer function magnitude becomes 5×0.2) before the closed-loop system becomes unstable.

The phase margin γ is defined as

$$\gamma = 180° + \angle\mathbf{G}(j\omega_g)\mathbf{H}(j\omega_g), \tag{13.23}$$

where ω_g is the frequency for which the magnitude of the open-loop transfer function is unity:

$$|\mathbf{G}(j\omega_g)\mathbf{H}(j\omega_g)| = 1. \tag{13.24}$$

Similar to the gain margin, the phase margin supplies information about how close the closed-loop system is to unstable behavior, but, unlike the gain margin, it provides no direct information about how much the open-loop gain can be increased and still ensure closed-loop system stability. Figure 13.8 presents a graphical interpretation of both phase and gain margins on a polar plot of a typical open-loop transfer function.

In typical control system applications, the lowest acceptable gain margin is usually between 1.2 and 1.5 (or 1.6 and 3.5 dB). The typical range for the acceptable phase margin is from $30°$ to $45°$.

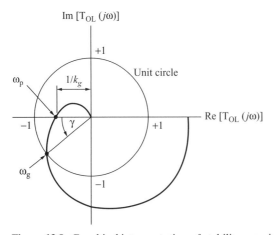

Figure 13.8. Graphical interpretation of stability margins.

EXAMPLE 13.6

Determine the open-loop gain necessary for the closed-loop system shown in Fig. 13.9 to be stable with the gain margin $k_g \geq 1.2$ and the phase margin $\gamma \geq 45°$.

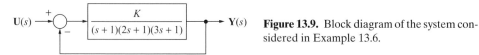

Figure 13.9. Block diagram of the system considered in Example 13.6.

SOLUTION

The open-loop system transfer function is

$$\mathbf{T}_{OL}(s) = \frac{K}{6s^3 + 11s^2 + 6s + 1}. \tag{13.25}$$

When $s = j\omega$, the sinusoidal transfer function is found to be

$$\mathbf{T}_{OL}(j\omega) = \frac{K}{(1 - 11\omega^2) + j6\omega(1 - \omega^2)}, \tag{13.26}$$

or, equally,

$$\mathbf{T}_{OL}(j\omega) = \frac{K[(1 - 11\omega^2) - j6\omega(1 - \omega^2)]}{[(1 - 11\omega^2)^2 + 36\omega^2(1 - \omega^2)^2]}. \tag{13.27}$$

The frequency ω_p for which the phase angle is $-180°$ is found by the solution of

$$\mathrm{Im}[\mathbf{T}_{OL}(j\omega)] = 0. \tag{13.28}$$

Substituting into Eq. (13.28) the expression for the imaginary part of the open-loop transfer function from Eq. (13.27) yields

$$\omega_p = 1 \, \mathrm{rad/s}. \tag{13.29}$$

The real part of the open-loop transfer function for this frequency is

$$\mathrm{Re}[\mathbf{T}_{OL}(j\omega_p)] = -\frac{K}{10}. \tag{13.30}$$

Substituting Eq. (13.30) into the condition for the gain margin, $k_g \geq 1.2$, yields

$$\frac{10}{K} \geq 1.2, \quad K \leq 8.3333.$$

Now, to satisfy the phase margin requirement, the phase angle of the open-loop transfer function for the frequency at which the magnitude is 1 should be

$$\angle \mathbf{T}_{OL}(j\omega_g) = -180° + 45° = -135°. \tag{13.31}$$

From the expression for the open-loop transfer function [Eq. (13.26)], the phase angle of $\mathbf{T}_{OL}(j\omega)$ for ω_g is

$$\angle \mathbf{T}_{OL}(j\omega_g) = \tan^{-1}(0) - \tan^{-1}\left[\frac{6\omega_g(1 - \omega_g^2)}{1 - 11\omega_g^2}\right]. \tag{13.32}$$

Comparing the right-hand sides of Eqs. (13.31) and (13.32) yields

$$-\tan^{-1}\left[\frac{6\omega_g(1-\omega_g^2)}{1-11\omega_g^2}\right] = -135°. \tag{13.33}$$

Taking the tangents of both sides of Eq. (13.33) yields a cubic equation for ω_g:

$$6\omega_g^3 + 11\omega_g^2 - 6\omega_g - 1 = 0, \tag{13.34}$$

which has the solution

$$\omega_g = 0.54776\,\text{rad/s}.$$

Now use Eq. (13.24) to obtain

$$\left|\mathbf{T}_{\text{OL}}(j\omega_g)\right| = 1. \tag{13.35}$$

The magnitude of $\mathbf{T}_{\text{OL}}(j\omega)$ for this system is

$$\left|\mathbf{T}_{\text{OL}}(j\omega_g)\right| = \frac{K}{\sqrt{(1-11\omega_g^2)^2 + 36\omega_g^2(1-\omega_g^2)^2}}. \tag{13.36}$$

Combine Eqs. (13.35) and (13.36) to obtain the solution for K that satisfies the phase condition:

$$K = 3.27.$$

This value is smaller than the value of K obtained from the gain condition: Therefore the open-loop gain must be equal to 3.27 (or less) to meet both the gain and the phase angle conditions.

13.6   ROOT-LOCUS METHOD

In Chap. 4 and also in Sections 13.2 and 13.3, a system transient performance and ultimately the system stability were shown to be governed by the locations of the roots of the system characteristic equation in the s plane. This fact has important implications for methods used in both analysis and design of dynamic systems. In the analysis of system dynamics, it is desirable to know the locations of the characteristic roots in order to be able to predict basic specifications of the transient performance. In the design process, the system parameters are selected to obtain the desired locations of the roots in the s plane. Knowledge of the locations of the characteristic roots and of the paths of their migration in the s plane as a result of variations in system parameters is therefore extremely important for a system engineer.

A very powerful and relatively simple technique, called the root-locus method, was developed by W. R. Evans[7,8] to assist in determining locations of roots of the characteristic equations of feedback systems. The graphs generated with the root-locus method show the migration paths of the characteristic roots in the s plane resulting from variations of selected system parameters. The parameter that is of particular interest in the analysis of feedback systems is an open-loop gain. When

---

[7] W. R. Evans, "Graphical analysis of control systems," *Trans. Am. Inst. Electr. Eng.* **67**, 547–51 (1948).
[8] W. R. Evans, *Control-System Dynamics* (McGraw-Hill, New York, 1954), pp. 96–121.

the open-loop gain increases, the system response may become faster, but if the increase is too great it may lead to very oscillatory or even unstable behavior.

Although the root-locus method can be used to determine the migration of roots caused by variation of any of the system parameters, it is most often used to examine the effect of varying the open-loop gain.

The significance of the root-locus method is illustrated in Example 13.7.

EXAMPLE 13.7

Consider the unity feedback system shown in Fig. 13.10. The system open-loop transfer function is

$$\mathbf{T}_{OL}(s) = \frac{K}{s(s+2)},$$ (13.37)

where $K \geq 0$ is the open-loop gain. The closed-loop transfer function is

$$\mathbf{T}_{CL}(s) = \frac{K}{(s^2 + 2s + K)}.$$ (13.38)

Hence the closed-loop system characteristic equation is

$$s^2 + 2s + K = 0,$$ (13.39)

and hence the roots are

$$s_1 = -1 - \sqrt{1 - K}, \quad s_2 = -1 + \sqrt{1 - K}$$ (13.40)

As the open-loop gain varies from zero to infinity, the roots change as follows:

- For $K = 0$: $s_1 = -2$ and $s_2 = 0$.
- For K increasing from 0 to 1: Both roots remain real, with s_1 moving from -2 to -1 and s_2 moving from 0 to -1.
- For K between 1 and infinity: The roots are complex conjugate, $s_1 = -1 - j\sqrt{K-1}$ and $s_2 = -1 + j\sqrt{K-1}$.

The migration of roots for $0 \leq K < \infty$ is shown in Fig. 13.11. Once the root locus is plotted, it is easy to determine the locations of the characteristic roots necessary for a desired system performance. For instance, if the desired damping ratio for the system considered in this example is $\zeta = 0.7$, the roots must be located at points A_1 and A_2, which correspond to

$$s_1 = -1 + j, \quad s_2 = -1 - j.$$

The characteristic equation for these roots takes the form

$$(s + 1 - j)(s + 1 + j) = 0$$

or

$$s^2 + 2s + 2 = 0.$$ (13.41)

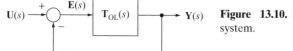

U(s)   +   E(s) $\mathbf{T}_{OL}(s)$ Y(s)

Figure 13.10. Block diagram of unity feedback system.

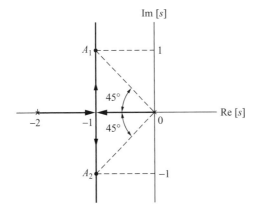

Figure 13.11. Root locus for $\mathbf{T}(s) = K/(s^2 + 2s + K)$.

When Eq. (13.41) is compared with the characteristic equation in terms of K [Eq. (13.39)], the required value of the open-loop gain is found to be

$$K = 2.$$

The direct procedure for plotting loci of characteristic roots used in Example 13.7 becomes impractical for higher-order systems. The method developed by Evans, consisting of several simple rules, greatly simplifies the process of plotting root loci. A block diagram of the system to which the method applies is shown in Fig. 13.10. The characteristic equation of this system is

$$1 + \mathbf{T}_{\mathrm{OL}}(s) = 0, \tag{13.42}$$

where $\mathbf{T}_{\mathrm{OL}}(s)$ is an open-loop transfer function that can be expressed as

$$\mathbf{T}_{\mathrm{OL}}(s) = \frac{K\mathbf{B}(s)}{\mathbf{A}(s)}, \tag{13.43}$$

where K is an open-loop gain and $\mathbf{A}(s)$ and $\mathbf{B}(s)$ are polynomials in s of nth and mth order, respectively. It should be noted that $m \le n$. The open-loop transfer function is, in general, a complex quantity, and therefore Eq. (13.42) is equivalent to two equations representing magnitude and phase angle conditions, i.e.,

$$|\mathbf{T}_{\mathrm{OL}}(s)| = 1 \tag{13.44}$$

and, for $K > 0$,

$$\angle\mathbf{T}_{\mathrm{OL}}(s) = \pm(2k+1)\pi \quad \text{for } k = 0, 1, 2, \ldots, \tag{13.45a}$$

or, for $K < 0$,

$$\angle\mathbf{T}_{\mathrm{OL}}(s) = \pm 2k\pi \quad \text{for } k = 0, 1, 2, \ldots. \tag{13.45b}$$

All roots of the characteristic equation, s_i, $i = 1, 2, \ldots, n$, must satisfy both the magnitude condition [Eq. (13.44)] and the phase angle condition [Eq. (13.45a,b)]. The following rules for plotting root loci are derived from these two conditions.

(1) The root locus is symmetric about the real axis of the s plane.

(2) The number of branches of the root loci is equal to the number of roots of the characteristic equation or, equally, to the order of $\mathbf{A}(s)$.

(3) The loci start at open-loop poles for $K = 0$ and terminate either at open-loop zeros (m branches) or at infinity along asymptotes for $K \to \infty$ ($n - m$ branches).

(4) The loci exist on section of the real axis between neighboring real poles and/or zeros, if the number of real poles and zeros to the right of this section is odd, for $K > 0$ (or, if the number is even, for $K < 0$).

(5) In accordance with rule (3), $n - m$ loci terminate at infinity along asymptotes. The angles between the asymptotes and the real axis are given, for $K > 0$, by

$$\alpha = \pm 180 \frac{2k+1}{n-m} \text{deg}, \quad k = 0, 1, 2, \ldots, \tag{13.46a}$$

and, for $K < 0$, by

$$\alpha = \pm 180 \frac{2k}{n-m} \text{deg}, \quad k = 0, 1, 2, \ldots. \tag{13.46b}$$

(6) All the asymptotes intersect the real axis at the same point. The abscissa of that point is

$$\sigma_a = \frac{\sum\limits^{n} \text{poles} - \sum\limits^{m} \text{zeros}}{n-m} \tag{13.47}$$

where the summations are over finite poles and zeros of $\mathbf{T}_{\text{OL}}(s)$.

(7) The loci depart the real axis at breakaway points and enter the real axis at break-in points. One can find the locations of these points by solving for "min–max" values of s by using[9]

$$\frac{d}{ds}\left[-\frac{\mathbf{A}(s)}{\mathbf{B}(s)}\right] = 0. \tag{13.48}$$

(8) The angles of departure from complex poles at $K = 0$ and the angles of arrival at complex zeros at $K \to \infty$ are found by application of the angle condition given by Eqs. (13.45a) and (13.45b) to a point infinitesimally close to the complex pole or zero in question.

(9) The points where the loci cross the imaginary axis in the s plane are determined by the solution of the system characteristic equation for $s = j\omega$ or by use of the Routh stability criterion.

Table 13.1 shows examples of root loci for common transfer functions for $K > 0$.

---

[9] R. C. Dorf and R. H. Bishop, *Modern Control Systems*, 7th ed. (Addison-Wesley, New York, 1995), pp. 325–28.

Table 13.1. Examples of root loci for common system transfer functions for $K > 0$

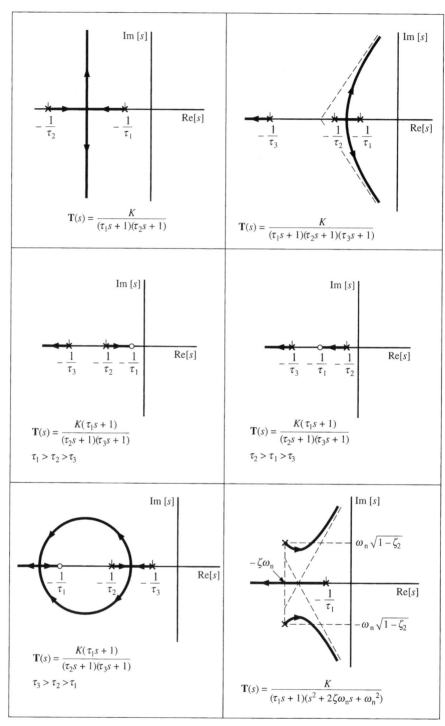

13.7 MATLAB TOOLS FOR SYSTEM STABILITY ANALYSIS

As with the frequency response techniques described in Chap. 12, MATLAB's Control System Toolbox provides powerful numerical tools to aid in the determination of relative stability of closed-loop systems. In particular, the function `margin` computes the phase and gain margin of a system in a unity feedback configuration. It also provides a graphical representation in the form of an annotated Bode diagram. The function `rlocus` automatically draws the loci of closed-loop poles given the transfer function of an open-loop system. The plot it generates is interactive and allows the user to easily and rapidly find the open-loop gain that corresponds to a given location of closed-loop poles.

13.7.1 Phase and Gain Margin Determination

Given a closed-loop feedback system of the form shown in Fig. 13.10, there are two different methods to call the routine that determines the gain and phase margins of the system:

```
>> [Gm, Pm, Wcg, Wcp] = margin(Tsys)
```

where `Tsys` is a system object (see description in Appendix 3), `Gm` is the gain margin (returned as a linear gain factor), `Pm` is the phase margin (in degrees), `Wcg` is the frequency at which the gain is unity, and `Wcp` is the frequency at which the phase is $-180°$.

Alternatively, invoking the function without any return variables,

```
>> margin(Tsys)
```

generates an annotated Bode diagram on which the margins and their crossover frequencies are indicated.

EXAMPLE 13.8

Given the open-loop transfer function in Example 13.6, find the gain margin and phase margin of the system for the case in which $K = 2.0$.

SOLUTION

First, create at transfer function object in the MATLAB environment:

```
>> tf138=tf([2], [6 11 6 1])
Transfer function:
      2
---------------------
6 s^3 + 11 s^2 + 6 s +1
```

Next, find the margins associated with this open-loop transfer function:

```
>> [gm pm wcg wcp]=margin(tf138)
```

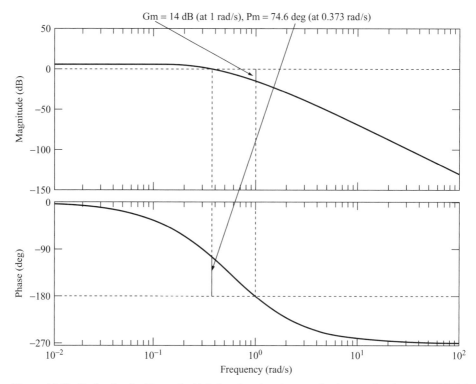

Figure 13.12. Bode plot for Example 13.8 showing the phase and gain margins (arrows added for emphasis).

which leads to the following results:

$$k_g = 5.0 \ (14 \text{ dB}),$$
$$\gamma = 74.6°,$$
$$\omega_{cg} = 1.0,$$
$$\omega_{cp} = 0.373.$$

To better interpret these results, consider the Bode plot that is generated by the alternative call of margin, Fig. 13.12.

13.7.2 Root-Locus Plots in MATLAB

Section 13.6 describes the method by which the approximate loci of closed-loop poles can be sketched given the location of the poles and zeros of the open-loop system. This method of sketching is very effective and should not be underestimated as a tool in developing the structure of control systems. However, fine-tuning the gain of the final closed-loop design often requires a level of accuracy that is difficult to achieve in hand-sketched plots. MATLAB's rlocus command is a fast and easy way to generate an accurate root-locus plot and to compute the gain corresponding to any set of poles along the loci. The format of the command is

```
>> rlocus(Tsys)
```

Example 13.9

Given the transfer function from Example 13.6, find the value of gain K that results in a pair of complex poles with equal real and imaginary parts (corresponding to a damping ratio of 0.707).

Again, start the process by entering the transfer function. In this case, enter the transfer function by using unity gain, knowing that the root-locus method will compute a gain factor that, when multiplied by the original open-loop gain, will result in a closed-loop system of desired properties:

```
>> tf139=tf([1],[6 11 6 1])
Transfer function:
              1
   _____

   6 s^3 + 11 s^2 + 6 s + 1
>> rlocus(tf139)
```

The resulting plot shows the three different loci (one for each of the system poles). However, this particular plot is interactive. If any of the loci is selected with a mouse click, a small black block appears indicating a pole location along with a callout showing the properties of that particular pole and the gain associated with that pole location. By dragging the block, any pole location along any of the loci can be found (and hence, the corresponding gain factor). Figure 13.13 shows a screen shot of the root-locus plot, with the pole indicator at the location desired by the problem statement. Therefore a closed-loop gain of 0.561 results in a pair of complex poles whose damping ratio is near 0.7071.

13.8 SYNOPSIS

Along with many beneficial effects that can be attributed to the presence of feedback in engineering systems, there are also some unwanted side effects of feedback on system performance. One such side effect is system instability, which is of utmost importance for design engineers. In this chapter, the conditions for stability of feedback systems were formulated. First, the stability conditions were stated in general, descriptive terms to enhance understanding of the problem. A more rigorous definition of stability for linear systems, developed by Lyapunov, was then presented. It was shown that for stability of linear systems it is necessary and sufficient that all roots of the closed-loop system characteristic equation lie strictly in the left half of the complex plane. Two algebraic stability criteria developed by Hurwitz and Routh were introduced. The Hurwitz criterion uses determinants built on coefficients of the characteristic equation to determine if there are any roots located in the right half of the complex plane. The Routh method leads to the same result and, in addition, gives the number of unstable roots. The Nyquist stability criterion was also presented. The practical importance of the Nyquist method lies in that it determines the stability of a closed-loop system based on the frequency response of the system components with the feedback loop open. Two quantitative measures of stability, gain margin and phase margin, derived from the Nyquist criterion, were introduced. Finally, a

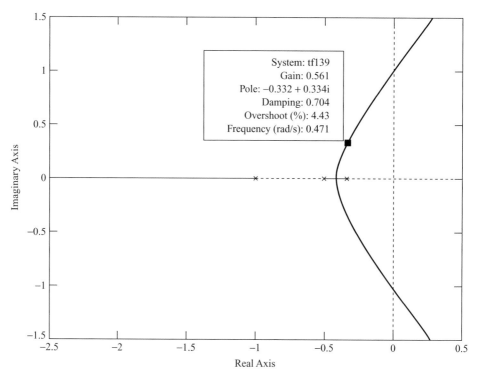

Figure 13.13. Root locus for the system considered in Example 13.9.

root-locus method for design of feedback systems, based on the migration paths of the system characteristic roots in the complex plane, was described.

PROBLEMS

13.1 A closed-loop transfer function of a dynamic system is

$$T_{CL}(s) = \frac{s + 10}{10s^4 + 10s^3 + 20s^2 + s + 1}.$$

Use the Hurwitz criterion to determine the stability of this system.

13.2 The transfer functions of the system represented by the block diagram shown in Fig. P13.2 are

$$G(s) = \frac{2s + 1}{3s^3 + 2s^2 + s + 1},$$

$$H(s) = 10.$$

(a) Determine the stability of the open-loop system.

(b) Determine the stability of the closed-loop system.

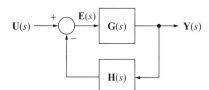

Figure P13.2. Block diagram of the feedback system considered in Problem 13.2.

13.3 Figure P13.3 shows a block diagram of a control system. The transfer functions of the controller $\mathbf{T}_C(s)$ and of the controlled process $\mathbf{T}_P(s)$ are

$$\mathbf{T}_C(s) = K\left(1 + \frac{1}{4s}\right),$$

$$\mathbf{T}_P(s) = \frac{5}{100s^2 + 20s + 1}.$$

Using the Hurwitz criterion, determine the stability conditions for the open-loop and closed-loop systems in terms of the controller gain K.

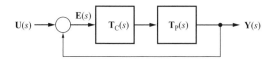

Figure P13.3. Block diagram of a control system.

13.4 The transfer functions of the system shown in Fig. P13.3 are

$$\mathbf{T}_C(s) = K,$$

$$\mathbf{T}_p(s) = \frac{2}{s(\tau s + 1)^2}.$$

Determine the stability condition for the closed-loop system in terms of the controller gain K and the process time constant τ. Show the area of the system stability in the (τ, K) coordinate system.

13.5 Examine the stability of the systems whose characteristic equations are subsequently listed. Determine the number of roots of the characteristic equation having positive real parts for unstable systems:

(a) $s^3 + 12s^2 + 41s + 42 = 0$,

(b) $400s^3 + 80s^2 + 44s + 10 = 0$,

(c) $s^4 + s^3 - 14s^2 + 26s - 20 = 0$.

13.6 The transfer functions of the system shown in Fig. P13.3 are

$$\mathbf{T}_C(s) = K,$$

$$\mathbf{T}_P(s) = \frac{1}{s(0.2s + 1)(0.08s + 1)}.$$

Sketch the polar plot of the open-loop system $\mathbf{T}_C\mathbf{T}_P$, and determine the stability condition for the closed-loop system in terms of the controller gain K by using the Nyquist criterion.

13.7 The block diagram for a control system has been developed as shown in Fig. P13.7. The system parameters are

$$k_C = 3.0 \, \text{v/v}, \quad k_P = 4.6 \, \text{m/V},$$
$$\tau_i = 3.5 \, \text{s}, \qquad \tau_P = 1.4 \, \text{s},$$
$$\tau_C = 0.1 \, \text{s}, \quad k_f = 1.0 \, \text{V/m}.$$

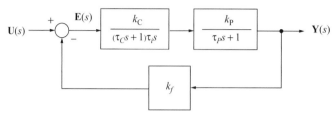

Figure P13.7. Block diagram of the system considered in Problem 13.7.

(a) Determine whether the system is stable or unstable.

(b) If the system is stable, find the stability gain and phase margins.

13.8 A simplified block diagram of a servomechanism used to control the angular position of an antenna dish, θ, is shown in Fig. P13.8. A potentiometer having gain k_p is used to produce a voltage signal proportional to the angular position of the antenna. This voltage signal is compared with voltage $u(t)$, which is proportional to the desired position of the antenna. The difference between the desired and actual positions is amplified to produce a driving signal for the electric dc motor.

(a) Determine the stability of the closed-loop system if the potentiometer gain is $k_p = 1.5 \, \text{V/rad}$.

(b) Find the potentiometer gain for which the closed-loop system will be stable with the gain margin $k_g = 1.2$.

(c) Find the potentiometer gain for which the closed-loop system will be stable with the phase margin $\gamma = 45°$.

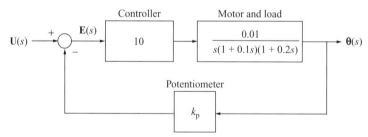

Figure P13.8. Simplified block diagram of the servomechanism considered in Problem 13.8.

13.9 Consider the feedback system represented by the block diagram shown in Fig. P13.3 with the following transfer functions:

$$\mathbf{T}_C(s) = K,$$
$$\mathbf{T}_P(s) = \frac{10}{(s+5)(s+0.2)}.$$

(a) Construct the root locus for this system.

(b) Determine the locations of the roots of the system characteristic equation required for 20 percent overshoot in the system step response. Find the value of K necessary for the roots to be at the desired locations. What will be the period of damped oscillations T_d in the system step response?

13.10 The transfer functions for the system represented by the block diagram shown in Fig. P13.3 are

$$\mathbf{T}_C(s) = K,$$

$$\mathbf{T}_P(s) = \frac{1}{(s+1)(s+2)(s+5)}.$$

(a) Construct the root locus for this system.

(b) Use the constructed root locus to determine the value of K for which the closed-loop system is marginally stable.

13.11 Subsection 13.3.1 describes the first four Hurwitz determinants used to determine stability of an nth-order system. Derive the general expression for the jth Hurwitz determinant of an nth-order system.

13.12 For the liquid-level control system in Problem 11.11, use the Routh criterion to find the limits for k_p to ensure the stability of the closed-loop system.

Control Systems

LEARNING OBJECTIVES FOR THIS CHAPTER

14–1 To characterize the steady-state behavior of a system through analysis of the system transfer function.

14–2 To evaluate the steady-state disturbance sensitivity of a system.

14–3 To understand the trade-off between transient and steady-state performance specifications in control system design.

14–4 To select gains of a proportional–integral–derivative controller based on open-loop system performance.

14–5 To design an appropriate cascade compensator based on steady-state and transient performance specifications.

14.1 INTRODUCTION

In Chap. 1, it was pointed out that negative feedback is present in nearly all existing engineering systems. In control systems, which are introduced in this chapter, negative feedback is included intentionally as a means of obtaining a specified performance of the system.

Control is an action undertaken to obtain a desired behavior of a system, and it can be applied in an open-loop or a closed-loop configuration. In an open-loop system, shown schematically in Fig. 14.1, a process is controlled in a certain prescribed manner regardless of the actual state of the process. A washing machine performing a predefined sequence of operations without any information and "with no concern" regarding the results of its operation is an example of an open-loop control system.

In a closed-loop control system, shown in Fig. 14.2, the controller produces a control signal based on the difference between the desired and the actual process output. The washing machine previously considered as an open-loop system would operate in a closed-loop mode if it were equipped with a measuring device capable of generating a signal related to the degree of cleanness of the laundry being washed.

Open-loop control systems are simpler and less expensive (at least by the cost of the measuring device necessary to produce the feedback signal); however, their performance can be satisfactory only in applications involving highly repeatable processes that have well-established characteristics and are not exposed to disturbances. The methods for analysis of open-loop control systems are the same as the methods

Input signal   Control signal   Output signal

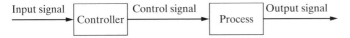

Controller   Process

Figure 14.1. Open-loop control system.

developed for analysis of dynamic systems in general, discussed in previous chapters. In this chapter, the basic characteristics of linear closed-loop systems are investigated.

In Chap. 13, the conditions under which a linear system is stable were derived. It was also shown how to design a linear closed-loop system for a specified gain and phase margin. Equally important to stability characteristics is the knowledge of the expected steady-state performance of the system. In Section 14.2, a control error (that is, the difference between the desired and the actual system output) at steady state is evaluated for various types of systems and various types of input signals. Another aspect of steady-state performance of control systems – namely, the sensitivity to disturbances – is discussed in Section 14.3. The problem of designing a control system that will meet specified steady-state and transient performance criteria is presented in Section 14.4. In Section 14.5, the most common algorithms used in industrial controllers are described. More specialized control devices, called compensators, are the subject of Section 14.6.

14.2   STEADY-STATE CONTROL ERROR

For most closed-loop control systems the primary goal is to produce an output signal that follows an input signal as closely as possible. The steady-state performance of a control system is therefore judged by the steady-state difference between the input and output signals – that is, the steady-state error. Any physical control system inherently suffers steady-state error in response to certain types of inputs as a result of inadequacies in the system components, such as insufficient gain, output limiting, static friction, amplifier drift, or aging. In general, the steady-state performance depends on not only the control system itself but also on the type of input signal applied. A system may have no steady-state error in response to a step input, yet the same system may exhibit nonzero steady-state error in response to a ramp input. This error can usually be reduced by an increase in the open-loop gain. Increasing the open-loop gain should, however, be done with care, because it usually has other effects on the system performance. Such effects may include an increase in the speed of response – that is, a reduction in the time to reach steady state, which is usually

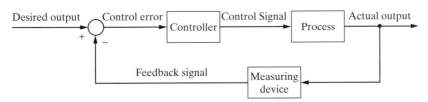

Desired output   Control error   Control Signal   Actual output

+   −   Controller   Process

Feedback signal   Measuring device

Figure 14.2. Closed-loop control system.

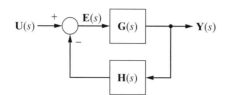

Figure 14.3. Block diagram of a closed-loop system.

welcome, and an increase in the system's tendency to oscillate, eventually reducing stability margins, which is unwelcome.

The steady-state error that occurs in control systems owing to their incapability of following particular types of inputs are now discussed.

Consider the closed-loop system shown in Fig. 14.3. The open-loop transfer function of this system, $\mathbf{G}(s)\mathbf{H}(s)$, is

$$\mathbf{G}(s)\,\mathbf{H}(s) = \frac{b_m s^m + \cdots + b_1 s + b_0}{s^r \left(a_n s^n + \cdots + a_1 s + a_0\right)}, \tag{14.1}$$

where m is the number of zeros, n is the number of nonzero poles, and r is the multiplicity of poles at the origin. The "error transfer function" of the system, relating the error signal $e(t)$ and the input signal $u(t)$ in the domain of complex variable s, is defined as

$$\mathbf{T}_E(s) = \frac{E(s)}{U(s)}. \tag{14.2}$$

For the system shown in Fig. 14.3, $\mathbf{T}_E(s)$ takes the form

$$\mathbf{T}_E(s) = \frac{U(s) - Y(s)\,H(s)}{U(s)}. \tag{14.3}$$

The closed-loop transfer function is

$$\frac{Y(s)}{U(s)} = \frac{G(s)}{1 + G(s)\,H(s)}. \tag{14.4}$$

Substituting Eq. (14.4) into Eq. (14.3) yields

$$\mathbf{T}_E(s) = \frac{1}{1 + G(s)\,H(s)}. \tag{14.5}$$

The error signal is thus given by

$$E(s) = \mathbf{T}_E(s)\,U(s) \tag{14.6}$$

or

$$E(s) = \frac{U(s)}{1 + G(s)\,H(s)}. \tag{14.7}$$

When the final-value theorem from Laplace transform theory (Appendix 2) is applied, the steady-state error e_{ss} can be calculated as

$$e_{ss} = \lim_{t \to \infty} e(t) = \lim_{s \to 0} s E(s) = \lim_{s \to 0} \frac{s U(s)}{1 + G(s)\,H(s)}. \tag{14.8}$$

This result will now be used in evaluating the steady-state error in response to step inputs and ramp inputs. The results obtained for these two types of input signals can be applied to more general cases involving linear systems in which actual inputs may be considered combinations of such inputs.

14.2.1 Unit Step Input, $u(t) = U_s(t)$

The unit step function in the domain of complex variable s is represented by

$$U(s) = \frac{1}{s}, \tag{14.9}$$

where $U(s) = \mathscr{L}[u(t)]$, where $u(t)$ is a unit step at $t = 0$. The steady-state error for a unit step is calculated with Eq. (14.8)[1]:

$$e_{ss} = \lim_{s \to 0} \frac{s(1/s)}{1 + G(s)H(s)} \tag{14.10}$$

or

$$e_{ss} = \frac{1}{1 + K_p}, \tag{14.11}$$

where K_p is the static-position error coefficient, defined as

$$K_p = \lim_{s \to 0} G(s)H(s) = G(0)H(0), \tag{14.12}$$

or, by use of Eq. (14.1),

$$K_p = \lim_{s \to 0} \frac{b_m s^m + \cdots + b_1 s + b_0}{s^r (a_n s^n + \cdots + a_1 s + a_0)}. \tag{14.13}$$

Now assume that the system has no poles at the origin, $r = 0$. Such systems are called type 0 systems. The static-position error coefficient for the type 0 system is

$$K_p = \frac{b_0}{a_0} = K \tag{14.14}$$

and the steady-state error is

$$e_{ss} = \frac{1}{1 + K}. \tag{14.15}$$

If there is at least one pole at the origin, $r \geq 1$, the static-position error coefficient K_p is infinite and the steady-state error is zero, $e_{ss} = 0$. Systems with $r = 1, 2, \ldots$, are called type 1, 2, ..., systems, respectively. Therefore it can be said that

$$e_{ss} = 0 \quad \text{for type 1 or higher systems.} \tag{14.16}$$

---

[1] Note that s cancels $1/s$ in Eq. (14.10). This leads to the notion of using a simplified final-value expression for a unit step input, i.e.,

$$e_{ss} = \lim_{s \to 0} T_E(s) = \lim_{s \to 0} \frac{1}{1 + G(s)H(s)}.$$

The use of this simplified final-value expression can be substantiated by use of

$$u(t) = \lim_{s \to 0} e^{st}$$

to represent a unit step and finding the system response as

$$y(t) = \lim_{s \to 0} Y(s)e^{st} = \lim_{s \to 0} T_E(s)e^{st}.$$

14.2.2 Unit Ramp Input, $u(t) = t$

The steady-state error for a unit ramp input[2] is given by

$$e_{ss} = \lim_{s \to 0} \frac{s \frac{1}{s^2}}{1 + G(s)H(s)} = \lim_{s \to 0} \frac{1}{sG(s)H(s)}. \tag{14.17}$$

The static-velocity error coefficient is defined as

$$K_v = \lim_{s \to 0} sG(s)H(s). \tag{14.18}$$

The steady-state error can thus be expressed as

$$e_{ss} = \frac{1}{K_v}. \tag{14.19}$$

For a type 0 system, $r = 0$, the static-velocity error coefficient is

$$K_v = \lim_{s \to 0} s \frac{b_m s^m + \cdots + b_1 s + b_0}{s^r (a_n s^n + \cdots + a_1 s + a_0)} = 0 \tag{14.20}$$

and the steady-state error[3] is

$$e_{ss} = \frac{1}{K_v} = \infty \text{ for type 0 systems.} \tag{14.21}$$

For a type 1 system, $r = 1$,

$$K_v = \lim_{s \to 0} s \frac{b_m s^m + \cdots + b_1 s + b_0}{s^r (a_n s^n + \cdots + a_1 s + a_0)} = \frac{b_0}{a_0} = K \tag{14.22}$$

and the steady-state error is

$$e_{ss} = \frac{1}{K_v} = \frac{1}{K} \quad \text{for type 1 systems.} \tag{14.23}$$

For a type 2 or higher system, $r \geq 2$,

$$K_v = \lim_{s \to 0} s \frac{b_m s^m + \cdots + b_1 s + b_0}{s^r (a_n s^n + \cdots + a_1 s + a_0)} = \infty \tag{14.24}$$

and the steady-state error is

$$e_{ss} = \frac{1}{K_v} = 0 \quad \text{for type 2 or higher systems.} \tag{14.25}$$

---

[2] Note again that a simplified final-value expression such as the one stated in the preceding footnote but for the case of a unit ramp input yields an identical result as obtained from Eq. (14.17), i.e.,

$$e_{ss} = \lim_{s \to 0} \left(\frac{1}{s}\right) T_E(s) = \lim_{s \to 0} \left(\frac{1}{s}\right) \frac{1}{1 + G(s)H(s)} = \lim_{s \to 0} \frac{1}{sG(s)H(s)}.$$

[3] Although the system does not reach a steady value for y, the time rate of change of y does reach a steady state if the system is otherwise stable.

Table 14.1. Values of the steady-state error, e_{ss}

Input signal	Type 0 system	Type 1 system	Type 2 system	...	Type N system
$U_s(t) = 1$	$1/(1 + K)$	0	0		0
$U_r(t) = t$	∞	$1/K$	0		0
t^2	∞	∞	$1/K$		0
$\vdots$	$\vdots$	$\vdots$	$\vdots$		$\vdots$
t^N	∞	∞	∞	...	$1/K$

Table 14.1 summarizes the steady-state errors for systems of types 0–N when they are subjected to inputs of order 0–N. When the order of the input signal (exponent to which the variable t is raised) is equal to the type of the system, the steady-state error is finite, as depicted by the values on the diagonal line in Table 14.1. When the input signal is of a higher order when compared with the system type number, the steady-state error approaches infinity. Finally, when the input is of an order that is smaller than the type number of the system, the steady-state error is zero.

14.3 STEADY-STATE DISTURBANCE SENSITIVITY

Another important aspect of a control system's performance is its sensitivity to disturbances. Disturbances are all those other inputs that are not directly controlled by feedback to a separate summing point in the system. Very often, disturbances are difficult to measure, and so their presence can be detected only by observations of variations in the process output signal that take place while the control input signal is unchanged. Some of the most common disturbances are variations in load force or torque in mechanical systems, variations in ambient temperature in thermal systems, and variations in load pressure or load flow rate in fluid systems. A control system should be capable of maintaining the process output signal at the desired level in the presence of disturbances. In fact, as pointed out at the beginning of this chapter, it is the presence of disturbances in the system environment that provides the rationale for many closed-loop control systems.

When a disturbance changes suddenly, the system output signal deviates at least temporarily from its desired value, even in the best-designed system. What is expected from a well-designed control system is that, after the transients die out, the output signal will return to its previous level. A system's ability to compensate for the steady-state effects of disturbances is determined quantitatively in terms of the steady-state disturbance sensitivity S_D, defined as the ratio of the change of the output Δy to the change of the disturbance Δv at steady state:

$$S_D = \frac{\Delta y_{ss}}{\Delta v_{ss}}. \tag{14.26}$$

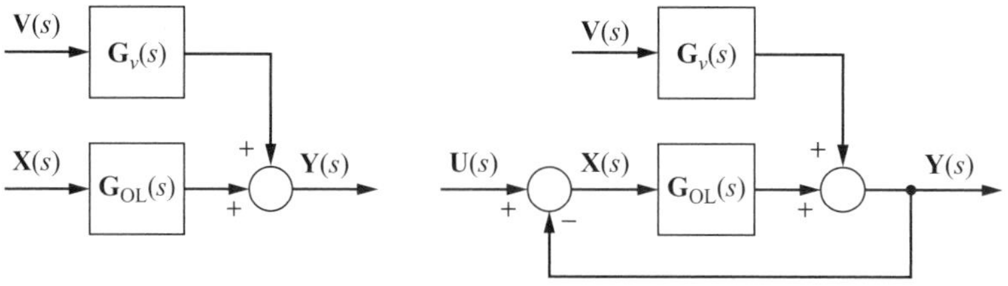

Figure 14.4. (a) Open-loop and (b) closed-loop systems subjected to disturbance $\mathbf{V}(s)$.

In the open-loop system shown in Fig. 14.4(a), the disturbance sensitivity can be calculated with the final-value theorem, assuming that $\mathbf{X}(s) = 0$:

$$S_{DO} = \frac{\lim\limits_{s \to 0} s \mathbf{Y}(s)}{\lim\limits_{s \to 0} s \mathbf{V}(s)}. \qquad (14.27)$$

Assuming that $v(t)$ is a step function, Eq. (14.27) leads to the simplified final-value expression:

$$S_{DO} = \frac{\lim\limits_{s \to 0} s \dfrac{1}{s} \mathbf{G}_v(s)}{\lim\limits_{s \to 0} s \dfrac{1}{s}} = \lim\limits_{s \to 0} \mathbf{G}_v(s) = \mathbf{G}_v(0) = K_D, \qquad (14.28)$$

where K_D is the static gain of $\mathbf{G}_v(s)$.

In the closed-loop system shown in Fig. 14.4(b), the output $\mathbf{Y}(s)$ for $\mathbf{U}(s) = 0$ is

$$\mathbf{Y}(s) = \frac{\mathbf{V}(s)\,\mathbf{G}_v(s)}{1 + \mathbf{G}_{OL}(s)}, \qquad (14.29)$$

and hence the sensitivity of the closed-loop system, S_{DC}, for a step disturbance is

$$S_{DC} = \frac{\mathbf{G}_v(0)}{1 + \mathbf{G}_{OL}(0)} = \frac{K_D}{1 + K}, \qquad (14.30)$$

where K is the steady-state gain of the open-loop transfer function $\mathbf{G}_{OL}(s)$. When Eqs. (14.28) and (14.30) are compared, it can be seen that the closed-loop system is less sensitive to disturbance, $S_{DC} < S_{DO}$, if the steady-state gain K is positive.

Example 14.1 illustrates the mathematical considerations just presented.

EXAMPLE 14.1

Consider a heating system for a one-room sealed-up house, shown schematically in Fig. 14.5. The house is modeled as a lumped system having contents of mass m and average specific heat c. The rate of heat supplied by an electric heater is $Q_{in}(t)$. The spatial average temperature inside the house is T_1, and the ambient air temperature is T_2. Determine the effect of variation of T_2 on T_1 at steady state.

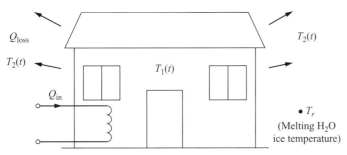

Figure 14.5. House heating system from Example 14.1.

The heat balance equation is

$$mc\frac{dT_{1r}(t)}{dt} = Q_{in}(t) - Q_{loss}(t). \tag{14.31}$$

The rate of heat losses, Q_{loss}, is given by

$$Q_{loss}(t) = U_o[T_{1r}(t) - T_{2r}(t)], \tag{14.32}$$

where U_o is a heat loss coefficient. The heat balance equation becomes

$$mc\frac{dT_{1r}(t)}{dt} = Q_{in}(t) - U_o[T_{1r}(t) - T_{2r}(t)]. \tag{14.33}$$

Transferring Equation (14.33) from the time domain into the domain of complex variable s yields

$$(mcs + U_o)\mathbf{T}_{1r}(s) = \mathbf{Q}_{in}(s) + U_o\mathbf{T}_{2r}(s)$$

or

$$\mathbf{T}_{1r}(s) = \frac{\mathbf{Q}_{in}(s) + U_o\mathbf{T}_{2r}(s)}{mcs + U_o}. \tag{14.34}$$

The block diagram of the system represented by Eq. (14.34) is shown in Fig. 14.6.

When Figs. 14.4(a) and 14.6 are compared, the disturbance and process transfer functions for the open-loop system can be identified as

$$\mathbf{G}_v(s) = \frac{1}{\dfrac{mc}{U_o}s + 1},$$

$$\mathbf{G}_{OL}(s) = \frac{\dfrac{1}{U_o}}{\dfrac{mc}{U_o}s + 1}.$$

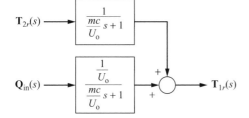

Figure 14.6. Block diagram of the open-loop thermal system considered in Example 14.1.

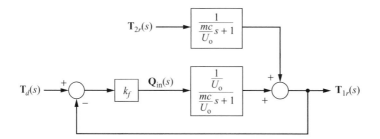

Figure 14.7. Block diagram of the closed-loop temperature control system considered in Example 14.1.

Using Eq. (14.28), one can calculate the steady-state disturbance sensitivity as

$$S_{DO} = \lim_{s \to 0} \frac{1}{\frac{mc}{U_o}s + 1} = 1. \tag{14.35}$$

Equation (14.35) indicates that the change of ambient temperature by ΔT will cause the change in the house temperature by the same value ΔT.

Now consider a closed-loop system in which the rate of heat supply is controlled to cause the house temperature to approach the desired level, T_d. The block diagram of the closed-loop temperature control system is shown in Fig. 14.7. The rate of heat supply $Q_{in}(t)$ is assumed to be proportional to the temperature deviation:

$$Q_{in}(t) = k_f [T_d(t) - T_{1r}(t)].$$

By use of Eq. (14.30), the disturbance sensitivity in this closed-loop system is found to be

$$S_{DC} = \frac{1}{1 + \frac{k_f}{U_o}},$$

where k_f/U_o is the open-loop gain of the system. Because both k_f and U_o are positive, the sensitivity of the closed-loop system to variations in ambient air temperature is smaller than that of the open-loop system. The greater the open-loop gain, the less sensitive the closed-loop system is to disturbances.

14.4   INTERRELATION OF STEADY-STATE AND TRANSIENT CONSIDERATIONS

In Sections 14.2 and 14.3, it was shown that, to improve the system steady-state performance, the open-loop gain has to be increased or an integration has to be added to the open-loop transfer function. Both remedies will, however, aggravate the stability problem. In general, the design of a system with more than two integrations in the feedforward path is very difficult. A compromise between steady-state and transient system characteristics is thus necessary. Example 14.2 illustrates this problem.

EXAMPLE 14.2

Examine the effect of the open-loop gain K on stability and steady-state performance of the system shown in Fig. 14.8, which is subjected to unit ramp input signals.

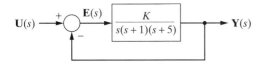

Figure 14.8. Block diagram of the system considered in Example 14.2.

SOLUTION

The open-loop sinusoidal transfer function is

$$\mathbf{G}(j\omega)\,\mathbf{H}(j\omega) = \mathbf{T}_{\text{OL}}(j\omega) = \frac{K}{j\omega(j\omega+1)(j\omega+5)}.$$

The real and imaginary parts of $\mathbf{T}_{\text{OL}}(j\omega)$ are

$$\text{Re}\left[\mathbf{T}_{\text{OL}}(j\omega)\right] = \frac{-6K}{(\omega^4 + 26\omega^2 + 25)},$$

$$\text{Im}\left[\mathbf{T}_{\text{OL}}(j\omega)\right] = \frac{K(\omega^2 - 5)}{(\omega^5 + 26\omega^3 + 25\omega)}.$$

The stability gain margin k_g was defined in Chap. 13 as

$$k_{g\text{dB}} = 20\log\frac{1}{\left|\mathbf{T}_{\text{OL}}(j\omega_p)\right|},$$

where ω_p is such that

$$\angle\mathbf{T}_{\text{OL}}(j\omega_p) = -180°, \quad \text{or} \quad \text{Im}\left[\mathbf{T}_{\text{OL}}(j\omega)\right] = 0.$$

For this system, $\omega_p = \sqrt{5}$ rad/s, and the gain margin in decibels is

$$k_{g\text{dB}} = 20\log\frac{30}{K}.$$

Now, to examine the steady-state performance of the system subjected to a unit ramp input $u(t) = t$, the static-velocity error coefficient must be determined. Using Equation (14.22) for a type 1 system yields

$$K_v = \lim_{s\to 0}\left[\frac{sK}{s(s+1)(s+5)}\right] = \frac{K}{5},$$

and hence the steady-state error is

$$e_{ss} = \frac{5}{K}.$$

Figure 14.9 shows the system gain margin $k_{g\text{dB}}$, static, velocity error coefficient K_v, and steady-state error e_{ss} as functions of the open-loop gain K. Note that the system is marginally stable for $K = 30$, at which the minimum steady-state error approaches 0.1667. Selecting values of K less than 30 will improve system stability at the cost of increasing steady-state error.

14.5 INDUSTRIAL CONTROLLERS

In most process control applications, standard "off-the-shelf" devices are used to obtain desired system performance. These devices, commonly called industrial controllers, compare the actual system output with the desired value and produce a

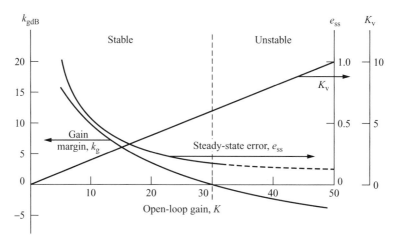

Figure 14.9. Effect of the open-loop gain on stability gain margin k_{gdB}, velocity error coefficient K_v, and steady-state error e_{ss}.

signal to reduce the output signal deviation to zero or to a small value (Fig. 14.2). The manner in which the controller produces the control signal in response to the control error is referred to as a control algorithm or a control law. Five of the most common control algorithms implemented by typical industrial controllers will now be described.

14.5.1 Two-Position or On–Off Control

The control algorithm of the two-position controller is

$$u(t) = \begin{cases} M_1 & \text{for} \quad e(t) > 0 \\ M_2 & \text{for} \quad e(t) < 0 \end{cases}. \tag{14.36}$$

The relationship between the control signal $u(t)$ and the error signal $e(t)$ is shown in Fig. 14.10. In many applications, the control signal parameters M_1 and M_2 correspond to "on" and "off" positions of the actuating device. In an on–off temperature control system, a heater is turned on or off depending on whether the process temperature is below or above the desired level. In an on–off liquid-level control system, the supply valve is either opened or closed depending on the sign of the control error. In every two-position control system, the process output oscillates as the control signal is being switched between its two values, M_1 and M_2. Figure 14.11 shows the output

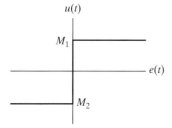

Figure 14.10. Control signal versus error signal in two-position control.

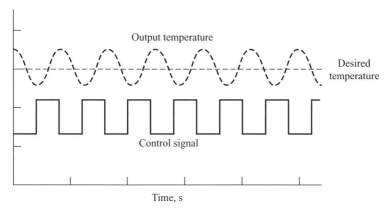

Figure 14.11. Output temperature and control signal in an on–off control system.

and control signal of an on–off temperature control system. However, this system is only piecewise linear.

14.5.2 Proportional Control

The signal produced by a proportional (P) controller is, as its name implies, proportional to the control error:

$$u(t) = k_p e(t). \tag{14.37}$$

The transfer function of the proportional controller is

$$T_C(s) = \frac{\mathbf{U}(s)}{\mathbf{E}(s)} = k_p. \tag{14.38}$$

The steady-state performance and speed of response of a system with a proportional controller improve with increasing gain k_P. Increasing the controller gain may, however, decrease stability margins.

14.5.3 Proportional–Integral Control

The steady-state performance can also be improved by adding an integral action to the control algorithm. The ideal proportional–integral (PI) controller produces a control signal defined by the equation

$$u(t) = k_p \left[e(t) + \frac{1}{T_i} \int_0^t e(\tau)\, d\tau \right]. \tag{14.39}$$

The ideal controller transfer function is

$$\mathbf{T}_C(s) = k_p \left(1 + \frac{1}{T_i s} \right). \tag{14.40}$$

Adding integral control, although improving the steady-state performance, may lead to oscillatory response (that is, reduced potential stability margin), which is usually undesirable.

14.5.4 Proportional–Derivative Control

The stability of a system can be improved by the addition of a derivative action to the control algorithm. The control signal produced by an ideal proportional–derivative (PD) controller is

$$u(t) = k_\mathrm{p} \left[e(t) + T_\mathrm{d} \frac{de\,(t)}{dt} \right]. \tag{14.41}$$

The controller transfer function is[4]

$$\mathbf{T}_C\,(s) = k_\mathrm{p}\,(1 + T_\mathrm{d}s)\,. \tag{14.42}$$

The derivative action provides an anticipatory effect that results in a damping of the system response. When the stability margin is increased in this way, it becomes possible to use a greater loop gain, thus improving speed of response and reducing steady-state error.

The most significant drawback to derivative control action is that it tends to amplify higher-frequency noise, which is often present in feedback signals because of limitations of transducer performance.

14.5.5 Proportional–Integral–Derivative Control

All three control actions are incorporated in a proportional–integral–derivative (PID) control algorithm. The control signal generated by an ideal PID controller is

$$u(t) = k_\mathrm{p} \left[e(t) + \frac{1}{T_\mathrm{i}} \int_0^t e(\tau)d\tau + T_\mathrm{d} \frac{de\,(t)}{dt} \right]. \tag{14.43}$$

The transfer function of the ideal PID controller is

$$\mathbf{T}_C\,(s) = k_\mathrm{p} \left(1 + \frac{1}{T_\mathrm{i}s} + T_\mathrm{d}s \right). \tag{14.44}$$

Determining optimal adjustments of the control parameters k_p, T_i, and T_d is one of the basic problems faced by control engineers. The tuning rules of Ziegler and Nichols[5] provide one of the simplest procedures developed for this purpose. There are two versions of this method: One is based on the process step response and the other on characteristics of sustained oscillations of the system under proportional

---

[4] The mathematical models given in Eqs. (14.41)–(14.44) for ideal controllers are not physically realizable. All real controllers have a transfer function incorporating at least a fast first-order parasitic lag term $(1 + \tau_\mathrm{p}s)$ in the denominator. Inclusion of this lag term becomes important for programming the controller for computer implementation.

[5] J. G. Ziegler and N. B. Nichols, "Optimum settings for automatic controllers," *Trans. ASME* **64**, 759 (1942).

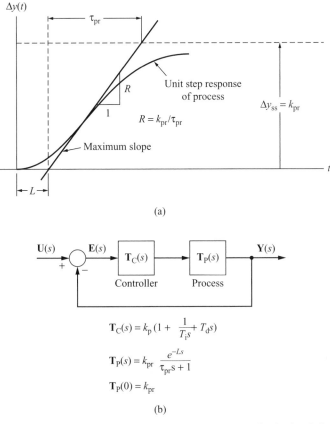

Figure 14.12. Determining controller parameters on the basis of the process step response: (a) process response graph, (b) transfer function block diagram of system.

control at the stability limit. The first method, based on a delay-lag model of the process, can be applied if process step response data are available in the form shown in Fig. 14.12. The transfer function e^{-Ls} for the time delay L is discussed in Chap. 15.

The controller parameters are calculated with the values of slope R and delay time L of the unit step response, as follows:

- For a proportional controller,

$$k_p = \frac{1}{RL}.$$

- For a PI controller,

$$k_p = \frac{0.9}{RL} \quad T_i = 3.3L.$$

- For a PID controller,

$$k_p = \frac{1.2}{RL} \quad T_i = 2L \quad T_d = 0.5L.$$

In the other method, the gain of the proportional controller in a closed-loop system test, shown in Fig. 14.13(a), is increased until a stability limit is reached with

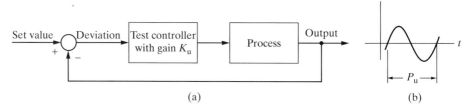

Figure 14.13. Determining controller parameters on the basis of stability limit oscillations: (a) block diagram of process with controller, (b) response with sustained oscillations.

a test controller gain K_u. The control parameters are then calculated on the basis of this critical value of gain K_u and the resulting period of sustained oscillations P_u by use of the following relations:

- For a proportional controller,

$$k_p = 0.5 K_u.$$

- For a PI controller,

$$k_p = 0.45 K_u \quad T_i = 0.83 P_u.$$

- For a PID controller,

$$k_p = 0.6 K_u \quad T_i = 0.5 P_u \quad T_d = 0.125 P_u.$$

It has to be emphasized that the rules of Ziegler and Nichols were developed empirically and that the control parameters provided by these rules are not optimal. However, they do give a good starting point from which further tuning can be performed to obtain satisfactory system performance.

More advanced industrial controllers available on the market today are capable of self-tuning – that is, of automatically adjusting the values of their control settings to obtain the best performance with a given process.[6]

EXAMPLE 14.3

Figure 14.14 shows a block diagram of a feedback control system in which the transfer function is the same as the one used in Examples 13.6 and 13.8 and is subsequently repeated. For the purposes of this example, assume that the transfer function represents the dynamics' large-capacity material-handling system in which the input signal $u(t)$ is the desired position and the output signal $y(t)$ is the actual, measured position of the payload, measured in meters:

$$\mathbf{T}_p(s) = \frac{2}{6s^2 + 11s^2 + 6s + 1}.$$

[6] K. J. Aström and T. Hagglund, "Automatic tuning of PID regulators," Instrument Society of America, Research Triangle Park, NC, 1988.

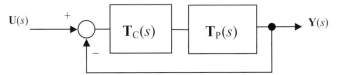

Figure 14.14. Feedback control structure for Example 14.3.

Design a controller to meet the following specifications:

- e_{ss} for step input $= 0$,
- no overshoot for step input,
- response to step input as rapid as possible.

The simplest controller is the proportional control described in Subsection 14.5.3:

$$\mathbf{T}_C (s) = k_p.$$

The open-loop transfer function, combining the controller and the system under control, is

$$\mathbf{G} (s) = \mathbf{T}_C (s) \, \mathbf{T}_P (s) = \frac{2k_p}{6s^3 + 11s^2 + 6s + 1}.$$

For a unit step input, the steady state error is determined according to Eq. (14.10):

$$e_{ss} = \lim_{s \to 0} \frac{1}{1 + \mathbf{G} (s)} = \frac{1}{1 + 2k_p}.$$

This result indicates that, regardless of the value of k_p, the specification for steady-state error will never be met. Therefore a different control structure is desired. In Subsection 14.5.3, the PI control law is presented. The main characteristic of this control law is that it increases the type of the system and hence improves steady-state performance. The original system is a type 0 system. If the controller is a PI controller, then the combined system will be type 1, and, according to Table 14.1, the steady-state error to a step input will be zero. To achieve the desired transient response, further analysis is required.

The controller transfer function is

$$\mathbf{T}_C (S) = k_p \left(1 + \frac{1}{T_i s} \right),$$

and the combined open-loop transfer function is

$$\mathbf{G} (s) = \frac{\frac{2k_p}{T_i} (T_i s + 1)}{s (6s^3 + 11s^2 + 6s + 1)}.$$

As a starting point for the values of k_p and T_i, the Ziegler–Nichols tuning rules are used. Figure 14.15 shows a step response of the original system, computed with MATLAB's step command. Indicated on the plot are the values of R and L for this response, as defined in Fig. 14.12.

From the plot in Fig. 14.15, the following characteristics can be measured:

$L = 1.5$ s,

$R = 0.27$ m/s.

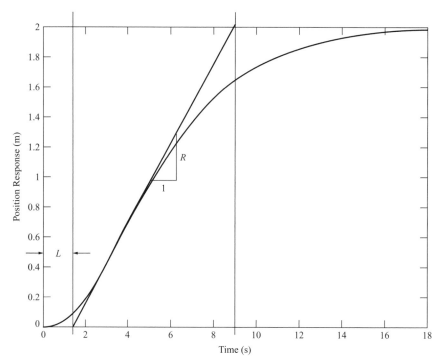

Figure 14.15. Step response of open-loop system for Example 14.3, showing the characteristics required for Ziegler–Nichols tuning rules.

The Ziegler–Nichols tuning rules for PI controllers suggest the following values for gains:

$$k_p = 2.2,$$
$$T_i = 4.95 \text{ s}.$$

To test whether or not the second design criterion is satisfied with these values, the following steps in MATLAB will generate a closed-loop step response:

```
>> kp = 2.2
>> ti = 4.95                          % set the gain values
>> num = 2*kp/ti*[ti 1];
>> den = [6 11 6 1 0];
>> tfol = tf(num, den)                % establish the open-loop
                                      %transfer function
>> tfcl = tfol/(1+tfol);              % compute the closed-loop
                                      %transfer function
>> tfcl = minreal(tfcl)               % cancel poles and zeros
                                      (see Section 11.7)
>> step(tfcl)                         % compute and plot step
                                      %response
```

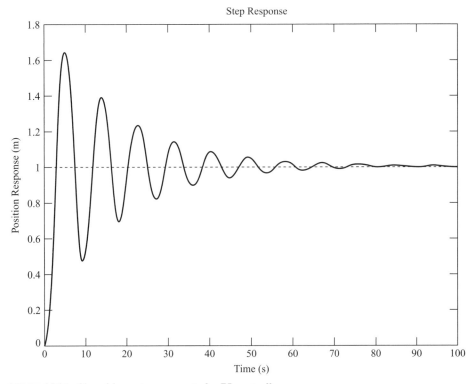

Figure 14.16. Closed-loop step response for PI controller.

Figure 14.16 shows the step response for the tuning parameters computed with the Zeigler–Nichols tuning rules.

The oscillatory behavior of the closed-loop system clearly violates the second design constraint, that there be no overshoot. Note, however, that the first design constraint, zero steady-state error to step input, is clearly met. Oscillatory behavior can be mitigated by lowering the proportional gain factor. Through trial and error and use of MATLAB's commands to reevaluate the closed-loop transfer function and step response for various gain values, it was discovered that when k_p is reduced to a value of 0.48, the step response shows no overshoot, as demonstrated in Fig. 14.17.

Finally, the third design criterion, that the response be as rapid as possible, is more difficult to assess. How can it be determined if the response can be any quicker? Figure 14.15, for example, shows that the open-loop system can respond to a step input in approximately 20 s. Experience with feedback control dictates that closed-loop response can nearly always be more rapid than open-loop. Therefore the response in Fig. 14.17 seems a bit sluggish.

The next step in the design process is to consider a PID control architecture, as described in Subsection 14.5.5. The new combined open-loop transfer function is

$$\mathbf{G}(s) = \frac{\frac{2k_p}{T_i}\left(T_i T_d s^2 + T_i s + 1\right)}{s\left(6s^3 + 11s^2 + 6s + 1\right)}.$$

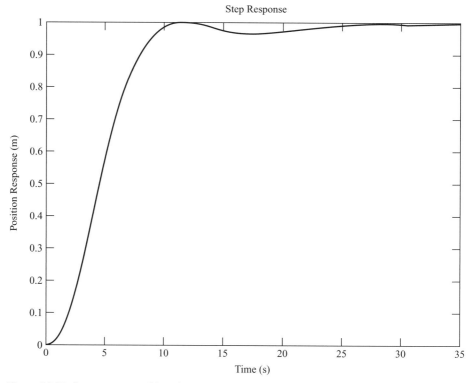

Figure 14.17. Step response with reduced proportional control gain.

The Ziegler–Nichols tuning rules suggest the following starting place for the PID parameters:

$$k_p = 2.96,$$
$$T_i = 3.0 \text{ s},$$
$$T_d = 0.75 \text{ s},$$

which lead to the closed-loop step response seen in Fig. 14.18.

Note that the system is somewhat less oscillatory than the first attempt with a PI controller, but the overshoot is very high. As with the PI controller, the response can be improved by adjustment of the three control parameters. The overall gain k_p must be reduced and the derivative term T_d increased to eliminate the overshoot. In addition, decreasing the value of T_i strengthens the integral action and helps drive the system to its steady-state value. Although there are literally an infinite number of parameter combinations that will bring this system in compliance with the first two design goals, the following set of parameters result in the response seen in Fig. 14.19:

$$k_p = 1.25,$$
$$T_i = 1.6 \text{ s},$$
$$T_d = 1.8 \text{ s}.$$

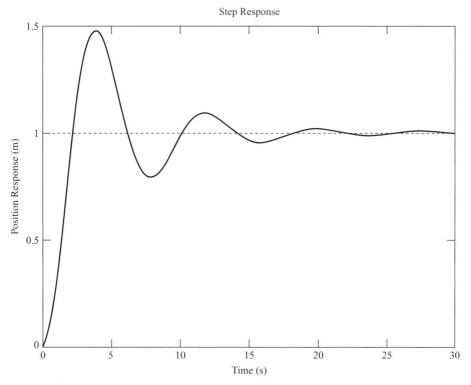

Figure 14.18. Step response with a PID controller.

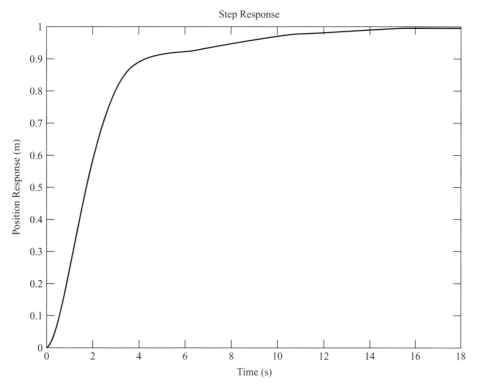

Figure 14.19. Step response with PID control after tuning.

In comparing the step responses obtained with the PI controller (Fig. 14.17) with those obtained with the PID controller, it is important to note the change in time scale on the two plots. The PID controller offers a far superior performance in the speed of response and hence is a better choice in light of the third design criterion.

Example 14.4

Although the previous example portrays the manner in which control systems are often designed in practice, the heuristic nature of the approach is unsatisfying for a number of reasons. In particular, because there are three different parameters to adjust, each of which has a unique contribution, the process of trial-and-error tuning is unlikely to arrive at the very best, or optimal, set of controller parameters.

In this example, the problem of Example 14.3 is solved by use of a PID controller, but a more directed approach is used to select the parameters. Recall that a PID controller gives rise to a second-order polynomial in the numerator of the open-loop transfer function:

$$\mathbf{G}(s) = \frac{\frac{2k_p}{T_i}\left(T_i T_d s^2 + T_i s + 1\right)}{s\left(6s^3 + 11s^2 + 6s + 1\right)}.$$

The second-order numerator means that there are two zeros of the open-loop transfer function. If those zeros are placed (by the appropriate selection of T_i and T_d) in such a way that they coincide with poles of the transfer function, those poles will, in effect, be canceled. As long as the poles that are canceled are not the fastest poles in the system, then overall dynamic performance will be improved. Further, as long as the pole at the origin is not the one canceled, then the appropriate steady-state performance will not be compromised as well. The poles of this transfer function are

$$0,$$
$$-0.3333,$$
$$-0.5,$$
$$-1.0.$$

Hence, if the controller parameters are chosen such that the zeros correspond with the middle two poles, dynamic performance will be improved without sacrificing steady-state behavior.

The second-order polynomial whose roots are $-1/3$ and $-1/2$ is

$$6s^2 + 5s + 1.$$

When the coefficients are equated with the coefficients of the second-order polynomial in the numerator of $\mathbf{G}(s)$, it is easy to solve for T_i and T_d:

$$T_i = 5.0 \text{ s},$$
$$T_d = 1.20 \text{ s}.$$

Now the only remaining task is to choose the proportional gain k_p. Again, a trial-and-error approach could be used, this time much more easily because there is only one parameter to vary and there is a high probability that one could achieve nearly optimal performance for these choices of T_i and T_d. However, the root-locus method provides a more systematic approach.

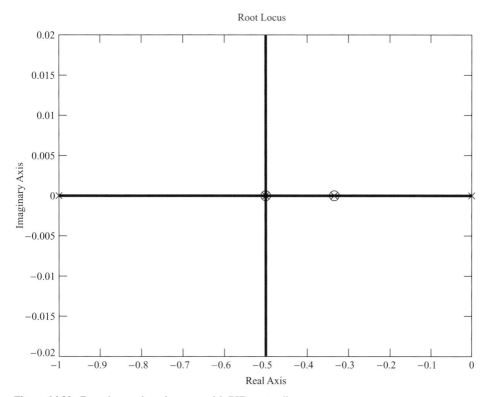

Figure 14.20. Root-locus plot of system with PID controller.

Begin with the new open-loop transfer function based on the new values of T_i and T_d:

$$\mathbf{G}(s) = \frac{0.4k_p \left(6s^2 + 5s + 1\right)}{s \left(6s^3 + 11s^2 + 6s + 1\right)}.$$

Now generate the root-locus plot for this transfer function and find the highest value of k_p for which the system shows no overshoot. In MATLAB, the root-locus plot of this transfer function is easily generated, as shown in Fig. 14.20.

The plot is a little difficult to interpret because so much of the loci sit on the axes. The $\times$'s represent the four open-loop poles (at 0, -0.333, -0.5, and -1). The circles represent the two zeros (chosen to cancel the poles at -0.333 and -0.5). The two loci begin at the other poles (at $k_p = 0$) and move toward each other until they meet at 0.5. At that point, the loci depart from the real axis and move out vertically as k_p approaches infinity. The point of departure corresponds to a maximum value of k_p for which there is no overshoot. The value of gain that corresponds to zero overshoot is 0.625. Summarizing, the PID parameters that should satisfy the design criteria are

$T_i = 5.0$ s,
$T_d = 1.2$ s,
$k_p = 0.625$.

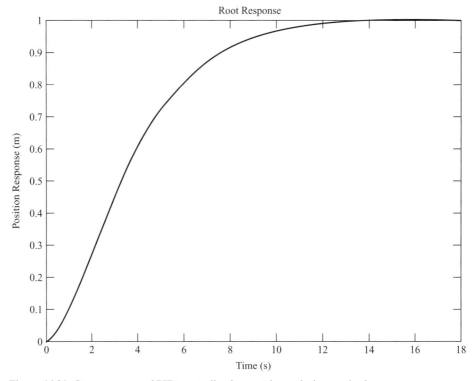

Figure 14.21. Step response of PID controller for root-locus design method.

Figure 14.21 shows the step response for the system obtained by use of these parameters. When compared with the step response of Fig. 14.19, the step response of Fig. 14.21 can be seen to be much smoother, and reaches steady state in approximately the same time.

One final note is in order before leaving this example. The astute reader will note that the much-touted Ziegler–Nichols tuning rules suggest a set of gains that appear quite far from those that satisfy our design requirements. The reason lies in the assumptions under which those tuning rules were developed. In particular, the system is assumed to be dominated by a time delay and a first-order response. The system in this example is a third-order system with no delay. At the time Ziegler and Nichols did their work, the chemical process industry dominated the field of industrial control systems and the model they assumed was a very good choice for a wide variety of systems under study. Finally, however, it is important to point out that the tuning rules provided gains that led to a stable response and provided a much better starting point for the design work than did a random combination of gains.

14.6 SYSTEM COMPENSATION

In some applications it may be very difficult or even impossible to obtain both desired transient and steady-state system performance by adjusting parameters of PID controllers. In such cases, additional devices are inserted into the system to modify the

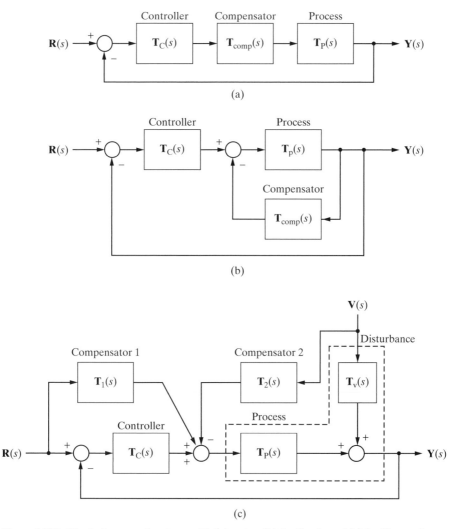

Figure 14.22. Block diagram of systems with (a) series, (b) feedback, and (c) feedforward compensation.

open-loop characteristics and enhance the system performance. This technique is called system compensation, and the additional devices inserted into the system are called compensators. Unlike controllers, compensators are usually designed for specific applications and their parameters are not adjustable.

There are several ways of inserting a compensator into a control system. Figure 14.22 shows block diagrams of control systems with series compensation [Fig. 14.22(a)], feedback compensation [Fig. 14.22(b)], and feedforward compensation [Fig. 14.22(c)]. The series structure is the most common and usually is the simplest to design. A typical example of feedback compensation is a velocity feedback in a position control system. The feedforward compensation is used to improve the system speed of response to disturbance when the disturbance is measurable.

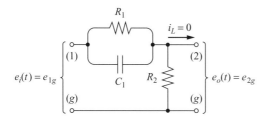

Figure 14.23. Electrical lead compensator.

There are two types of series compensators: lead compensators and lag compensators. The transfer function of the lead compensator is

$$\mathbf{T}_{\text{lead}}(s) = \alpha \frac{\tau s + 1}{\alpha \tau s + 1}, \quad \alpha < 1. \tag{14.45}$$

Compensators can be made of mechanical, electrical, and fluid components. An electrical lead compensator is shown in Fig. 14.23. The transfer function relating the output voltage $\mathbf{E}_{2g}(s)$ to the input voltage $\mathbf{E}_{1g}(s)$ for this circuit when $i_L = 0$ is

$$\mathbf{T}(s) = \frac{\mathbf{E}_{2g}(s)}{\mathbf{E}_{1g}(s)} = \frac{R_2}{R_1 + R_2} \frac{R_1 C_1 s + 1}{\frac{R_1 R_2}{R_1 + R_2} C_1 s + 1}. \tag{14.46}$$

When Eqs. (14.45) and (14.46) are compared, the parameters α and τ are found to be

$$\alpha = \frac{R_2}{R_1 + R_2}, \tag{14.47}$$

$$\tau = R_1 C_1. \tag{14.48}$$

It can be noted that the value of α given by Eq. (14.47) is always smaller than 1 because $R_2 < (R_1 + R_2)$. The lead compensator is used primarily to improve system stability. As its name indicates, this type of compensator adds a positive phase angle (phase lead) to the open-loop system frequency characteristics in a critical range of frequencies and thus increases the potential stability phase margin. Also, by increasing the potential stability margin, the lead compensator allows for further increasing of the open-loop gain to achieve good dynamic and steady-state performance.[7]

The transfer function of the lag compensator when $i_1 = 0$ is

$$\mathbf{T}_{\text{lag}}(s) = \frac{\tau s + 1}{\beta \tau s + 1}, \quad \beta > 1. \tag{14.49}$$

Figure 14.24 shows an electrical lag compensator. The transfer function of this circuit is

$$\mathbf{T}(s) = \frac{\mathbf{E}_{2g}(s)}{\mathbf{E}_{1g}(s)} = \frac{R_2 C_2 s + 1}{(R_1 + R_2) C_2 s + 1}, \tag{14.50}$$

---

[7] Thus, using a lead compensator accomplishes somewhat the same effect as using a PD controller, and therefore the lead compensator is sometimes referred to as the "poor man's" PD controller.

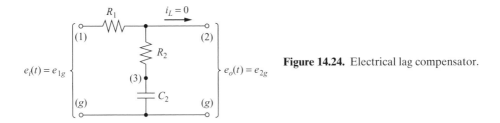

Figure 14.24. Electrical lag compensator.

or, equally,

$$T(s) = \frac{R_2 C_2 s + 1}{\frac{R_1+R_2}{R_2} R_2 C_2 s + 1}.$$ (14.51)

When Eqs. (14.49) and (14.51) are compared, the compensator parameters are found to be

$$\beta = \frac{R_1 + R_2}{R_2},$$ (14.52)

$$\tau = R_2 C_2,$$ (14.53)

where β is always greater than 1 because $(R_1 + R_2) > R_2$.

   The lag compensator improves steady-state performance of the system. System stability may, however, be seriously degraded by the lag compensator.[8]

   The advantages of each of the two types of compensators are combined in a lag–lead compensator. An electrical lag–lead compensator is shown in Fig. 14.25. The transfer function of this circuit for zero load, $i_L = 0$, is

$$T_{\text{lag–lead}}(s) = \frac{E_{2g}(s)}{E_{1g}(s)}$$

$$= \frac{(R_1 C_1 s + 1)(R_2 C_2 s + 1)}{[(R_1 C_1 s + 1)(R_2 C_2 s + 1) + s R_1 C_2]},$$ (14.54)

or, equally,[9]

$$T_{\text{lag–lead}}(s) = \left(\frac{\tau_1 s + 1}{\tau_2 s + 1}\right)\left(\frac{\gamma \tau_2 s + 1}{\gamma \tau_1 s + 1}\right),$$ (14.55)

where

$$\tau_1 = R_1 C_1,$$ (14.56)

$$\tau_2 = R_2 C_2,$$ (14.57)

$$\gamma = 1 + \frac{R_1 C_2}{\tau_1 - \tau_2}, \tau_1 > \tau_2.$$ (14.58)

---

[8] Thus, using a lag compensator, especially when $\beta\tau$ is very large, accomplishes somewhat the same effect as using a PI controller, and therefore a properly designed lag compensator is sometimes referred to as a "poor man's" PI controller.

[9] F. H. Raven, *Automatic Control Engineering* (McGraw-Hill, New York, 1987), pp. 530–2.

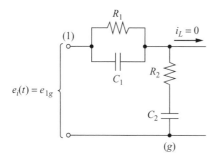

Figure 14.25. Electrical lag–lead compensator.

The lag–lead compensator is designed to improve both transient and steady-state performances, and its characteristics are similar to those of the PID controller.

The passive electrical networks shown in Figs. 14.23–14.25 are inexpensive implementations of lead and lag compensators. However, the assumption that the load current is zero ($i_L = 0$) is often limiting. If the load current is nonzero, then the compensator performance becomes degraded, or worse, unstable.

To avoid this problem, lead and lag compensation can be implemented with operational amplifiers. Op-amps, as described in Chap. 7, are versatile electronic devices that are easily incorporated into electronic designs. Figure 14.26 shows a generic lead or lag compensator constructed from two op-amps, four resistors and two capacitors. It is left as an exercise for the student to verify that the transfer function for this circuit is

$$\frac{E_o(s)}{E_i(s)} = \frac{R_4}{R_1} \frac{R_1 Cs + 1}{R_2 Cs + 1}. \tag{14.59}$$

To make a lead compensator, R_1 has to be chosen greater than R_2, and R_4 has to be adjusted to achieve the appropriate overall gain of the compensator. Lag compensators can be implemented by the choice of R_2 to be larger than R_1. A combined lead–lag compensator can also be designed by use of just two op-amps. Again, it is left as an exercise for the student to develop an appropriate design.

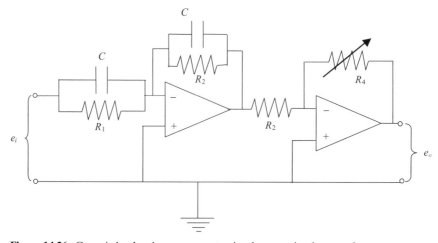

Figure 14.26. Generic lead or lag compensator implementation by use of op-amps.

14.7 SYNOPSIS

In previous chapters, the analysis of system transient performance, including stability, was emphasized. In this chapter, two important aspects of the steady-state performance were addressed. First, the steady-state control error was considered. It was shown that the steady-state error depends on both the system transfer function and the type of input signal.

A system is said to be of type r if there are r poles of the system open-loop transfer function located at the origin of the complex plane, $s = 0$. A type r system will produce zero steady-state errors if the input signal is a function of time of the order of less than r, $u(t) = t^p$, where $p < r$. If $r = p$, the steady-state error has a finite value that depends on the gain of the open-loop transfer function; the greater the gain, the smaller the steady-state error. If the input signal is proportional to time raised to a power higher than the type of the system, $p < r$, the steady-state error is infinity.

The second aspect of steady-state performance addressed in this chapter was steady-state sensitivity to disturbances. It was shown that feedback reduces the effect of external disturbances on the system output at steady state.

The steady-state performance and the speed of response of feedback systems both usually improve when the system open-loop gain increases. On the other hand, the system stability margins usually decrease and the system may eventually become unstable when the open-loop gain is increased. A compromise is therefore necessary in designing feedback systems to ensure satisfactory steady-state and transient performances at the same time.

The most common algorithms used with industrial controllers (on–off, PI, PD, and PID), and their basic characteristics, were presented. In general, the integral action improves steady-state performance, whereas the derivative action improves transient performance of the control system. Compensators, custom-designed control devices that complement the typical controllers to enhance the control system performance, were also introduced.

PROBLEMS

14.1 The open-loop transfer function of a system was found to be

$$\mathbf{T}_{\text{OL}}(s) = \frac{K}{(s+5)(s+2)^2}.$$

Determine the range of K for which the closed-loop system meets the following performance requirements: (1) the steady-state error for a unit step input is less than 10% of the input signal, and (2) the system is stable.

14.2 A simplified block diagram of the engine speed control system known as a flyball governor, invented by James Watt in the 18th century, is shown in Fig. P14.2.

(a) Develop the closed-loop transfer function $\mathbf{T}_{\text{CL}}(s) = \Omega_o(s)/\Omega_d(s)$ and write the differential equation relating the actual speed of the engine $\Omega_o(t)$ to the desired speed $\Omega_d(t)$ in the time domain.

(b) Find the gain of the hydraulic servo necessary for the steady-state value of the error in the system, $e(t) = \Omega_o(t) - \Omega_d(t)$, to be less than 1% of the magnitude of the step input.

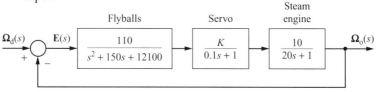

Figure P14.2. Block diagram of steam engine speed control system.

14.3 The block diagram of the system designed to control angular velocity Ω of a motor shaft is shown in Fig. P14.3. The system parameters are

$$L = 100\,\text{mH}, \qquad R = 12\,\Omega \qquad\qquad \alpha = 68\,\text{V s/rad (or N m/amp)},$$
$$J = 4\,\text{N m s}^2, \qquad B = 15\,\text{N m s},$$

where L and R are the series inductance and resistance of the armature of a dc motor, respectively, J is the combined motor and load inertia, B is the combined motor and load friction coefficient, and α is the electromechanical coupling coefficient.

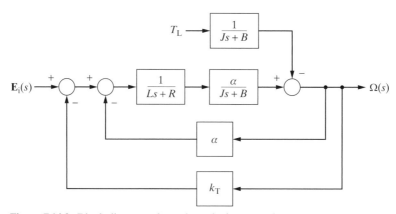

Figure P14.3. Block diagram of angular velocity control system.

(a) Determine the value of the tachometer gain k_T for which the damping ratio of the closed-loop system is greater than 0.5 and the system sensitivity to the load torque T_L is less than 2.0×10^{-3} rad/s N m.

(b) Find the steady-state control error for a unit step change in the input voltage, $e_i(t) = U_s(t)$, using the value of k_T calculated in part (a).

14.4 The block diagram of the control system developed for a thermal process is shown in Fig. P14.4.

(a) Determine the gain of the proportional controller k_P necessary for the stability gain margin $k_g = 1.2$.

(b) Find the steady-state control error in the system when the input temperature changes suddenly by 10 °C, $T_i(t) = 10U_s(t)$ by using the value of the proportional gain obtained from the stability requirement in part (a).

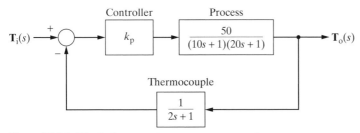

Figure P14.4. Block diagram of temperature control system.

14.5 A system open-loop transfer function is

$$\mathbf{T}_{OL}(s) = \frac{k}{s^2(\tau s + 1)}.$$

Find the steady-state control error in the closed-loop system subjected to input $u(t) = t^2$. Express the steady-state error in terms of the static-acceleration error coefficient K_a, defined as

$$K_a = \lim_{s \to 0} s^2 \mathbf{T}_{OL}(s).$$

14.6 The block diagram of a control system is shown in Fig. P14.6. The process transfer function, $\mathbf{T}_P(s)$, is

$$\mathbf{T}_P(s) = \frac{1}{10s + 1}.$$

Compare the performance of the control system with proportional and PI controllers. The controller transfer functions are

$$\mathbf{T}_C(s) = 9$$

for the proportional controller and

$$\mathbf{T}_C(s) = 9\left(1 + \frac{1}{1.8s}\right)$$

for the PI controller. In particular, compare the percentage of overshoot of the step responses and the steady-state errors for a unit step input obtained with the two controllers.

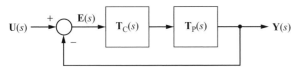

Figure P14.6. Block diagram of control system.

14.7 The process transfer function of the control system shown in Fig. P14.6 has been found to be

$$\mathbf{T}_P(s) = \frac{k}{(\tau_1 s + 1)(\tau_2 s + 1)}.$$

Compare the damping ratios and the steady-state errors for a step input obtained in this system with proportional and PD controllers. The controller transfer functions are

$$\mathbf{T}_C(s) = k_p$$

for the proportional controller and

$$\mathbf{T}_C(s) = k_p(1 + T_d s)$$

for the ideal PD controller.

14.8 A position control system, shown in Fig. P14.8(a), is being considered for a large turntable. The turntable is to be driven by a "torque motor" that provides an output torque that is proportional to its input signal, using power supplied from a dc source to achieve its inherent power amplification. One special requirement for the control system is to minimize the effects of an external load torque that may act from time to time on the turntable. The system has been modeled as shown in Fig. P14.8(b) to investigate the use of a proportional controller for this task. The values of the system parameters are

$$k_m = 1.0 \text{ N m/V} \qquad k_a = 2.0 \text{ V/rad,}$$
$$J = 3.75 \text{ N m s}^2, \qquad B = 1.25 \text{ N m s.}$$

The general requirements for the steady-state performance of this system are as follows:

(a) The steady-state error after a step input must be zero.

(b) The steady-state load sensitivity resulting from a step change in the load torque T_L is to be less than 0.1 rad/N m.

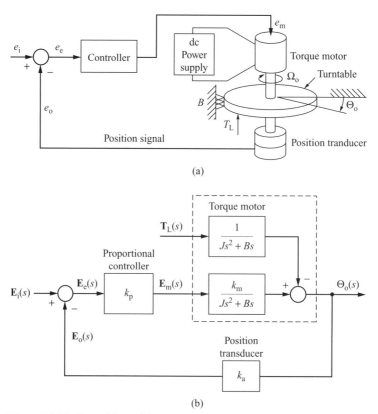

(a)

(b)

Figure P14.8. Turntable position control system with position feedback.

   Find the value of the controller gain k_p required for achieving the desired steady-state performance. Calculate the damping ratio and the natural frequency of the system with this value of k_p.

14.9 This problem is a continuation of Problem 14.8, which must be solved first. Consider again the turntable control system shown in Fig. P14.8(a). To improve the degree of damping of this system with the parameter values found in Problem 14.8, it is now proposed to use the velocity transducer of gain k_v. Figure P14.9 shows a block diagram of the system developed to investigate the performance attainable with a proportional controller augmented with an inner loop that is closed by velocity feedback to the torque motor. In addition to the steady-state performance requirements stated in Problem 14.8, the system must be stable, having a damping ratio of at least 0.5. Using the value of the controller gain k_p obtained in Problem 14.8. find the value of k_v needed to achieve the required damping ratio of 0.5. Check if all the steady-state and transient performance requirements specified in Problem 14.8 and in this problem are satisfied in the redesigned system. Also, compare the speeds of response obtained with the systems designed in these two problems.

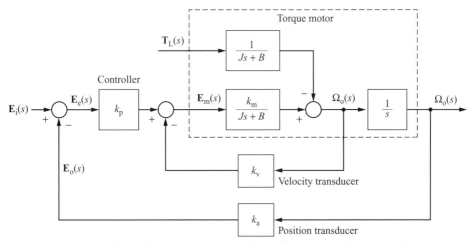

Figure P14.9. Turntable position control system with position and velocity feedback.

14.10 Derive the transfer function for the compensator circuit shown in Fig. 14.26.

14.11 Starting with the design of the lead or lag compensator shown in Fig. 14.26, design a lead–lag compensator with the addition of two capacitors. Derive the transfer function for that lead–lag compensator.

14.12 For the third-order system in Examples 14.3 and 14.4, design a lead-lag compensator to meet the same design goals.

14.13 Consider the system you modeled in Problem 4.16, shown in Fig. P14.13. Assume that the dynamics of the motor is very slow compared with the dynamics of the mechanical system (usually a very good assumption). We are going to develop a system to control the position of the large inertial mass (J_2). Assume that we have a transducer on the large mass, measuring θ_2, and that the gain of that transducer is unity (1 V implies 1 rad). Assume also that we can generate a voltage to command the system with the same interpretation (1 V implies a desired position of 1 rad.)

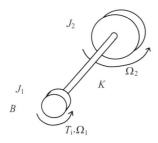

Figure P14.13. Schematic representation of a drive system.

We will implement proportional control according to the following equation:

$$T_i = K_p \left(\theta_d - \theta_2 \right).$$

The following table lists the appropriate parameter values.

Parameter	Value	Units
J_1	2.0	kg m^2
J_2	5.0	kg m^2
B	10	N s/m
K	20,000	N m/rad

Answer the following questions about the system:

(a) What is the gain margin of the system? What is the maximum value of K_p that ensures closed-loop stability?

(b) What value of K_p will give us a phase margin of 30°?

(c) If the damping factor of the original system (B) was 10 times as high (100), what would the be the answers to parts (a) and (b)?

(d) Using Simulink to simulate this system in closed loop and by observing the response to a step input, choose the value of K_p that you think is best for a fast response but minimal overshoot/oscillation.

15

Analysis of Discrete-Time Systems

LEARNING OBJECTIVES FOR THIS CHAPTER

15–1 To use the finite-difference approximation of a derivative to develop an approximate discrete-time model corresponding to a continuous input–output model.

15–2 To derive discrete-time state models for linear dynamic systems.

15–3 To develop block diagrams of a digital control system including sampling and holding devices.

15–4 To use the z transform to develop pulse transfer functions of discrete-time systems.

15.1 INTRODUCTION

In almost all existing engineering systems, the system variables (input, output, state) are continuous functions of time. The first 14 chapters of this book deal with this category of systems, classified in Chap. 1 as continuous dynamic systems. The last two chapters are devoted to discrete-time systems in which, according to the definition given in Chap. 1, the system variables are defined only at distinct instants of time. It may seem that there are not many such systems, and, indeed, very few examples of intrinsically discrete engineering systems come to mind. There are, however, many systems involving continuous subsystems that are classified as discrete because of the discrete-time elements used to monitor and control the continuous processes. Any system in which a continuous process is measured and/or controlled by a digital computer is considered discrete. Although some variables in such systems are continuous functions of time, they are known only at distinct instants of time determined by the computer sampling frequency, and therefore they are treated as discrete-time variables.

The number of digital computer applications in data acquisition and control of continuous processes has been growing rapidly over the past two decades; thus the knowledge of basic methods available for analysis and design of discrete-time systems is an increasingly important element of engineering education. This chapter and the next provide introductory material on analysis and control of linear discrete-time systems.

In Section 15.2, a problem of mathematical modeling of discrete-time systems is presented. Both input–output and state forms of system models are introduced. The process of discretization of continuous systems as a result of sampling at discrete-time intervals is described in Section 15.3. Theoretical and practical criteria for selecting the sampling frequency to ensure that no information relevant to the dynamics of the continuous process is lost as a result of sampling are discussed. In Section 15.4, the z transform is introduced. The concept of the pulse transfer function of discrete-time systems defined in the domain of complex variable z is presented in Section 15.5. A procedure for calculation of a response of a discrete-time system to an arbitrary input is also outlined.

15.2 MATHEMATICAL MODELING

A discretized model of a continuous system uses a sequence of values of each continuous variable taken only at carefully chosen distinct increments of time[1] (usually equal increments). A continuous variable $x(t)$, for instance, is represented in a discretized model by a sequence $\{x(k)\}$, $k = 0,\ 1,\ 2, \ldots$, consisting of the values $x(0), x(T), x(2T)$, or, simply, $x(0), x(1), x(2), \ldots$. Furthermore, the amplitude of a signal in a discrete-time system may be quantized; that is, it may take only a finite number of values, and in such a case the signal is called a digital signal. If the signal amplitude is not quantized, such a signal is referred to as a sampled-data signal. In this introductory treatment of discrete-time systems, the effect of quantization is neglected; thus no distinction will be made between digital and sampled-data signals.

Mathematical discrete-time models can be derived in either the form of an input–output equation or in a state form. Just as in the case of continuous systems, both forms of mathematical models of discrete-time systems are equivalent in terms of the information incorporated in them. The decision of which form should be used in modeling discrete-time systems depends on a particular application. The input–output form is usually preferred in modeling low-order linear systems. The state-variable form is used primarily in modeling higher-order systems and in solving optimal control problems.

15.2.1 Input–Output Models

As mentioned in the previous section, discrete-time models are often derived by discretizing continuous models. It may therefore be expected that a certain correspondence exists between the two types of models. The connection between the corresponding continuous and discrete-time models can best be illustrated by consideration of simple first- and second-order models.

A linear first-order continuous input–output model equation is

$$a_1 \frac{dy}{dt} + a_0 y = b_1 \frac{du}{dt} + b_0 u. \tag{15.1}$$

A frequently used discrete-time model results from an approximate discretization based on a finite-difference approximation of the state-variable time derivatives.

---

[1] The choice of T will be discussed later in this chapter and in Chap. 16. See also discussion of choice of Δt in Chap. 5.

Equation (15.1) can be discretized if the continuous derivatives of input and output variables are replaced with appropriate approximating quotients at the selected time increments. A backward-difference approximation scheme for a derivative of a continuous variable $x(t)$ is

$$\left.\frac{dx}{dt}\right|_{t=kT} \approx \frac{x(kT) - x[(k-1)T]}{T}. \tag{15.2}$$

A simplified notation, i.e., $x(k)$ instead of $x(kT)$, will be used from now on for all discrete-time variables. With this notation, approximation (15.2) can be rewritten as

$$\left.\frac{dx}{dt}\right|_{t=kT} \approx \frac{x(k) - x(k-1)}{T}. \tag{15.3}$$

When the backward-difference approximation defined by approximation (15.3) is applied to Eq. (15.1), the following first-order difference equation is obtained, which will be used here as an approximate discretized model:

$$y(k) = \frac{a_1}{a_1 + a_0 T} y(k-1) + \frac{b_1 + b_0 T}{a_1 + a_0 T} u(k) - \frac{b_1}{a_1 + a_0 T} u(k-1), \tag{15.4}$$

or, simply,

$$y(k) + g_1 y(k-1) = h_0 u(k) + h_1 u(k-1), \tag{15.5}$$

where g_1, h_0, and h_1 are the parameters of this discrete-time model. By comparison of Eqs. (15.4) and (15.5), these parameters can be expressed in terms of the parameters of the continuous model:

$$g_1 = \frac{-a_1}{a_1 + a_0 T},$$

$$h_0 = \frac{b_1 + b_0 T}{a_1 + a_0 T}, \tag{15.6}$$

$$h_1 = \frac{-b_1}{a_1 + a_0 T}.$$

A similar procedure can be applied to a linear second-order model of a continuous system described by the differential equation

$$a_2 \frac{d^2 y}{dt^2} + a_1 \frac{dy}{dt} + a_0 y = b_2 \frac{d^2 u}{dt^2} + b_1 \frac{du}{dt} + b_0 u. \tag{15.7}$$

A discrete approximation of a second derivative of a continuous variable x is

$$\left.\frac{d^2 x}{dt^2}\right|_{t=kT} \approx \frac{x(k) - 2x(k-1) + x(k-2)}{T^2}, \tag{15.8}$$

which leads to the following discrete-time model:

$$y(k) + g_1 y(k-1) + g_2 y(k-2) = h_0 u(k) + h_1 u(k-1) + h_2 u(k-2), \tag{15.9}$$

where

$$g_1 = \frac{-(2a_2 + a_1 T)}{a_2 + a_1 T + a_0 T^2},$$

$$g_2 = \frac{a_2}{a_2 + a_1 T + a_0 T^2},$$

$$h_0 = \frac{b_2 + b_1 T + b_0 T^2}{a_2 + a_1 T + a_0 T^2}, \tag{15.10}$$

$$h_1 = \frac{-(2b_2 + b_1 T)}{a_2 + a_1 T + a_0 T^2},$$

$$h_2 = \frac{b_2}{a_2 + a_1 T + a_0 T^2}.$$

From Equations (15.5) and (15.9), it can be deduced that an nth-order approximate discrete-time model can be presented in the form

$$\begin{aligned} y(k) + g_1 y(k-1) + \cdots + g_n y(k-n) \\ = h_0 u(k) + h_1 u(k-1) + \cdots + h_n u(k-n), \end{aligned} \tag{15.11}$$

where some of the coefficients g_1 ($i = 1, 2, \ldots, n$) and h_j ($j = 0, 1, \ldots, n$) may be equal to zero.

It should be realized that the connection between the corresponding continuous and discrete-time models is rather symbolic. Although the parameters of the discrete-time model can be expressed in terms of the parameters of the corresponding continuous model for a given discretization method, that relationship is not unique. By choice of different values of the sampling time T, different sets of discrete-time model parameters are obtained, and each of these approximate discrete-time models can be considered to be "corresponding" to the continuous model. The selection of time T is not uniquely determined either and is usually based on a rule of thumb. One such rule states that T should be smaller than one-fourth of the smallest time constant of the continuous model. Another rule of thumb, developed for models producing oscillatory step responses, recommends that the value of T be smaller than about one-sixth of the period of the highest-frequency oscillation of interest. Obviously, none of these rules is very precise in determining the value of the sampling time.

On the other hand, the order of the model is the same, regardless of whether the modeling is performed in a continuous or in a discrete-time domain. And an output variable of a discrete-time system at time $t = kT$, $y(kT)$, can be expressed as a function of n previous values of the output and $m + 1$ present and past values of the input variable u, which can be written mathematically as

$$\begin{aligned} y(k) = f[y(k-1), y(k-2), \ldots, y(k-n), u(k), \\ u(k-1), \ldots, u(k-m)]. \end{aligned} \tag{15.12}$$

For linear stationary systems, the input–output model takes the form of a linear difference equation [Eq. (15.11)]. A recursive solution of Eq. (15.11) can be obtained for given initial conditions and a specified input sequence, $u(k)$, $k = 0, 1, \ldots$, as illustrated in Example 15.1.

EXAMPLE 15.1

Find the solution of the following input–output difference equation for $k = 0, 1, \ldots, 10$:

$$y(k) - 0.6y(k-1) + 0.05y(k-2) = 0.25u(k-1) + 0.2u(k-2).$$

The input signal $u(k)$ is a unit step sequence given by

$$u(k) = \begin{cases} 0 & \text{for} \quad k < 0 \\ 1 & \text{for} \quad k = 0, 1, 2, \ldots, \end{cases}$$

The output sequence $y(k)$ is initially zero:

$$y(k) = 0 \quad \text{for} \quad k < 0.$$

SOLUTION

The recursive solution of the given difference equation can be obtained in a step-by-step manner, starting at $k = 0$ and progressing toward the final value of $k = 10$, as follows:

$$\text{For } k = 0, \quad y(0) = 0.6y(-1) - 0.05y(-2) + 0.25u(-1)$$
$$+ 0.2u(-2) = 0,$$
$$\text{For } k = 1, \quad y(1) = 0.6y(0) - 0.05y(-1) + 0.25u(0)$$
$$+ 0.2u(-1) = 0.25.$$

In a similar manner, the corresponding values of $y(k)$ are calculated from the equation

$$y(k) = 0.6y(k-1) - 0.05(k-2) + 0.25u(k-1)$$
$$+ 0.2u(k-2) \text{ for } k = 1, 2, \ldots, 10.$$

A listing of a very simple MATLAB script for solving the difference equation considered in this example is as follows:

```
% y(k)-output
% u(k)-input
% t(k)-time index
% Set up the initial conditions
y(1) = 0.0;
y(2) = 0.25;
%
u(1) = 1.0; % input is the unit step
u(2) = 1.0;
%
t(1) = 0;
t(2) = 1;
%
% set up loop to solve the difference equation for the next
% nine time steps
%
for k = 3:11
u(k) = 1.0;
t(k) = k-1;
```

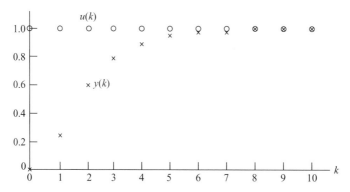

Figure 15.1. Solution of the difference equation considered in Example 15.1.

```
%
y(k) = 0.6*y(k − 1) − 0.05*y(k − 2) + 0.25*u(k − 1) + 0.2*u(k − 2);
%
end;
%
plot (t,y, 'x', t, u, 'o')
```

The plots of $u(k)$ and $y(k)$ are shown in Fig. 15.1.

15.2.2 State Models

The basic concept and definitions associated with state models of discrete-time systems are the same as those used in modeling continous systems.

To derive a state model for a linear system described by an nth-order input–output equation [Eq. (15.11)], first an auxiliary discrete-time variable x is introduced; this satisfies a simplified input–output equation in which all coefficients on the right-hand side except h_0 are assumed to be zero, that is,

$$x(k+n) + g_1 x(k+n-1) + \cdots + g_n x(k) = u(k). \tag{15.13}$$

The following set of n discrete-time state variables is then selected:

$$
\begin{aligned}
q_1(k) &= x(k), \\
q_2(k) &= x(k+1) = q_1(k+1), \\
q_3(k) &= x(k+2) = q_2(k+1).
\end{aligned}
\tag{15.14}
$$

Equations (15.14) yield $n - 1$ state equations:

$$
\begin{aligned}
q_1(k+1) &= q_2(k), \\
q_2(k+1) &= q_3(k) \\
&\vdots \\
q_{n-1}(k+1) &= q_n(k).
\end{aligned}
\tag{15.15}
$$

To obtain the nth state equation, first note that the last of Eqs. (15.14) for $k = k + 1$ becomes

$$q_n(k+1) = x(k+n). \tag{15.16}$$

Substituting $x(k + n)$ from Eq. (15.13) gives the nth state equation:

$$q_n(k+1) = -g_1 q_n(k) - g_2 q_{n-1}(k) - \cdots - g_n q_1(k) + u(k). \tag{15.17}$$

Hence a complete state model for a discrete system is

$$\begin{bmatrix} q_1(k+1) \\ q_2(k+1) \\ \vdots \\ q_n(k+1) \end{bmatrix} = \begin{bmatrix} 0 & 1 & 0 & \cdots & 0 \\ 0 & 0 & 1 & \cdots & 0 \\ & & & & \\ -g_n & -g_{n-1} & & \cdots & -g_1 \end{bmatrix} \begin{bmatrix} q_1(k) \\ q_2(k) \\ \vdots \\ q_n(k) \end{bmatrix} + \begin{bmatrix} 0 \\ 0 \\ \vdots \\ 1 \end{bmatrix} u(k), \tag{15.18}$$

or, in a more compact form,

$$\mathbf{q}(k+1) = \mathbb{G}\mathbf{q}(k) + \mathbf{h}u(k), \tag{15.19}$$

where $\mathbf{q}$ is the state vector, $\mathbb{G}$ is the system matrix, $\mathbf{h}$ is the input vector, and $u(k)$ is an input signal.

The preceding state model was derived with all but one of the terms on the right-hand side of Eq. (15.11) neglected. To incorporate these terms into the system representation, an output model is derived relating the output variable $y(k)$ to the discrete-time state vector $\mathbf{q}(k)$ and the input variable $u(k)$. The procedure followed in the derivation is very similar to that used in Chap. 3 to derive the output model for continuous systems described by input–output equations that involve derivatives of input variables. The resulting output equation for a discrete-time model is

$$y(k) = [(h_n - h_0 g_n)\,(h_{n-1} - h_0 g_{n-1}) \cdots (h_1 - h_0 g_1)] \begin{bmatrix} q_1(k) \\ q_2(k) \\ \vdots \\ q_n(k) \end{bmatrix} + h_0 u(k). \tag{15.20}$$

A complete state model of a single-input, single-output discrete system, described by Eq. (15.11), can now be presented in the following form:

$$\begin{aligned} \mathbf{q}(k+1) &= \mathbb{G}\mathbf{q}(k) + \mathbf{h}u(k), \\ y(k) &= \mathbf{p}^T\mathbf{q}(k) + ru(k), \end{aligned} \tag{15.21}$$

where $r = h_0$ and column vector $\mathbf{p}$ is

$$\mathbf{p} = \begin{bmatrix} (h_n - h_0 g_n) \\ (h_{n-1} - h_0 g_{n-1}) \\ \vdots \\ (h_1 - h_0 g_1) \end{bmatrix}. \tag{15.22}$$

It is interesting to note that the set of discrete-time state variables is *not* the same as the set of discretized continuous system state variables. The discrete-time state variables are successively advanced versions of the output $y(k)$, whereas the continuous system state variables are successively differentiated versions of the output $y(t)$. Corresponding solution techniques thus involve successive time delays between the discrete-time variables and successive integrations between the continuous system variables chosen in this way. The procedure just presented is illustrated in Example 15.2.

EXAMPLE 15.2

Derive a state model for the discrete system considered in Example 15.1.

SOLUTION

The input–output equation of the system can be written in the form

$$y(k+2) - 0.6y(k+1) + 0.05y(k) = 0.25u(k+1) + 0.2u(k).$$

Define the state variables as

$$q_1(k) = y(k),$$
$$q_2(k) = y(k+1).$$

An auxiliary variable $x(k)$ is introduced such that

$$x(k+2) - 0.6x(k+1) + 0.05x(k) = u(k).$$

The state equations take the form

$$\begin{bmatrix} q_1(k+1) \\ q_2(k+1) \end{bmatrix} = \begin{bmatrix} 0 & 1.0 \\ -0.05 & 0.6 \end{bmatrix} \begin{bmatrix} q_1(k) \\ q_2(k) \end{bmatrix} + \begin{bmatrix} 0 \\ 1 \end{bmatrix} u(k).$$

The output equation is

$$y(k) = [0.2 \; 0.25] \begin{bmatrix} q_1(k) \\ q_2(k) \end{bmatrix}.$$

15.3 SAMPLING AND HOLDING DEVICES

Most systems classified as discrete-time systems involve discrete as well as continuous components: An example of such a system is a digital control system, shown in block-diagram form in Fig. 15.2. In this system, a digital device is used to control a continuous process.

Figure 15.3 shows the signals appearing in the system using the digital PID algorithm to control a third-order process.

The digital controller generates a discrete-time control signal and accepts only discrete input signals. The continuous process produces a continuous output signal

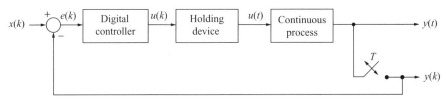

Figure 15.2. Block diagram of a digital control system with a continuous process.

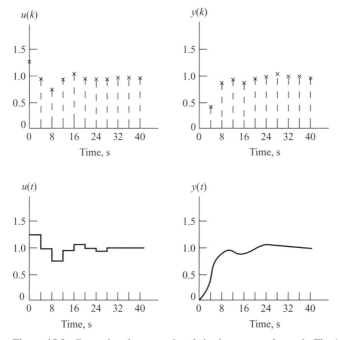

Figure 15.3. Control and output signals in the system shown in Fig. 15.2.

and can be effectively manipulated by a continuous input signal. Interface devices capable of transforming continuous signals into discrete signals and vice versa are therefore necessary to create compatibility between a digital controller and a continuous process. The two devices, a sampling device that converts a continuous signal into a discrete-time signal and a holding device that performs the opposite signal conversion, are now described.

The *sampling device* allows the continuous input signal to pass through at distinct instants of time. In an actual sampler, the path for the input signal remains open for a finite period of time Δ, as illustrated in Fig. 15.4(a). It is usually assumed that Δ is much smaller than the sampling period T; as a result, the output signal from a sampling device is presented by the strengths of a sequence of impulses, as shown in Fig. 15.4(b). The strength of an impulse is defined here as the area of the impulse.

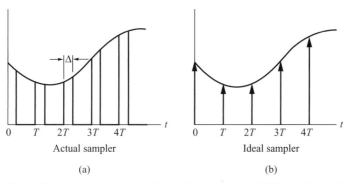

Figure 15.4. Input and output signals in (a) actual and (b) ideal sampling devices.

Think of the sampled signal $x^*(t)$ obtained from a continuous signal $x(t)$ as the result of modulation of $x(t)$ with a gating function $G_s(t)$. Then

$$x^*(t) = G_s(t)x(t) \tag{15.23}$$

when the function $G_s(t)$ is equal to 1 at $t = 0, T, 2T, \ldots$, and zero elsewhere.

Some authors[2] use Dirac's delta function to describe $x^*(t)$ mathematically. The mathematical formula representing an idealized sampled signal [Fig. 15.4(b)] is

$$x^*(t) = \sum_{k=0}^{\infty} x(kT)U_i(t - kT), \tag{15.24}$$

where Dirac's delta function, introduced in Section 4.3, is defined as a unit impulse having area of unity:

$$U_i(t - kT) = \begin{cases} 0 & \text{for} \quad t \neq kT \\ \infty & \text{for} \quad t = kT \end{cases}, \tag{15.25a}$$

$$\lim_{\Delta \to 0} \int_{kT}^{kT+\Delta} U_i(t - kT)dt = 1. \tag{15.25b}$$

Equation (15.24) describes a sampled signal $x^*(t)$ in terms of a series of impulses such that the strength of each impulse is equal to the magnitude of the continuous signal at the corresponding instant of time, $t = kT, k = 0, 1, 2, \ldots$, that is,

$$\int_{-\infty}^{+\infty} x(t)U_i(t - kT)dt = x(kT). \tag{15.26}$$

Remember that, as pointed out in Chap. 4, although Dirac's delta function represents a useful mathematical idealization, it cannot be generated physically.

Intuitively, it seems that a certain loss of information must occur when a continuous signal is replaced with a discrete signal. However, according to Shannon's theorem,[3] a continuous, band-limited signal of maximum frequency ω_{max} can be recovered from a sample signal if the sampling frequency ω_s is greater than twice the maximum signal band frequency, that is, if

$$\omega_s > 2\omega_{max}, \tag{15.27}$$

or, in terms of the sampling period T, if

$$T < \pi/\omega_{max}. \tag{15.28}$$

In digital control practice, the value of T should be less than $1/(2\,\omega_{max})$. Half of the frequency that satisfies inequality (15.27), $\omega_s/2$, is often referred to as the Nyquist frequency.[4] Selecting sampling frequency on the basis of the condition imposed by inequality (15.27) is difficult in practice because real signals in engineering systems have unlimited frequency spectra; hence, determining the value of ω_{max} involves some uncertainty. Note also that the sampling frequency used in digital control systems

[2] G. F. Franklin, J. D. Powell and M. Workman, *Digital Control of Dynamic Systems*, 3rd ed. (Addison Wesley Longman, Menlo Park, CA, 1998).
[3] C. E. Shannon and W. Weaver, *The Mathematical Theory of Communication* (University of Illinois Press, Urbana, IL, 1949).
[4] Franklin *et al.*, *op cit.*, p. 163.

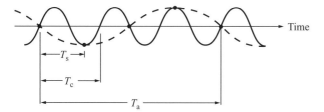

Figure 15.5. Aliasing as a result of sampling frequency $\omega_s < 2\omega_c$.

can be higher than the frequency determined from Shannon's theorem because its selection is based on different criteria. The problem of selecting a sampling frequency in digital control systems will be discussed in more detail in Chap. 16.

A serious problem, called aliasing, occurs as a result of sampling if the sampling frequency is too small. Aliasing is manifested by the presence of harmonic components in the sampled signal that are not present in the original continuous signal. The frequency of the aliasing harmonics is

$$\omega_a = \omega_s - \omega_c \tag{15.29}$$

where ω_c is a frequency of the continuous signal. Aliasing is illustrated in Fig. 15.5, where a sinusoidal signal of frequency ω_c (and period T_c) is sampled with frequency $\omega_s = 4/3\,\omega_c$, which corresponds to period $T_s = (3/4)T_c$ and is less than the sampling frequency required by the Shannon theorem. As a result, an alias sinusoidal signal is generated with a frequency of $\omega_a = \omega_s - \omega_c = 1/3\,\omega_c$ and a period of $T_a = 3T_c$.

To prevent aliasing from affecting the sampled signal, the sampling frequency should be high enough, which is easier to accomplish with a continuous signal if it is processed by a low-pass filter before sampling. The bandwidth of such an antialiasing low-pass filter should be higher than the bandwidth of the sampled signal.

A *holding device* is used to convert sampled values (such as a control sequence generated by a digital controller) into signals that can be applied to continuous systems. Operation of a holding device can be thought of as an extrapolation of past values of the discrete-time signal over the next sampling period. Mathematically, for nth-order extrapolation, the output signal from a holding device, $x_h(t)$ for time t such that $kT < t < (k+1)T$, can be expressed as

$$x_h(t) = x(kT) + a_1\tau + a_2\tau^2 + \cdots + a_n\tau^n, \tag{15.30}$$

where $0 < \tau < T$. The coefficients $a_1, a_2, \ldots, a_n$ have to be estimated, by use of n past values of $x(k)$. In digital control practice, the simplest holding device – the one that maintains the last value over the next sampling period – is used. This device, a zero-order hold (ZOH), is described by

$$x_h(kT + \tau) = x(kT) \quad \text{for} \quad 0 \le \tau < T. \tag{15.31}$$

In general, a staircase signal $x_h(t)$ that has the same form as $u(t)$ shown in Fig. 15.3 can be expressed mathematically as the summation

$$x_h(t) = \sum_{k=0}^{\infty} x(kT)\{U_s(t - kT) - U_s[(k+1)T]\}, \tag{15.32}$$

where U_s is a unit step function. The transfer function of the ZOH can be found by use of summation (15.32):

$$\mathbf{T_h}(s) = \frac{\mathbf{X_h}(s)}{\mathbf{X}^*(s)} = \frac{1 - e^{-sT}}{s}, \tag{15.33}$$

where $\mathbf{X_h}(s)$ and $\mathbf{X}^*(s)$ are Laplace transforms of output and input signals of the ZOH, respectively. It can be seen in Eq. (15.33) that the ZOH involves nontrivial dynamics that may have a significant effect on performance of the discrete system incorporating a continuous process. Very often, the dynamics of the holding device and the dynamics of the process are combined to obtain a single-block "equivalent" process.

15.4 THE z TRANSFORM

As discussed in Section 15.2, difference equations describing discrete-time systems are usually solved recursively – that is, in a step-by-step manner – starting with the initial conditions and progressing toward the final time. This method, although quite effective when applied with a computer, does not produce a closed-form solution, which is often needed in analysis of system dynamics. The z transform provides a useful tool that allows difference equations derived in the time domain to be transformed into equivalent algebraic equations in the domain of a complex variable z. The algebraic equations in the z domain are usually much easier to solve than the original difference equations. By application of the inverse z transform, closed-form solutions of the model difference equations can be obtained. Moreover, there are many powerful methods for analysis of discrete-time systems in the z domain. Most of these methods, by the way, are analogous to corresponding methods for analysis of continuous systems in the s domain.

15.4.1 Definition and Basic z Transforms

The z transform of sequence $x(kT)$ such that $x(kT) = 0$ for $k < 0$ is defined as follows:

$$\mathbf{X}(z) = \mathcal{Z}\{x(kT)\} = \sum_{k=0}^{\infty} x(kT)z^{-k}. \tag{15.34}$$

Because of the assumption that $x(kT)$ is equal to zero for negative k, the z transform defined in Eq. (15.34) is referred to as the one-sided z transform. It can be noted that $\mathbf{X}(z)$ is a power series of z^{-1}:

$$\mathbf{X}(z) = x(0) + x(T)z^{-1} + x(2T)z^{-2} + \cdots + . \tag{15.35}$$

Example 15.3 demonstrates how Eq. (15.34) can be used to find z transforms of functions representing typical input and output signals in linear discrete-time systems.

EXAMPLE 15.3

Find z transforms of the following functions.

(a) A unit step function defined as

$$x(k) = U_s(kT) = \begin{cases} 0 & \text{for} \quad k < 0 \\ 1 & \text{for} \quad k > 0 \end{cases}.$$

Using the definition of the z transform given in Eq. (15.34), we obtain

$$\mathbf{X}(z) = \sum_{k=0}^{\infty} 1\,(z^{-k}) = \frac{1}{1 - z^{-1}} = \frac{z}{z - 1}.$$

Note that $\mathbf{X}(z)$ converges if $|z| > 1$. In general, in calculating z transforms it is not necessary to determine the region of convergence of $\mathbf{X}(z)$; it is sufficient to know that such a region exists.

(b) A unit ramp function defined as

$$x(k) = \begin{cases} 0 & \text{for} \quad k < 0 \\ kT & \text{for} \quad k \geq 0 \end{cases}.$$

Using Eq. (15.34), we obtain

$$\mathbf{X}(z) = \sum_{k=0}^{\infty} kTz^{-k} = T(z^{-1} + 2z^{-2} + 3z^{-3} + \cdots +$$

$$= Tz^{-1}(1 + 2z^{-1} + 3z^{-2} + \cdots +$$

$$= \frac{Tz^1}{(1 - z^{-1})^2} = \frac{Tz}{(z - 1)^2}.$$

(c) A Kronecker delta function defined as

$$x(k) = \begin{cases} 1 & \text{for} \quad k = 0 \\ 0 & \text{for} \quad k \neq 0 \end{cases}.$$

Using Eq. (15.34), we obtain

$$\mathbf{X}(z) = (1 \cdot z^{-0}) + (0 \cdot z^{-1}) + (0 \cdot z^{-2}) + \cdots = 1.$$

(d) A power function defined as

$$x(k) = \begin{cases} 0 & \text{for} \quad k < 0 \\ a^{kT} & \text{for} \quad k \geq 0 \end{cases}.$$

Again, using Eq. (15.34), we find the z transform to be

$$\mathbf{X}(z) = \sum_{k=0}^{\infty} a^{kT} z^{-k} = \sum_{k=0}^{\infty} (a^{-T} z)^{-k} = \frac{1}{1 - a^T z^{-1}} = \frac{z}{z - a^T}.$$

A special case of the power function is the exponential function defined as

$$x(k) = \begin{cases} 0 & \text{for} \quad k < 0 \\ e^{-bkT} & \text{for} \quad k \geq 0 \end{cases},$$

for which the z transform is

$$\mathbf{X}(z) = \frac{1}{1 - e^{-bT} z^{-1}} = \frac{z}{z - e^{-bT}}.$$

More z transforms of selected discrete functions can be found in Appendix 2.

15.4.2 z-Transform Theorems

Several basic theorems of the z transform – those that are most useful in analysis of discrete-time systems – are introduced in this subsection. For proofs of these theorems, the reader is referred to the textbook by Ogata.[5]

(a) Linearity:

$$\mathcal{Z}\{a_1 f_1(kT) + a_2 f_2(kT)\} = a_1 \mathcal{Z}\{f_1(kT)\} + a_2 \mathcal{Z}\{f_2(kT)\}. \tag{15.36}$$

(b) Delay of argument:

$$\mathcal{Z}\{f(k-n)T\} = z^{-n}\mathcal{Z}\{f(kT)\} \quad \text{for } n > 0. \tag{15.37}$$

(c) Advance of argument:

$$\mathcal{Z}\{f(k+n)T\} = z^n \left[\mathcal{Z}\{f(kT)\} - \sum_{k=0}^{n-1} f(kT)z^{-k} \right]. \tag{15.38}$$

(d) Initial-value theorem:

$$f(0^+) = \lim_{z \to \infty} \mathbf{F}(z), \tag{15.39}$$

provided $\lim_{z \to \infty} \mathbf{F}(z)$ exists.

(e) Final-value theorem:

$$\lim_{k \to \infty} f(kT) = \lim_{z \to 1}(z-1)\mathbf{F}(z), \tag{15.40}$$

provided $f(kT)$ remains finite for $k = 0, 1, 2, \ldots$.

15.4.3 Inverse z Transform

It was mentioned earlier that one of the main applications of the z transform is in solving difference equations. Once the closed-form solution in the z domain is found, it must be transformed back into the discrete-time domain by use of the inverse z transform. The following notation is used for the inverse z transform:

$$f(k) = f(kT) = \mathcal{Z}^{-1}\{\mathbf{F}(z)\} \tag{15.41}$$

In analysis of discrete-time systems, the function to be inverted is usually in the form of the ratio of polynomials in z^{-1}:

$$\mathbf{F}(z) = \frac{b_0 + b_1 z^{-1} + \cdots + b_m z^{-m}}{1 + a_1 z^{-1} + \cdots + a_n z^{-n}}. \tag{15.42}$$

---

[5] K. Ogata, *Discrete-Time Control Systems* (Prentice-Hall, New York, 1995).

By direct division of the numerator and denominator polynomials, a series is obtained:

$$\mathbf{C}(z) = c_0 + c_1 z^{-1} + c_2 z^{-2} + \cdots + . \qquad (15.43)$$

When this form is compared with the definition of the z transform, Eq. (15.34), the values of the sequence $f(k)$ can be found:

$$f(0) = c_0, \quad f(1) = c_1, \quad f(2) = c_2, \ldots . \qquad (15.44)$$

In general, the direct method does not yield a closed-form solution and is practical only if no more than the first several terms of the sequence $f(k)$ are to be found.

The most powerful method for calculation of inverse z transforms is the partial fraction expansion method. In this method, which is parallel to the method used in the inverse Laplace transformation, function $\mathbf{F}(z)$ is expanded into a sum of simple terms, which are usually included in tables of common z transforms. Because of linearity of the z transform, the corresponding function $f(k)$ is obtained as a sum of the inverse z transforms of the simple terms resulting from the partial fraction expansion. The use of the partial fraction expansion method is demonstrated in Examples 15.4 and 15.5.

EXAMPLE 15.4

Find the inverse z transform of the function

$$\mathbf{F}(z) = \frac{z(z+1)}{(z^2 - 1.4z + 0.48)(z - 1)}. \qquad (15.45)$$

First, the denominator of $\mathbf{F}(z)$ must be factored. The roots of the quadratic term in the denominator are 0.6 and 0.8; hence the factored form is

$$\mathbf{F}(z) = \frac{z(z+1)}{(z - 0.6)(z - 0.8)(z - 1)}. \qquad (15.46)$$

When $\mathbf{F}(z)$ has a zero at the origin, $z = 0$, it is convenient to find a partial fraction expansion for $\mathbf{F}(z)/z$. In this case,

$$\frac{\mathbf{F}(z)}{z} = \frac{z+1}{(z - 0.6)(z - 0.8)(z - 1)}, \qquad (15.47)$$

and the expanded forms is

$$\frac{\mathbf{F}(z)}{z} = \frac{c_1}{z - 0.6} + \frac{c_2}{z - 0.8} + \frac{c_3}{z - 1}. \qquad (15.48)$$

If all poles of $\mathbf{F}(z)$, Eq. (15.46), are of multiplicity 1, the constants c_i are calculated as

$$c_i = \left. \frac{(z - z_i)\mathbf{F}(z)}{z} \right|_{z=z_i}, \qquad (15.49)$$

where z_i are the poles of $\mathbf{F}(z)$. By use of Eq. (15.49), the expanded form of $\mathbf{F}(z)/z$ is found to be

$$\frac{\mathbf{F}(z)}{z} = \frac{20}{z - 0.6} - \frac{45}{z - 0.8} + \frac{25}{z - 1}. \qquad (15.50)$$

Multiplying both sides of Eq. (15.50) by z yields

$$\mathbf{F}(z) = \frac{20z}{z - 0.6} - \frac{45z}{z - 0.8} + \frac{25}{z - 1}. \tag{15.51}$$

The inverse z transforms of each of the three terms on the right-hand side of Eq. (15.51) can be found easily to obtain the solution in the discrete-time domain:

$$f(k) = (20 \times 0.6^k) - (45 \times 0.8^k) + 25. \tag{15.52}$$

The final-value theorem can be used to verify the solution for k approaching infinity. When k approaches infinity, the first two terms on the right-hand side of Eq. (15.52) approach zero, and thus the final value of $f(k)$ is 25. Applying the final-value theorem to Eq. (15.45) produces

$$\lim_{k \to \infty} f(k) = \lim_{z \to 1} (z - 1)\mathbf{F}(z) = \lim_{z \to 1} \frac{z(z + 1)}{z^2 - 1.4z + 0.48} = 25, \tag{15.53}$$

which verifies the final value of the solution.

EXAMPLE 15.5

Find the inverse z transform of the function

$$\mathbf{F}(z) = \frac{z(z + 2)}{(z - 1)^2}.$$

Because $\mathbf{F}(z)$ has a zero at $z = 0$, it is convenient to expand $\mathbf{F}(z)/z$ rather than $\mathbf{F}(z)$. The expanded form is

$$\frac{\mathbf{F}(z)}{z} = \frac{c_1}{z - 1} + \frac{c_2}{(z - 1)^2}, \tag{15.54}$$

where the constants c_1 and c_2 are

$$c_1 = \left[\frac{d}{dz} \frac{(z - 1)^2 \mathbf{F}(z)}{z} \right]_{z=1} = 1,$$

$$c_2 = \left[\frac{(z - 1)^2 \mathbf{F}(z)}{z} \right]_{z=1} = 3.$$

Thus the expanded form of $\mathbf{F}(z)$ is

$$\mathbf{F}(z) = \frac{z}{(z - 1)} + \frac{3z}{(z - 1)^2}$$

or

$$\mathbf{F}(z) = \frac{1}{1 - z^1} + \frac{3z^{-1}}{(1 - z^{-1})^2}. \tag{15.55}$$

The inverse transforms of the terms on the right-hand side of Eq. (15.55) are found in the table of z transforms given in Appendix 2 and yield the solution for $f(k)$:

$$f(k) = 1(k) + 3k. \tag{15.56}$$

The final-value theorem cannot be applied in this case to verify the solution because $\mathbf{F}(z)$ in Eq. (15.53) has a double pole at $z = 1$ and thus $f(k)$ does not remain finite for $k = 0$, $1, 2, \ldots$. As will be shown in Chap. 16, for $f(k)$ to remain finite for $k = 0, 1, 2, \ldots$, it is necessary that all poles of $\mathbf{F}(z)$ lie inside the unit circle in the z plane, with the possible exception of a single pole at $z = 1$.

15.5 PULSE TRANSFER FUNCTION

Another form of mathematical model of a linear discrete-time system, in addition to the input–output and state models introduced in Section 15.2, is a pulse transfer function. For a system with input $u(k)$ and output $y(k)$, the pulse transfer function is defined as the ratio of z transforms of $y(k)$ and $u(k)$ for zero initial conditions:

$$\mathbf{T}(z) = \frac{\mathbf{Y}(z)}{\mathbf{U}(z)}, \tag{15.57}$$

where $\mathbf{U}(z)$ and $\mathbf{Y}(z)$ are z transforms of $u(k)$, and $y(k)$, respectively. The pulse transfer function for a system described by Eq. (15.11) is

$$\mathbf{T}(z) = \frac{h_0 + h_1 z^{-1} + \cdots + h_m z^{-m}}{1 + g_1 z^{-1} + \cdots + g_n z^{-n}}. \tag{15.58}$$

In general, the pulse transfer function for an engineering discrete-time system takes the form of the ratio of polynomials in z^{-1}. Equation (15.58) can be rewritten as

$$\mathbf{T}(z) = \frac{\mathbf{H}(z^{-1})}{\mathbf{G}(z^{-1})}, \tag{15.59}$$

where

$$\mathbf{G}(z^{-1}) = 1 + g_1 z^{-1} + \cdots + g_n z^{-n}, \tag{15.60}$$

$$\mathbf{H}(z^{-1}) = h_0 + h_1 z^{-1} + \cdots + h_m z^{-m}. \tag{15.61}$$

The pulse transfer function is obtained by application of the z transform to the system input–output equation, as depicted in Fig. 15.6. Note that the form of the pulse transfer function is not affected by the shifting of the argument of the system input–output equation, provided the system is initially at rest. In particular, the transfer function obtained from Eq. (15.11) is the same as the transfer function derived from

$$y(k+n) + g_1 y(k+n-1) + \cdots + g_n y(k)$$
$$= h_0 u(k+n) + h_1 u(k+n-1) + \cdots + h_m u(k+n-m), \tag{15.62}$$

provided both $u(k)$ and $y(k)$ are zero for $k < 0$. This property of the pulse transfer function is illustrated in the following simple example.

EXAMPLE 15.6

Consider the system in Example 15.1. Find the pulse transfer function $\mathbf{T}(z)$ for the system having the input–output equation

$$y(k) - 0.6y(k-1) + 0.05y(k-2) = 0.25u(k-1) + 0.2u(k-2).$$

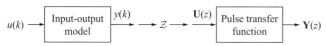

Figure 15.6. Obtaining the pulse transfer function from input–output model.

SOLUTION

Taking the z transform of both sides of this equation yields

$$\mathbf{Y}(z) - 0.6z^{-1}\mathbf{Y}(z) + 0.05z^{-2}\mathbf{Y}(z) = 0.25z^{-1}\mathbf{U}(z) + 0.2z^{-2}\mathbf{U}(z). \qquad (15.63)$$

Hence the system pulse transfer function is

$$\mathbf{T}(z) = \frac{\mathbf{Y}(z)}{\mathbf{U}(z)} = \frac{0.25z^{-1} + 0.2z^{-2}}{1 - 0.6z^{-1} + 0.05z^{-2}}. \qquad (15.64)$$

Now, shift the argument of the original input–output equation by two steps to yield

$$y(k+2) - 0.6y(k-1) + 0.05y(k-2) = 0.25u(k+1) + 0.2u(k). \qquad (15.65)$$

Transforming Eq. (15.65) into the z domain produces

$$z^2\mathbf{Y}(z) - z^2 y(0) - y(1)z - 0.6z\mathbf{Y}(z) + 0.6zy(0) + 0.05\mathbf{Y}(z)$$
$$= 0.25z\mathbf{U}(z) - 0.25zu(0) + 0.2\mathbf{U}(z). \qquad (15.66)$$

To determine $y(0)$ and $y(1)$, substitute first $k = -2$ and then $k = -1$ into Eq. (15.65). For $k = -2$,

$$y(0) - 0.6y(-1) + 0.05y(-2) = 0.25u(-1) + 0.2u(-2).$$

It is assumed here that both $u(k)$ and $y(k)$ are zero for $k < 0$, which yields

$$y(0) = 0.$$

For $k = -1$, Eq. (15.65) is

$$y(1) - 0.6y(0) + 0.05y(-1) = 0.25u(0) + 0.2u(-1),$$

and hence

$$y(1) = 0.25u(0) = 0.25.$$

Now substitute $y(0) = 0$, $y(1) = 0.25$, and $u(0) = 1$ into Eq. (15.66) to obtain

$$z^2\mathbf{Y}(z) - (z^2 \times 0) - 0.25z - 0.6z\mathbf{Y}(z) + (0.6 \times z \times 0) + 0.05\mathbf{Y}(z)$$
$$= 0.25z\mathbf{U}(z) - (0.25 \times z \times 1) + 0.2\mathbf{U}(z)$$

The resulting transfer function is

$$\mathbf{T}(z) = \frac{\mathbf{Y}(z)}{\mathbf{U}(z)} = \frac{0.25z + 0.2}{z^2 - 0.6z + 0.05}. \qquad (15.67)$$

Multiplying the numerator and denominator of Eq. (15.67) by z^{-2} gives

$$\mathbf{T}(z) = \frac{0.25z^{-1} + 0.2z^{-1}}{1 - 0.6z^{-1} + 0.05z^{-2}},$$

which is the same as Eq. (15.64), obtained from the original input–output equation.

A response of a linear discrete-time system to a discrete impulse function as defined in Example 15.3(c) is called a weighting sequence. The z transform of the discrete impulse function is equal to 1, and thus the z transform of the weighting sequence is

$$\mathbf{Y}(z) = \mathbf{T}(z) \cdot 1 = \mathbf{T}(z). \qquad (15.68)$$

The weighting sequence is thus given by an inverse z transform of the system pulse transfer function $\mathbf{T}(z)$:

$$\mathcal{Z}^{-1}\{\mathbf{T}(z)\} = w(k). \tag{15.69}$$

The weighting sequence of a discrete-time system, as do most of the other terms introduced in this chapter, has its analogous term in the area of continuous systems – the impulse response. Although this analogy between various aspects of continuous and discrete-time systems is, in most cases, clearly drawn and very useful, it should be taken with caution. One such example is an analogy between the relationships involving a continuous function of time $f(t)$ and its Laplace transform $\mathbf{F}(s)$ versus a discrete function $f(k)$ and its z transform $\mathbf{F}(z)$: By applying an inverse Laplace transformation to $\mathbf{F}(s)$, the same continuous function $f(t)$ is obtained. A discrete function $f(k)$ obtained by sampling a continuous function $f(t)$ having sampling period T is transformed into $\mathbf{F}(z)$ in the domain of complex variable z. Application of the inverse z transform to $\mathbf{F}(z)$ will result in the same discrete function $f(k)$; however, there is no basis for considering $f(k)$ as a sampled version of any specific continuous function of time. In other words, the function $f(k)$ obtained from the inverse z transform is defined only at distinct instants of time $0, T, 2T, \ldots$, and it would be entirely meaningless to deduce what values it might take between the sampling instants of time. After a continuous function of time is sampled, there is no unique transformation that will allow for a return from the discrete-time domain to the original function.

15.6 SYNOPSIS

In this chapter, basic methods for modeling and analysis of discrete-time systems were introduced. Although most engineering systems are continuous, more and more of those systems are observed and/or controlled by digital computers, which results in overall systems that are considered discrete. In such situations the continuous system variables are known only at distinct, usually equal, increments of time. As in the modeling of systems that include continuous elements only, input–output and state models are used in modeling discrete-time systems. Difference equations describing discrete-time models can be solved with simple computer programs based on recursive algorithms or with the z transform, which leads to closed-form solutions. General forms of the discrete-time models were presented and compared with the corresponding continuous system models. It was shown that the correspondence between continuous and discrete-time models is rather elusive. A continuous system can be approximated by many different discrete-time models resulting from different discretization methods or from different time intervals selected for the discrete-time approximation. In general, there are many similarities between continuous and discrete-time systems that are very helpful for those who have had considerable experience in the area of continuous systems in their introductory studies of discrete-time systems. However, as your knowledge of discrete-time systems progresses, you will notice many distinct characteristics of these systems that open new and attractive opportunities for analysis and, more importantly, for applications in process control, robotics, and so forth.

15.1 A linear discrete-time model is described by the input–output equation

$$y(k) - 1.2y(k - 1) + 0.6y(k - 2) = u(k - 1) + u(k - 2).$$

Find the output sequence, $y(k)$, $k = 0, 1, \ldots, 25$, assuming that $y(k) = 0$ for $k < 0$, for the following input signals:

(a) $u(k) = \begin{cases} 0 & \text{for} \quad k < 0 \\ 1 & \text{for} \quad k \geq 0 \end{cases}$,

(b) $u(k) = \begin{cases} 1 & \text{for} \quad k = 0 \\ 0 & \text{for} \quad k \neq 0 \end{cases}$.

15.2 Select the sampling frequency for a digital data-acquisition system measuring the velocity v of mass m in the mechanical system considered in Example 4.1. The mechanical system parameters are $m = 5$ kg and $b = 2$ N s/m.

15.3 Determine the sampling frequency for digital measurement of the position of mass m in the mechanical system shown in Fig. 4.16(a) and described by the input–output equation

$$9\ddot{x} + 4\dot{x} + 4x = F(t),$$

where $F(t)$ varies in a stepwise manner.

15.4 A sinusoidal signal $y(t) = \sin 200t$ is to be recorded with a computer data-acquisition system. It is expected that the measuring signal may be contaminated by an electric noise of frequency 60 Hz. Select the sampling frequency and determine the value of the time constant of the guard filter to prevent aliasing. The transfer function of the filter is $\mathbf{T}_f(s) = 1/(\tau_f s + 1)$.

15.5 Find z transforms of the following discrete functions of time defined for $k = 0, 1, 2, \ldots$:

(a) $a(1 - e^{-bkT})$,
(b) $(1 - akT)e^{-bkT}$,
(c) $e^{-bkT} \sin \omega kT$,
(d) $\cos \omega(k - 2)T$.

15.6 Obtain z transforms of the sequences $x(k)$ subsequently described. Express the solutions as ratios of polynomials in z:

(a) $x(k) = \begin{cases} 0 & \text{for} \quad k \leq 1 \\ 0.5 & \text{for} \quad k = 2, \\ 1 & \text{for} \quad k \geq 3 \end{cases}$

(b) $x(k) = \begin{cases} 0 & \text{for} \quad k < 0 \\ e^{0.5k} + U_s(k - 2) & \text{for} \quad k \geq 0 \end{cases}$.

15.7 Find a closed-form solution for a unit step response of the system having pulse transfer function $\mathbf{T}(z) = (z + 1)/(z^2 - 1.1z + 0.28)$. Verify the steady-state solution using the final-value theorem.

15.8 A weighting sequence of a linear discrete-time system was measured at equally spaced instants of time, and the results are given in the table. After time $10T$, the measured output signal was zero.

Time	0	T	$2T$	$3T$	$4T$	$5T$	$6T$	$7T$	$8T$	$9T$	$10T$
$w(kT)$	1.0	0.5	0.25	0.125	0.063	0.031	0.016	0.008	0.004	0.002	0.001

Find the response of this system to a unit step input $U_s(k)$.

15.9 Obtain the weighting sequence for the pulse transfer function $\mathbf{T}(z) = (z^2 + z)/[(z^2 - 0.988z + 0.49)(z - 0.6)]$.

15.10 Obtain the weighting sequence for the system represented by the input–output equation

$$y(k+2) + 0.7y(k+1) + 0.1y(k) = u(k+1).$$

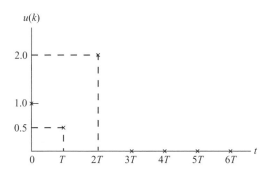

Figure P15.11. Input sequence used in Problem 15.11(b).

15.11 Find the output sequences for the discrete-time system of the transfer function $\mathbf{T}(z) = z/(z^2 - 1.125z + 0.125)$ for the following input signals:

(a) Kronecker delta function defined as

$$u(k) = \begin{cases} 1 & \text{for} \quad k = 0 \\ 0 & \text{for} \quad k \neq 0 \end{cases}.$$

(b) Sequence $u(k)$, shown in Fig. P15.11.

16

Digital Control Systems

LEARNING OBJECTIVES FOR THIS CHAPTER

16–1 To develop open-loop and closed-loop transfer functions in the z domain for simple digital control systems.

16–2 To evaluate stability and transient performance of linear discrete-time systems.

16–3 To assess steady-state performance of discrete-time systems.

16–4 To implement a discrete-time equivalent of a PID controller.

16.1 INTRODUCTION

Unprecedented advances in electronics have revolutionized control technology in recent years. Digital controllers, built around microcomputer chips as stand-alone units or implemented with ubiquitous personal computers, have dominated modern industrial process control applications. The computational power and operational speed of digital controllers allow for performance of much more sophisticated algorithms than were possible with analog controllers. Even for relatively simple control tasks, digital controllers are superior to analog controllers by virtue of their improved flexibility, greater reliability, and, more and more often, lower cost.

The main objective of this chapter is to introduce the very basic concepts of analysis of digital control systems. Only linear, stationary models are considered. In Section 16.2, a pulse transfer function block diagram of a single-loop digital control system is presented. Section 16.3 deals with transient characteristics determined by the locations of roots of the system characteristic equation; methods for determining stability are also briefly discussed. In Section 16.4, steady-state performance characteristics of digital control systems are reviewed. Section 16.5 provides introductory material on digital control algorithms. A digital version of the PID controller is given special attention because of its popularity in industrial process control applications.

16.2 SINGLE-LOOP CONTROL SYSTEMS

At the time of the first digital process control applications in the late 1950s, the cost of computers used to perform control functions was relatively high. To make these systems economical and to obtain reasonable payback time, at least 100 or more individual control loops had to be included in a single installation. Hundreds of measuring signals were transmitted from the process to the computer, often over very long distances. The control signals were sent back from the central computer to the process over the same long distances. As a result, an excessive network of wire and tubing was necessary to transmit electrical and pneumatic signals back and forth between the process and the computer. In addition to the obviously high cost of such installations, they were also very vulnerable to damage and interference from all kinds of industrial disturbances. Moreover, every failure of the central computer affected the entire process being controlled, which caused serious reliability problems.

In the 1970s, when microprocessors became available, distributed digital control systems were introduced. In these systems, controllers built around microprocessors are responsible for only local portions of the process, and so each digital controller is required to handle only one or a few control loops. Thus the controllers may be located in close proximity to the process, reducing cable and tubing cost in comparison with the centralized systems. The local controllers can be connected with a supervisory controller through a data bus, as illustrated in Fig. 16.1. During start-up and shutdown of complex multiloop process control systems, it is often advantageous to maintain autonomy of local control loops under the supervision of experienced personnel in order to achieve the transition between dead-start and normal operation. A failure of any of the local controllers or even of the supervisory controller has a limited impact on the performance of the rest of the system, which results in a much greater reliability than was possible in centralized systems. Also, it is usually easier and more economical to provide redundant digital controllers than to provide redundant continuous controllers when the need for reliability is very great. Moreover, the modular structure of distributed systems allows for gradual (piece-by-piece, controller-by-controller) implementation of new systems and easier expansion of existing systems. In summary, distributed control systems have proven to be the most efficient and reliable structures for industrial process control today. The material presented in this chapter is limited to single-loop systems such as those implemented at the lowest level of distributed control systems.

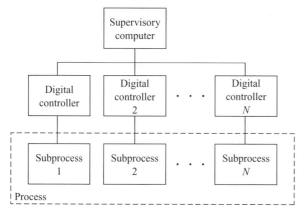

Figure 16.1. Distributed digital control system.

A block diagram of a single-loop digital control system is shown in Fig. 16.2. The controller pulse transfer function is $\mathbf{T}_C(z)$. The other block, $\mathbf{T}_P(z)$, represents a controlled process together with a preceding ZOH and can be expressed as

$$\mathbf{T}_P(z) = \mathcal{Z}\{\mathbf{T}_h(s)\mathbf{T}_P(s)\}, \tag{16.1}$$

where $\mathbf{T}_h(s)$ is the transfer function of the ZOH and $\mathbf{T}_P(s)$ represents a continuous model of the process. By use of Eq. (15.33), $\mathbf{T}_P(z)$ can be expressed as

$$\mathbf{T}_P(z) = \mathcal{Z}\left\{\frac{1 - e^{-sT}}{s}\mathbf{T}_P(s)\right\}, \tag{16.2}$$

and, hence,

$$\mathbf{T}_P(z) = (1 - z^{-1})\mathcal{Z}\left\{\frac{\mathbf{T}_P(s)}{s}\right\}. \tag{16.3}$$

A closed-loop pulse transfer function for the digital control system shown in Fig. 16.2 is

$$\mathbf{T}_{CL}(z) = \frac{\mathbf{T}_C(z)\mathbf{T}_P(z)}{[1 + \mathbf{T}_C(z)\mathbf{T}_P(z)]}. \tag{16.4}$$

In the next two sections, basic transient and steady-state characteristics of linear single-loop digital control systems are examined.

16.3 TRANSIENT PERFORMANCE

Just as in the case of continuous systems, transient performance of discrete-time systems is determined by location of the poles of the system transfer function. The poles of the transfer function are the roots of the system characteristic equation, which in the case of the system shown in Fig. 16.2 takes the form

$$1 + \mathbf{T}_C(z)\mathbf{T}_P(z) = 0. \tag{16.5}$$

For linear systems represented by input–output equation (15.11), the transfer function is of the form

$$\mathbf{T}(z) = \frac{h_0 + h_1 z^{-1} + \cdots + h_n z^{-n}}{1 + g_1 z^{-1} + \cdots + g_n z^{-n}}, \tag{16.6}$$

or, equally,

$$\mathbf{T}(z) = \frac{h_0 z^n + h_1 z^{n-1} + \cdots + h_n}{z^n + g_1 z^{n-1} + \cdots + g_n}. \tag{16.7}$$

The characteristic equation is thus an nth-order algebraic equation in z:

$$z^n + g_1 z^{n-1} + \cdots + g_n = 0. \tag{16.8}$$

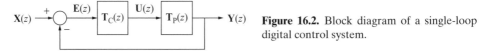

Figure 16.2. Block diagram of a single-loop digital control system.

This equation can be written in factored form as

$$(z - p_1)(z - p_2) \cdots (z - p_n) = 0, \tag{16.9}$$

where $p_1, p_2, \ldots, p_n$ are the poles of the transfer function. To investigate the effect of the locations of the poles in the domain of complex variable z on the system transient performance, a first-order system will be considered. The system pulse transfer function is

$$\mathbf{T}(z) = \frac{1}{1 + g_1 z^{-1}} \tag{16.10}$$

or, equally,

$$\mathbf{T}(z) = \frac{1}{1 + p_1 z^{-1}}, \tag{16.11}$$

where p_1 is a single real pole. The corresponding input–output equation is

$$y(k) - p_1 y(k - 1) = u(k). \tag{16.12}$$

The homogeneous equation is

$$y(k) - p_1 y(k - 1) = 0. \tag{16.13}$$

For a nonzero initial condition, $y(0) \neq 0$, the output sequence is

$$y(k) = y(0) p_1^k. \tag{16.14}$$

Plots of the sequence $y(k)$, for $k = 0, 1, 2, \ldots$, given by Eq. (16.14), for different values of p_1 are shown in Table 16.1. Note that the free response of the system, represented by the solution of the homogeneous input–output equation, converges to zero only if the absolute value of p_1 is less than unity, $|p_1| < 1$.

Discrete-time systems of higher than first order may have real as well as complex poles, which occur in pairs of complex-conjugate numbers just as they do in continuous systems. Moreover, the relation between the s-plane locations of continuous system poles and the z-plane locations of poles of a corresponding discrete-time system with sampling interval T is given by

$$z = e^{sT}, \tag{16.15}$$

or, equally,

$$s = \frac{1}{T} \ln z. \tag{16.16}$$

Equations (16.15) and (16.16) represent a mapping between the s plane and the z plane that applies to all poles of the system transfer function, not just the complex ones.[1] The term "corresponding" used here means the relation between a continuous system and a discrete-time system involving the original continuous system together with a ZOH and an ideal sampler, as shown in Fig. 16.3.

---

[1] G. F. Franklin, J. D. Powell, and M. Workman, *Digital Control of Dynamic Systems*, 3rd ed. (Addison Wesley Longman, Menlo Park, CA), 1998.

Table 16.1. Locations of the pole in the z plane and corresponding free response sequences for a first-order system

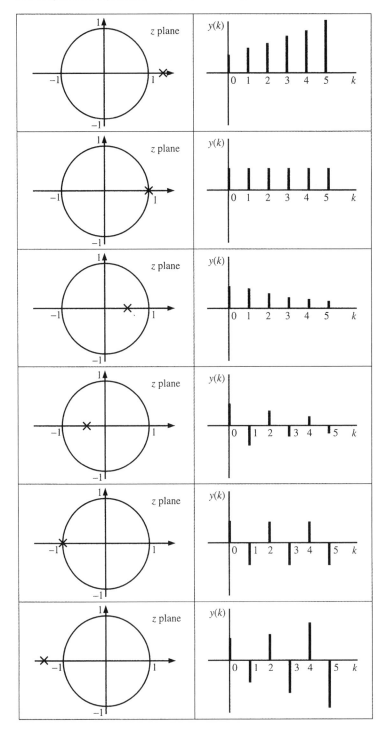

Figure 16.3. Illustration of mapping between the s and z planes.

In Chap. 4, the transient performance specifications of continuous systems were discussed in detail. The mapping given by Eq. (16.15) can be used to transform those specifications from continuous time to the discrete-time domain. A complex pole in the s plane can be expressed, from Eq. (4.81), as

$$s = -\zeta\omega_n + j\omega_d. \tag{16.17}$$

By use of mapping defined by Eq. (16.15), the corresponding pole in the z plane is

$$z = e^{(-\zeta\omega_n + j\omega_d)T} = e^{-\zeta\omega_n T}e^{j\omega_d T}. \tag{16.18}$$

The real decreasing exponential term on the right-hand side of Eq. (16.18) represents the distance d between the pole and the origin of the z-plane coordinate system:

$$d = |z| = e^{-\zeta\omega_n T}, \tag{16.19}$$

whereas the complex exponential factor represents the phase angle associated with the pole:

$$\varphi = \angle z = \omega_d T. \tag{16.20}$$

Replacing the sampling period T with the sampling frequency ω_s yields

$$\varphi = \frac{2\pi\omega_d}{\omega_s}. \tag{16.21}$$

For a specified sampling frequency ω_s and a constant damped frequency ω_d, a constant value of is obtained. The loci of the constant frequency ratio ω_d/ω_s are therefore straight lines crossing the origin of the coordinate system at the angle given by the right-hand side of Eq. (16.21) with respect to the positive real axis in the z plane.

One can find loci of another important parameter associated with complex poles, the damping ratio ζ, by combining Eqs. (16.19) and (16.20), which yields

$$d = e^{-(\varphi\zeta/\sqrt{1-\zeta^2})}. \tag{16.22}$$

Equation (16.22) describes spiral curves in the z plane with a constant value of the damping ratio along each curve. Figure 16.4 shows loci of constant frequency ratio given by Eq. (16.21) and constant damping ratio, Eq. (16.22). The loci are symmetrical with respect to the real axis; however, only the upper half of the z plane shown in Fig. 16.4 is of practical significance because it represents the area where the sampling frequency satisfies the Shannon theorem condition, Eq. (15.27).

The principal requirement regarding system transient performance is its stability. It was observed earlier that the free response of a system with real poles converges to zero (which indicates asymptotic stability according to the definition stated in Chap. 13) only if the absolute value of each real pole is less than unity. To determine the stability condition for discrete-time systems having both real and complex poles,

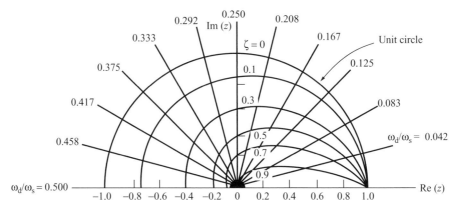

Figure 16.4. Loci of constant ω_d/ω_s and constant damping ratio.

the mapping defined by Eq. (16.15) can be used. When this mapping is used, the left-hand side of the s plane is transformed into the inside of a unit circle in the z plane. It can thus be concluded that a linear discrete-time system is asymptotically stable if all poles of the system transfer function lie inside the unit circle in the plane of complex variable z, that is, if

$$|p_i| < 1 \text{ for } i = 1, 2, \ldots, n.$$

The area inside the unit circle in the plane of complex variable z plays, therefore, the same role in analysis of stability of discrete-time systems as the left-hand side of the s plane does in analysis of stability of continuous systems.

A direct method for analysis of stability of a discrete-time system involves calculation of all roots of the system characteristic equation to determine their locations with respect to the unit circle in the z plane. This method, which is efficient only in analysis of low-order systems, is illustrated in Example 16.1.

EXAMPLE 16.1

Determine the stability condition for a digital control system with a proportional controller, shown in Fig. 16.5. The s-domain transfer function of the continuous process is

$$\mathbf{T}_P(s) = \frac{k_P}{\tau s + 1}.$$

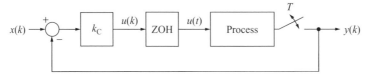

Figure 16.5. Digital control system considered in Example 16.1.

SOLUTION

First, the equivalent process transfer function representing the dynamics of the continuous process together with the zero-order hold can be found with Eq. (16.3), together with the partial fraction expansion for $T_P(s)/s$, $k_P(\frac{1}{s} - \frac{1}{s+1/\tau})$ and Table A2.1 in Appendix 2:

$$T_P(z) = (1 - z^{-1})\mathcal{Z}\left\{\frac{T_P(s)}{s}\right\} = \frac{k_P z^{-1}(1 - e^{-T/\tau})}{1 - e^{-T/\tau}z^{-1}},$$

where T is the sampling time.

The transfer function for the proportional controller is

$$T_C(z) = k_C.$$

The closed-loop system transfer function is

$$T_{CL}(z) = \frac{k_C k_P z^{-1}(1 - e^{-T/\tau})}{1 - e^{-T/\tau}z^{-1} + k_C K_P z^{-1}(1 - e^{-T/\tau})},$$

and the system characteristic equation is

$$1 - e^{-T/\tau}z^{-1} + k_C k_P z^{-1}(1 - e^{-T/\tau}) = 0.$$

The single real pole of the closed-loop system is

$$p = e^{-T/\tau} - k_C k_P(1 - e^{-T/\tau}).$$

To ensure stability, the pole must lie between -1 and $+1$ on the real axis in the z plane. The stability condition can thus be written as

$$-1 < [e^{-T/\tau} - k_C k_P(1 - e^{-T/\tau})] < 1,$$

which yields the admissible range of values for positive controller gain:

$$0 < k_C < \frac{1 + e^{-T/\tau}}{k_P(1 - e^{-T/\tau})}.$$

Several observations can be made in regard to Example 16.1. First, it can be seen that a simple discrete-time system, including a first-order process and a proportional controller, can be unstable if the open-loop gain is too high, whereas a first-order continuous control system is always stable. This difference is caused by the presence of the zero-order hold in the discrete-time system. Second, from the stability condition found for the system, it can be seen that the maximum admissible value of k_C increases when the sampling time T decreases. In fact, as T approaches zero, the system becomes unconditionally stable, that is, stable for all values of gain k_C from zero to infinity. Third, note that the relation between the continuous process pole $s = -1/\tau$ and the pole of the corresponding discrete-time pole $z = e^{-T/\tau}$ is indeed as given by Eq. (16.15).

Of course, the direct method for analysis of stability is not the most efficient. One of the more efficient methods is based on a bilinear transformation defined by

$$w = \frac{z+1}{z-1} \tag{16.23}$$

or, equally,

$$z = \frac{w+1}{w-1}. \tag{16.24}$$

The mapping defined by these equations transforms the unit circle in the z plane into the imaginary axis of the w plane and the inside of the unit circle in the z plane into the left-hand side of the w plane. Substituting the right-hand side of Eq. (16.24) for z in the system characteristic equation [Eq. (16.8)] yields an nth-order equation in variable w. If all roots of this equation lie in the left-hand side of the w plane, the discrete-time system is stable. To determine if the roots of the transformed equation satisfy this condition, the same methods can be used as those developed for analysis of stability of linear continuous systems, such as the Hurwitz and the Routh criteria presented in Chap. 13.

16.4 STEADY-STATE PERFORMANCE

In Section 14.2, the steady-state performance of continuous control systems was evaluated. The two primary criteria used in this evaluation, the steady-state control error and the steady-state sensitivity to disturbances, can also be applied to discrete-time systems.

The steady-state value of the error signal $e(k)$ in a single-loop control system (see Fig. 16.2) is defined as

$$e_{ss} = \lim_{k \to \infty} e(k). \tag{16.25}$$

When the final-value theorem is applied, e_{ss} can also be expressed as

$$e_{ss} = \lim_{z \to 1}(1 - z^{-1})\mathbf{E}(z), \tag{16.26}$$

where $\mathbf{E}(z)$ is a z transform of $e(k)$. An algebraic equation for the summing point in Fig. 16.2 is

$$\mathbf{E}(z) = \mathbf{X}(z) - \mathbf{Y}(z). \tag{16.27}$$

The output $\mathbf{Y}(z)$ is

$$\mathbf{Y}(z) = \mathbf{E}(z)\mathbf{T}_C(z)\mathbf{T}_P(z). \tag{16.28}$$

Substituting Eq. (16.28) into Eq. (16.27) yields

$$\mathbf{E}(z) = \mathbf{X}(z) - \mathbf{E}(z)\mathbf{T}_C(z)\mathbf{T}_P(z). \tag{16.29}$$

Hence the error pulse transfer function $\mathbf{T}_E(z)$ is defined as

$$\mathbf{T}_E(z) = \frac{\mathbf{E}(z)}{\mathbf{X}(z)} = \frac{1}{1 + \mathbf{T}_C(z)\mathbf{T}_P(z)}. \tag{16.30}$$

The steady-state error can now be expressed in terms of the error transfer function $\mathbf{T}_E(z)$ and the z transform of the input signal $\mathbf{X}(z)$:

$$e_{ss} = \lim_{z \to 1}(1 - z^{-1})\mathbf{X}(z)\mathbf{T}_E(z). \tag{16.31}$$

As indicated by Eq. (16.31), the steady-state performance of discrete-time control systems depends not only on the system characteristics, represented by the error transfer function $\mathbf{T}_E(z)$, but also on the type of input signal, $\mathbf{X}(z)$. This again is true for continuous control systems also. A steady-state response to unit step and unit ramp inputs will now be derived.

16.4.1 Unit Step Input

The z transform of the unit step function $U_s(kT)$ is

$$\mathbf{X}(z) = \mathcal{Z}\{U_s(kT)\} = \frac{1}{1 - z^{-1}}. \tag{16.32}$$

Substituting Eq. (16.32) into Eq. (16.31) gives the steady-state error for a unit step input:

$$e_{ss} = \lim_{z \to 1} \mathbf{T}_E(z) = \lim_{z \to 1} \frac{1}{1 + \mathbf{T}_C(z)\mathbf{T}_P(z)}. \tag{16.33}$$

A static-position error coefficient is defined as a limit of the open-loop pulse transfer function for z approaching unity:

$$K_P = \lim_{z \to 1} \mathbf{T}_C(z)\mathbf{T}_P(z). \tag{16.34}$$

The steady-state error can now be expressed as

$$e_{ss} = \frac{1}{1 + K_P}. \tag{16.35}$$

This result shows that the steady-state error in response to a unit step input will be zero only if the static-position error coefficient is infinity. From Eq. (16.34) it can be seen that K_P will approach infinity if the open-loop transfer function, $\mathbf{T}_C(z)\mathbf{T}_P(z)$, has at least one pole at $z = 1$. In general, $\mathbf{T}_C(z)\mathbf{T}_P(z)$ can be presented as

$$\mathbf{T}_C(z)\mathbf{T}_P(z) = \frac{k(z - z_1)(z - z_2)\ldots(z - z_m)}{(z - 1)^r(z - p_1)(z - p_2)\ldots(z - p_n)}, \tag{16.36}$$

where $z_1, z_2, \ldots, z_m$ are the open-loop zeros other than unity and $p_1, p_2, \ldots, p_n$ are the open-loop poles, none of which is equal to 1. There are also r open-loop poles at $z = 1$. The multiplicity of the open-loop poles at $z = 1$ determines the type of the system. As shown earlier, the steady-state error in response to a unit step input is zero for systems of type 1 or higher.

16.4.2 Unit Ramp Input

The z transform of the unit ramp signal was found in Example 15.3 to be

$$\mathbf{X}(z) = \mathcal{Z}\{x(kT)\} = \frac{Tz^{-1}}{(1 - z^{-1})^2}, \tag{16.37}$$

where $x(kT)$ is

$$x(kT) = \begin{cases} 0 & \text{for} \quad k < 0 \\ kT & \text{for} \quad k \geq 0 \end{cases}.$$

From Eq. (16.31), the steady-state error is

$$e_{ss} = \lim_{z \to 1}(1 - z^{-1})\frac{\dfrac{Tz^{-1}}{(1 - z^{-1})^2}}{1 + \mathbf{T}_C(z)\mathbf{T}_P(z)} \tag{16.38}$$

$$= \lim_{z \to 1}\frac{T}{(1 - z^{-1})\mathbf{T}_C(z)\mathbf{T}_P(z)}.$$

A static-velocity error coefficient K_v is defined as

$$K_v = \lim_{z \to 1}(1 - z^{-1})\mathbf{T}_C(z)\mathbf{T}_P(z)/T. \tag{16.39}$$

Comparing Eqs. (16.38) and (16.39) yields

$$e_{ss} = \frac{1}{K_v}. \tag{16.40}$$

From Equation (16.40), the steady-state error in response to a ramp signal can be seen to equal zero only if the system static-velocity error coefficient is infinity, which occurs when the open-loop transfer function has at least a double pole at $z = 1$. In other words, using the terminology introduced earlier in this section, the steady-state error in response to a ramp input is zero if the system is of type 2 or higher.

The results derived here for unit step and unit ramp inputs can be extended to parabolic and higher-order input signals.

Another important aspect of performance of control systems is their ability to eliminate steady-state effects of disturbances on the output variable. The parameter introduced in Chap. 14 for evaluation of continuous systems, a steady-state disturbance sensitivity, can also be applied to discrete-time systems. According to the definition formulated in Chap. 14, the steady-state disturbance sensitivity S_D is

$$S_D = \frac{\Delta y_{ss}}{\Delta v_{ss}}. \tag{16.41}$$

Figure 16.6 shows a digital control system subjected to a disturbance signal having z transform $\mathbf{V}(z)$. The output $\mathbf{Y}(z)$ in this system consists of two components – one caused by the input signal $\mathbf{X}(z)$, and the other produced by the disturbance $\mathbf{V}(z)$; that is,

$$\mathbf{Y}(z) = \mathbf{T}_{CL}(z)\mathbf{X}(z) + \mathbf{T}_D(z)\mathbf{V}(z), \tag{16.42}$$

where $\mathbf{T}_{CL}(z)$ is the system closed-loop pulse transfer function and $\mathbf{T}_D(z)$ is the disturbance pulse transfer function, defined as

$$\mathbf{T}_D(z) = \frac{\mathbf{Y}(z)}{\mathbf{V}(z)}\bigg|_{\mathbf{X}(z)=0}. \tag{16.43}$$

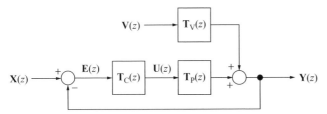

Figure 16.6. Block diagram of a digital control system subjected to disturbance $\mathbf{V}(z)$.

By use of block-diagram algebra, the disturbance transfer function is found to be

$$\mathbf{T}_{\mathrm{D}}(z) = \frac{\mathbf{T}_{\mathrm{V}}(z)}{1 + \mathbf{T}_{\mathrm{P}}(z)\mathbf{T}_{\mathrm{C}}(z)}. \tag{16.44}$$

Assuming that Δv_{ss} is unity, which does not restrict the generality of these considerations, the disturbance sensitivity can be expressed as

$$S_{\mathrm{D}} = \Delta y_{\mathrm{ss}} = \lim_{z \to 1}(1 - z^{-1})\mathbf{V}(z)\mathbf{T}_{\mathrm{D}}(z) = \lim_{z \to 1}\mathbf{T}_{\mathrm{D}}(z). \tag{16.45}$$

In general, to reduce the system disturbance sensitivity, the open-loop gain has to be high, but the extent to which the open-loop gain can be increased is usually limited by the system transient performance requirements.

16.5 DIGITAL CONTROLLERS

One of the greatest advantages of digital controllers is their flexibility. A control algorithm performed by a digital controller is introduced in the form of a computer code, in some cases just a few lines of BASIC or C programming. All it takes to change the control algorithm is to rewrite those few lines of programming. In addition, there are very few limitations, especially when compared with those of analog controllers, on what kinds of control actions can be encoded. Yet, in spite of the great flexibility and ease of implementing various control algorithms, over 90% of industrial digital controllers perform a classical PID algorithm. The main reason for the popularity of PID controllers is an extensive theoretical and practical knowledge of many aspects of their performance carried over from the years of analog controllers. The presentation of digital controllers in this book is limited to the PID control algorithm.

The continuous PID control law was given by Eq. (14.43):

$$u(t) = k_{\mathrm{p}} \left[e(t) + \frac{1}{T_{\mathrm{i}}} \int_0^t e(t_{\mathrm{D}})dt_{\mathrm{D}} + T_{\mathrm{d}}\frac{de(t)}{dt} \right].$$

In a discrete-time system, the integral and derivative terms are approximated by use of discretized models. Various discretization methods can be used to generate different versions of the digital PID algorithm. The simplest method for approximating an integral is a rectangular (staircase) backward-difference approximation described as

$$\int_0^t e(t_{\mathrm{D}})dt_{\mathrm{D}} \approx \sum_{i=0}^{k-1} e(i)T, \tag{16.46}$$

where T is the sampling time. The derivative of the control error is approximated by a backward-difference quotient:

$$\frac{de(t)}{dt} \approx \frac{e(k) - e(k-1)}{T}. \tag{16.47}$$

Substituting approximating expressions of Eqs. (16.46) and (16.47) into Eq. (14.43) and replacing $u(t)$ and $e(t)$ with $u(k)$ and $e(k)$, respectively, yields

$$u(k) = K_P e(k) + K_i \sum_{i=0}^{k-1} e(i) + K_d[e(k) - e(k-1)], \tag{16.48}$$

where K_p is the digital proportional gain and K_i and K_d are the digital integral and derivative gains, respectively, given by

$$K_p = k_p, \tag{16.49}$$

$$K_i = k_p T / T_i, \tag{16.50}$$

$$K_d = k_p T_d / T. \tag{16.51}$$

Equations (16.49)–(16.51) are provided only to show the relationship of a discrete model to a continuous model of a PID controller. Thus they do not imply the existence of more than nominal correspondence between discrete-model coefficients and the continuous-model coefficients derived in Section 14.5 from Ziegler and Nichols.

Equation (16.48) is referred to as a position PID algorithm. Another very widely used form of digital PID is a velocity algorithm. To derive the velocity algorithm, first consider the control signal defined by Eq. (16.48) at the $(k-1)T$ instant of time:

$$u(k-1) = K_p e(k-1) + K_i \sum_{i=0}^{k-2} e(i) + K_d[e(k-1) - e(k-2)]. \tag{16.52}$$

Subtracting Equation (16.52) from Equation (16.48) yields

$$\begin{aligned} \Delta u(k) = u(k) - u(k-1) &= K_p[e(k) - e(k-1)] \\ &+ K_i e(k-1) + K_d[e(k) - 2e(k-1) + e(k-2)], \end{aligned} \tag{16.53a}$$

or, in more compact form,

$$\Delta u(k) = K_0 e(k) + K_1 e(k-1) + K_2 e(k-2), \tag{16.53b}$$

where

$$\begin{aligned} K_0 &= K_p + K_d, \\ K_1 &= K_i - 2K_d - K_p, \\ K_2 &= K_d. \end{aligned} \tag{16.54}$$

The velocity algorithm is usually preferred over the position algorithm because it is computationally simpler, it is safer [in case of controller failure, the control signal remains unchanged, $\Delta u(k) = 0$], and it better handles "wind-up." A control error wind-up occurs within the controller after a control actuator – a control valve, for instance – hits a stop. When this happens, the error signal to the controller persists and the integrator output continues to increase, producing a "wind-up" phenomenon.

The controller gains K_p, K_i, and K_d are selected to meet specified process performance requirements and must be adjusted according to the process transient and steady-state characteristics. A set of tuning rules, derived from the Ziegler–Nichols rules for analog controllers presented in Chap. 14, can be used to adjust the controller gains on the basis of the process step response. The Ziegler–Nichols rules for a digital PID controller are

$$K_i = \frac{0.6T}{R(L+0.5T)^2},$$

$$K_p = \frac{1.2}{R(L+T)},$$

$$K_d = \frac{0.5}{RT}. \tag{16.55}$$

The values of R and L are determined from the step response curve, as shown in Fig. 14.12. It should be stressed that, just as in the case of analog controllers, the Ziegler–Nichols rules do not guarantee optimal settings for the digital controller gains. In most cases, however, the values obtained with Eqs. (16.55) provide a good starting point for further fine-tuning of the controller gains based on the system on-line performance.

As shown by Eqs. (16.55), the values of the control settings of a digital controller are also dependent on the sampling time T. Selecting a proper sampling time for a digital controller is a complex matter involving many factors. Some of the more important system characteristics affected by the sampling time are stability, bandwidth, sensitivity to disturbances, and sensitivity to parameter variations. All of these characteristics improve when the sampling interval is decreased. However, the cost of the digital system increases when a small sampling time or, equally, a high sampling frequency is required, simply because faster computers and faster process interface devices are more expensive than slower ones. Moreover, the initial magnitude of a control signal produced by a digital PID controller in response to a step change of an input signal is greater if the sampling time is smaller for most control algorithms. Thus, selecting the sampling time for a digital control system is a compromise between performance and cost (including the control effort), a familiar dilemma for a system designer.[2]

16.6   SYNOPSIS

Digital control is a broad and still rapidly growing field. In this chapter, an attempt was made to present in simple terms some of the basic problems associated with analysis and design of digital control systems. The discussion was limited to single-loop systems, the most common structure in industrial digital control, implemented either as simple stand-alone systems or as parts of more complex distributed systems. In spite

---

[2] For more information on selection of sampling time for digital control systems, see Franklin *et al.*, *op. cit.*, pp. 449–78.

of availability of many new and improved control algorithms developed specifically for discrete-time systems, industrial digital controllers are usually programmed to perform conventional PID algorithms in discrete-time form. Two versions of the digital PID algorithms were introduced, and tuning rules derived from Ziegler–Nichols rules for analog PID controllers were discussed.

The design of digital control systems, like the design of continuous control systems, is an art of compromise between transient and steady-state performance. A method of bilinear transformation for determining stability of discrete-time systems was described. The relation between location of poles of the closed-loop transfer function and the system transient performance was examined. Two basic criteria of steady-state performance – steady-state control error and sensitivity to disturbances – were shown to depend on the system open-loop gain and the type of input signal in a manner similar to that observed in continuous feedback systems.

Although digital control offers many advantages to both designers and users of closed-loop control systems, it should be noted that the best performance achievable with a digital PID controller is dependent on choice of a sufficiently small value of the sampling time interval T. Unless this value can become infinitesimal, the closed-loop performance of a given system with digital PID control will not be as good as the closed-loop performance of the same system attainable with continuous (analog) PID control, because of the pure delay time T inherent in the sample-and-hold process of the digital control system.

Furthermore, the presence of the derivative term in the discrete-model PID controller generates a derivative component of the controller output signal that varies inversely with T, often resulting in controller output signals that are too large for the process to use (signal saturation at the process input), with the result that the beneficial effects of the derivative action are lost. Hence this tends to limit the smallness of T. The same saturation effect also occurs with continuous PID control, unless a time lag is also present in the controller, but the continuous control system does not suffer from the presence of the finite delay time T.

For the engineer who has accumulated a body of experience working with continuous control systems, the design of a conventional digital controller is readily accomplished by using characteristics (PID, compensation, and so forth) needed for continuous control in the s domain and then finding the corresponding characteristics of a digital control system. This can be accomplished through the use of one of a number of possible transform methods, as discussed in some detail by Ogata,[3] for instance.

PROBLEMS

16.1 The shaded area in Fig. P16.1 represents a region of desirable locations of poles for a continuous dynamic system. The damping ratio associated with the complex-conjugate poles located in this region is greater than 0.7, and the damped natural frequency is smaller than 3.93 rad/s. Find a corresponding region in the z plane, assuming that the sampling period is 0.2 s.

---

[3] K. Ogata, *Discrete-Time Control Systems* (Prentice-Hall, New York, 1987), pp. 306–478.

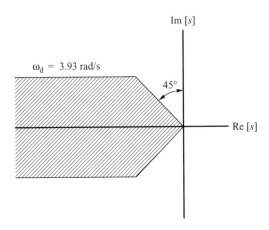

Figure P16.1. Region of desirable locations of poles for a continuous system.

16.2 In certain applications of continuous control systems, it is desirable that the closed-loop poles of the system transfer function be located far enough to the left of the imaginary axis in the s plane. Mathematically, this requirement can be written as

$$\text{Re}[p_i] \le a_{min}, \qquad i = 1, 2, \ldots, n,$$

where a_{min} is a negative number and p_i are closed-loop poles. This inequality is satisfied in the shaded area shown in Fig. P16.2. Find a corresponding region in the z plane.

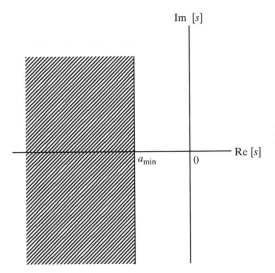

Figure P16.2. The region $\text{Re}[p_i] \le a_{min}$ in the s plane.

16.3 Obtain the z transform of the digital PID control algorithm given by Eqs. (16.53). Find a pulse transfer function of the PID controller and arrange it into a ratio of polynomials in z.

16.4 A control algorithm used in a digital control system, shown in Fig. P16.4, is given by the difference equation

$$u(k) = u(k-1) + K_p e(k) - 0.9512 K_p e(k-1).$$

The sampling time is 0.1 s, and the continuous process transfer function is

$$\mathbf{T}_p(s) = \frac{0.25}{s+1}.$$

(a) Determine the maximum value of the controller gain K_p for which the closed-loop system, consisting of the controller, the ZOH, and the continuous process $\mathbf{T}_p(s)$, remains stable.

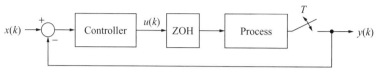

$x(k) \longrightarrow$ Controller $\xrightarrow{u(k)}$ ZOH $\longrightarrow$ Process $\xrightarrow{T} \longrightarrow y(k)$

Figure P16.4. Digital control system considered in Problem 16.4.

(b) Derive an expression for the steady-state error of the system in response to a unit ramp input kT, $k = 0, 1, \ldots$. Find the value of the steady-state error for the value of the controller gain found in part (a).

16.5 A pulse transfer function of a continuous process with ZOH was found to be

$$\mathbf{T}_p(z) = (z - 1)\mathcal{Z}\left\{\frac{\mathbf{T}_p(s)}{s}\right\} = \frac{0.025z^2 + 0.06z + 0.008}{z^3 - 1.6z^2 + 0.73z - 0.1}.$$

Determine the stability condition for a proportional controller of gain K_p in a digital control system with the process and ZOH represented by $\mathbf{T}_p(z)$.

16.6 A digital PD controller is to be designed for control of a continuous process having transfer function $\mathbf{T}_p(s) = 1/s(s+2)$. The design specifications for the closed-loop system include a damping ratio of 0.7 and a period of step response oscillation equal to 5 s.

(a) Find the required control parameters k_p and T_d for a continuous PD controller that meets the design specifications.

(b) Find the proportional and derivative gains of a corresponding digital PD controller, assuming a sampling time of $T = 0.2$ s, and write the discrete control algorithm in velocity form.

(c) Obtain the pulse transfer function of the closed-loop system, including the digital PD controller and the continuous process preceded by a ZOH.

(d) Find a difference equation relating output and input variables of the closed-loop system.

(e) Calculate the first 15 values of the closed-loop system unit step response.

Fourier Series and the Fourier Transform

When the input to a dynamic system is periodic – i.e., a continuously repeating function of time, having period T such as the function shown in Fig. A1.1 – it is often useful to describe this function in terms of an infinite series of pure sinusoids known as a Fourier series.

One form of such an infinite series is

$$x(t) = \frac{a_0}{2} + \sum_{k=1}^{\infty} a_k \cos k\omega_1 t + \sum_{k=1}^{\infty} b_k \sin k\omega_1 t, \qquad (A1.1)$$

where $a_0/2 = (1/T) \int_{t_0}^{t_0+T} x(t)\, dt$ is the average, or constant, value of the function, $\omega_1 = 2\pi/T$ is the radian frequency of the lowest-frequency component, and the amplitudes of the series of component sinusoids at succeeding frequencies $k\omega_1$ are given by

$$a_k = \frac{2}{T} \int_{t_0}^{t_0+T} x(t) \cos k\omega_1 t\, dt, \qquad (A1.2)$$

$$b_k = \frac{2}{T} \int_{t_0}^{t_0+T} x(t) \sin k\omega_1 t\, dt. \qquad (A1.3)$$

Alternatively, this function may be expressed as a series of only sine waves or only cosine waves by use of

$$x(t) = X_0 + \sum_{k=1}^{\infty} X_k \sin(k\omega_1 t + \phi_{ks}) \qquad (A1.4)$$

or

$$x(t) = X_0 + \sum_{k=1}^{\infty} X_k \cos(k\omega_1 t + \phi_{kc}), \qquad (A1.5)$$

where

$$X_0 = \frac{a_0}{2} = \frac{1}{T} \int_{t_0}^{t_0+T} x(t)\, dt,$$

$$X_k = \sqrt{a_k^2 + b_k^2},$$

$$\phi_{ks} = \tan^{-1} \frac{b_k}{a_k},$$

$$\phi_{kc} = -\tan^{-1} \frac{a_k}{b_k}.$$

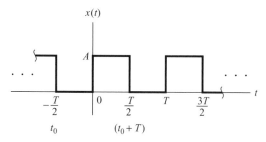

Figure A1.1. A typical periodic function.

The steady response of a dynamic system (i.e., the response remaining after all transients have decayed to zero) to an infinite series of sine waves may also be expressed as an infinite series:

$$y(t) = Y_o + \sum_{k=1}^{\infty} Y_k \sin(k\omega_1 t + \phi_{k_o}) \tag{A1.6}$$

where Y_o is the constant component of output and the coefficients Y_k are the amplitudes of the successive sine waves having frequencies $k\omega_1$ and phase angles ϕ_{k_o}. The values of Y_o, Y_k, and ϕ_{k_o} at each frequency are obtained by the methods developed in Chap. 12.

Alternatively, the periodic function $x(t)$ may be expressed in terms of an infinite series of exponentials having the form $e^{jk\omega_1 t}$ through the use of Euler's equations:

$$\sin \omega t = \frac{1}{2j} \left(e^{j\omega t} - e^{-j\omega t} \right), \tag{A1.7}$$

$$\cos \omega t = \frac{1}{2} \left(e^{j\omega t} + e^{-j\omega t} \right). \tag{A1.8}$$

Substituting these expressions into Eq. (A1.1) yields

$$x(t) = \frac{a_0}{2} + \sum_{k=1}^{\infty} \frac{1}{2} \left[(a_k - jb_k) e^{jk\omega_1 t} + (a_k + jb_k) e^{-jk\omega_1 t} \right]$$

or

$$x(t) = X_0 + \sum_{k=1}^{\infty} \left(\mathbf{X}_k e^{jk\omega_1 t} + \overline{\mathbf{X}}_k e^{-jk\omega_1 t} \right), \tag{A1.9}$$

where

$$X_0 = \frac{a_0}{2} = \frac{1}{T} \int_{t_0}^{t_0+T} x(t)\, dt, \tag{A1.10}$$

$$\mathbf{X}_k = \frac{a_k - jb_k}{2}, \tag{A1.11}$$

$$\overline{\mathbf{X}}_k = \frac{a_k + jb_k}{2}. \tag{A1.12}$$

Noting that

$$\sum_{k=1}^{\infty} \overline{\mathbf{X}}_k e^{-jk\omega_1 t} = \sum_{k=-1}^{-\infty} \mathbf{X}_k e^{jk\omega_1 t},$$

one may simplify the expression for $x(t)$ and then combine it with the expression for X_0 to form the single summation

$$x(t) = \sum_{n=-\infty}^{\infty} \mathbf{X}_n e^{jn\omega_1 t}, \tag{A1.13}$$

where

$$\mathbf{X}_n = \frac{a_n - jb_n \frac{n}{|n|}}{2} = \frac{1}{T} \int_{t_0}^{t_0+T} x(t) e^{-jn\omega_1 t} dt, \tag{A1.14}$$

and n represents the complete set of integers from negative infinity to positive infinity, *including zero*.

The complex coefficients $\mathbf{X}_n$ (i.e., the conjugate pairs $\mathbf{X}_k$ and $\overline{\mathbf{X}}_k$) have magnitudes and phase angles given by

$$|\mathbf{X}_n| = \frac{1}{2}\sqrt{a_n^2 + b_n^2}, \tag{A1.15}$$

$$\angle \mathbf{X}_n = -\tan^{-1} \frac{n}{|n|} \frac{b_n}{a}. \tag{A1.16}$$

Because the exponential components of the series in Eq. (A1.13) act in conjugate pairs for each value of $|n|$, the magnitude of $\mathbf{X}_n$ is one-half the amplitude of the corresponding wave in the sine wave series in Eq. (A1.4), and the phase angle of $\mathbf{X}_n$ is the negative of the phase angle ϕ_{ks} of the corresponding wave in the sine-wave series.

As a prelude to the discussion of Laplace transforms in Appendix 2, it can be seen that if the period T is allowed to increase and approach infinity, the frequency ω_0 approaches zero and the frequency interval between successive values of $n\omega_0$ becomes infinitesimal. In the limit, Eq. (A1.14) becomes the expression for the Fourier transform:

$$\mathbf{X}(j\omega) = \mathcal{F}\{x(t)\} = \int_{-\infty}^{\infty} x(t) e^{-j\omega t} dt. \tag{A1.17}$$

The Fourier transform is closely related[1] to the Laplace transform, which uses the exponential e^{-st} instead of the exponential $e^{-j\omega t}$ in the transformation integral.

EXAMPLE A1.1

Find the expression for the complex coefficients $\mathbf{X}_n$ of the infinite series of Fourier exponentials to represent the periodic function shown in Fig. A1.1 and develop the terms of the exponential series into a series of sines and/or cosines. Sketch roughly to scale each of the first two sinusoidal components (i.e., for $n = 1, 3$) on a graph, together with the square-wave periodic function.

---

[1] See D. Rowell and D. N. Wormley, *System Dynamics: An Introduction* (Prentice-Hall, Upper Saddle River, NJ, 1997), pp. 539–42.

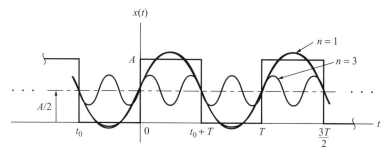

Figure A1.2. Sketch showing sine waves for $n = 1, 3$ superposed on square-wave function.

SOLUTION

From Eq. (A1.14), $\mathbf{X}_n$ is found to be

$$\mathbf{X}_n = \frac{1}{T} \int_{-T/2}^{+T/2} x(t) e^{-jn\omega_1 t} dt. \tag{A1.18}$$

Integrating from $-T/2$ to 0 and from 0 to $T/2$ and substituting the corresponding values of $x(t)$ yield

$$\mathbf{X}_n = \frac{1}{T} \left(\int_{-T/2}^{0} 0 e^{-jn\omega_1 t} dt + \int_{0}^{+T/2} A e^{-jn\omega_1 t} dt \right)$$

$$= \frac{1}{T} \left(0 - \frac{A}{jn\omega_1} e^{-jn\omega_1 t} \Big|_{0}^{+T/2} \right) = \frac{jA}{n\omega_1 T} \left(e^{-jn\omega_1 T/2} - 1 \right). \tag{A1.19}$$

Substituting the trigonometric form for the complex exponential in Eq. (A1.19) and using $\omega_1 = 2\pi/T$, we obtain

$$\mathbf{X}_n = \frac{jA}{2n\pi} (\cos n\pi - j \sin n\pi - 1) = \frac{jA}{2n\pi} (\cos n\pi - 1). \tag{A1.20}$$

Thus, for even values of n, $\mathbf{X}_{n_{\text{even}}} = 0$, and for odd values of n,

$$\mathbf{X}_{n_{\text{odd}}} = \frac{-jA}{n\pi}. \tag{A1.21}$$

For $n = 0$,

$$X_0 = \frac{A}{2}. \tag{A1.22}$$

Substituting Eqs. (A1.21) and (A1.22) into Equation (A1.13) yields

$$x(t) = \frac{A}{2} + \frac{2A}{\pi} \frac{1}{2j} \left(e^{j\omega_1 t} - e^{j\omega_1 t} \right)$$

$$+ \frac{2A}{3\pi} \frac{1}{2j} \left(e^{j3\omega_1 t} - e^{j3\omega_1 t} \right) + \cdots +$$

$$+ \frac{2A}{n\pi} \frac{1}{2j} \left(e^{jn\omega_1 t} - e^{jn\omega_1 t} \right). \tag{A1.23}$$

Then, using Euler's identity for $\sin \omega t$ yields

$$x(t) = \frac{A}{2} + \frac{2A}{\pi} \sin(\omega_1 t) + \frac{2A}{3\pi} \sin(3\omega_1 t)$$
$$+ \cdots + \frac{2A}{n\pi} \sin(n\omega_1 t). \qquad (A1.24)$$

The sine waves for $n = 1$ and $n = 3$ are shown superposed on the square-wave function in Fig. A1.2.

Laplace Transforms

A2.1 DEFINITION OF LAPLACE TRANSFORM

A Laplace transform is a mapping between the time domain and the domain of complex variable s defined by

$$\mathbf{F}(s) = \mathscr{L}\{f(t)\} = \int_0^\infty f(t)e^{-st}\,dt, \qquad (A2.1)$$

where s is a complex variable, $s = \sigma + j\omega$, and $f(t)$ is a sectionally continuous function of time. Function $f(t)$ is also assumed to be equal to zero for $t < 0$. With this assumption, the transform defined by Eq. (A2.1) is called a one-sided Laplace transform.

The condition for the existence of a Laplace transform of $f(t)$ is that the integral in Eq. (A2.1) exists, which in turn requires that there exist real numbers, A and b, such that $|f(t)| < Ae^{bt}$. Most functions of time encountered in engineering systems are Laplace transformable.

Laplace transforms are commonly used in solving linear differential equations. By application of the Laplace transform, the differential equations involving variables of time t are transformed into algebraic equations in the domain of complex variable s. The solutions of the algebraic equations, which are usually much easier to obtain than the solutions of the original differential equations, are then transformed back to the time domain by use of the inverse Laplace transform.

A2.2 INVERSE LAPLACE TRANSFORM

The inverse Laplace transform is defined by the Riemann integral:

$$f(t) = \mathscr{L}^{-1}\{\mathbf{F}(s)\} = \frac{1}{2\pi j}\int_{c-j\infty}^{c+j\infty} \mathbf{F}(s)e^{st}\,ds. \qquad (A2.2)$$

The Riemann integral is rarely used in practice. The most common practical method used in inverse Laplace transformation is the method of partial fraction expansion, which will be described later.

A short list of Laplace transforms and z transforms, defined in Chap. 15, of the most common functions of time is given in Table A2.1.

A2.3 BASIC PROPERTIES OF THE LAPLACE TRANSFORM

What follow are useful basic properties of the Laplace transform:

(1) Linearity:

$$\mathcal{L}\{a_1 f_1(t) + a_2 f_2(t)\} = a_1 \mathbf{F}_1(s) + a_2 \mathbf{F}_2(s). \tag{A2.3}$$

(2) Integration:

$$\mathcal{L}\left\{ \int_0^{-t} f(\tau) d\tau \right\} = \frac{\mathbf{F}(s)}{s}. \tag{A2.4}$$

(3) Differentiation:

$$\mathcal{L}\left\{ \frac{d^n f(t)}{dt^n} \right\} = s^n \mathbf{F}(s) - \sum_{k=0}^{n-1} s^{n-k-1} \left[\frac{d^k f(t)}{dt^k} \right]\Bigg|_{t=0^-}. \tag{A2.5}$$

(4) Shifting argument in the time domain (multiplication by an exponential in the s domain):

$$\mathcal{L}\{ f(t - a)\} = e^{-as} \mathbf{F}(s). \tag{A2.6}$$

(5) Shifting argument in the s domain (multiplication by an exponential in the time domain):

$$\mathcal{L}\{ f(t)e^{-at}\} = \mathbf{F}(s + a). \tag{A2.7}$$

In addition to these properties, two theorems are very useful in evaluating initial and steady-state values of function $f(t)$. The initial-value theorem is

$$f(0^+) = \lim_{s \to \infty} s\mathbf{F}(s). \tag{A2.8}$$

If the limit of function $f(t)$ as time approaches infinity exists, then the steady-state value of function $f(t)$ is given by the final-value theorem,

$$\lim_{t \to \infty} f(t) = \lim_{s \to \infty} s\mathbf{F}(s). \tag{A2.9}$$

A2.4 PARTIAL FRACTION EXPANSION METHOD

The method of partial fraction expansion is most commonly used in obtaining inverse Laplace transforms of functions that have a form of ratios of polynomials in s, such as

$$\mathbf{F}(s) = \frac{\mathbf{B}(s)}{\mathbf{A}(s)} = \frac{b_0 + b_1 s + \cdots + b_m s^m}{a_0 + a_1 s + \cdots + a_n s^n}, \tag{A2.10}$$

where $m \leq n$. Transfer functions of linear systems often take such form. If the roots of the characteristic equation, $\mathbf{A}(s) = 0$, are $s_1, s_2, \ldots, s_q$, the expansion of $\mathbf{F}(s)$ takes the form

$$\mathbf{F}(s) = \frac{C_{11}}{s - s_1} + \frac{C_{12}}{(s - s_1)^2} + \cdots + \frac{C_{1p_1}}{(s - s_1)^{p_1}} + \cdots + \frac{C_{qp_q}}{(s - s_q)^{p_q}}, \tag{A2.11}$$

where q is the number of roots, some of which may be multiple roots, and p_i is a multiplicity of the ith root. Note that if all roots are distinct, $q = n$. Equation (A2.11) can be rewritten in the more compact form

$$\mathbf{F}(s) = \sum_{i=1}^{q} \sum_{j=1}^{p_i} \frac{C_{ij}}{(s - s_i)^j}. \tag{A2.12}$$

The constants C_{ij} are given by

$$C_{i1} = \lim_{s \to s_i} [\mathbf{F}(s)(s - s_i)^{p_i}],$$

$$C_{i2} = \lim_{s \to s_i} \left\{ \frac{d[\mathbf{F}(s)(s - s_i)^{p_i}]}{ds} \right\},$$

$$C_{ij} = \frac{1}{(p_i - j)!} \lim_{s \to s_i} \frac{d^{p_i - j}[\mathbf{F}(s)(s - s_i)^{p_i - j}]}{ds^{p_i - j}}. \tag{A2.13}$$

In Examples A2.1 and A2.2, the partial fraction expansion method is used to find inverse Laplace transforms. In Example A2.3, Laplace transformation is used to obtain a transfer function and a step response for a mass–spring–dashpot system.

EXAMPLE A2.1

Find an inverse Laplace transform of function $\mathbf{F}(s)$:

$$\mathbf{F}(s) = \frac{2s + 4}{s^3 + 7s^2 + 15s + 9}.$$

SOLUTION

The first step is to find the poles of $\mathbf{F}(s)$. There is a single pole at $s = -1$ and a pole of multiplicity 2 at $s = -3$. $\mathbf{F}(s)$ can thus be rewritten as

$$\mathbf{F}(s) = \frac{2s + 4}{(s + 1)(s + 3)^2}.$$

Hence the expanded form of $\mathbf{F}(s)$ is

$$\mathbf{F}(s) = \frac{C_{11}}{s + 1} + \frac{C_{21}}{s + 3} + \frac{C_{22}}{(s + 3)^2}.$$

By use of Eqs. (A2.13), the constants C_{11}, C_{21}, and C_{22} are found to be

$$C_{11} = 0.5, \quad C_{21} = -0.5, \quad C_{22} = 1.$$

The partial fraction expansion of $\mathbf{F}(s)$ becomes

$$\mathbf{F}(s) = \frac{0.5}{s + 1} - \frac{0.5}{s + 3} + \frac{1}{(s + 3)^2}.$$

The inverse Laplace transforms of the three simple terms on the right-hand side of the preceding equation can be found in Table A2.1, yielding the solution

$$f(t) = 0.5e^{-t} - 0.5e^{-3t} + te^{-3t}.$$

Table A2.1. Laplace and z transforms of most common functions of time

Continuous	Discrete	Laplace transform	z Transform
$U_i(t)$	$U_i(kT)$	1	1
$U_s(t)$	$U_s(kT)$	$1/s$	$z/(z-1)$
t	kT	$1/s^2$	$Tz/(z-1)^2$
t^2	$(kT)^2$	$2/s^3$	$T^2 z(z+1)/(z-1)^3$
e^{-at}	e^{-akT}	$1/(s+a)$	$z/(z-e^{-at})$
te^{-at}	kTe^{-akT}	$1/(s+a)^2$	$Te^{-aT}z/(z-e^{-at})^2$
$\sin \omega t$	$\sin \omega kT$	$\omega/(s^2+\omega^2)$	$z \sin \omega T/(z^2 - 2z \cos \omega T + 1)$
$\cos \omega t$	$\cos \omega kT$	$s/(s^2+\omega^2)$	$z(z-\cos \omega T)/(z^2 - 2z \cos \omega T + 1)$

EXAMPLE A2.2

Find the inverse Laplace transform of

$$F(s) = \frac{4}{s(s^2 + 2.4s + 4)}.$$

SOLUTION

In addition to the pole $s = 0$, $F(s)$ has two complex-conjugate poles, $s_2 = -1.2 - j1.6$ and $s_3 = -1.2 + j1.6$. The factored form of $F(s)$ is

$$F(s) = \frac{4}{s(s + 1.2 + j1.6)(s + 1.2 - j1.6)}.$$

The partial fraction expansion in this case is

$$F(s) = \frac{C_{11}}{s} + \frac{C_{21}}{(s + 1.2 + j1.6)} + \frac{C_{31}}{(s + 1.2 - j1.6)}.$$

The constants C_{11}, C_{21}, and C_{31} are obtained by use of Eqs. (A2.13):

$$C_{11} = 1, \quad C_{21} = 0.625e^{-j0.6435}, \quad C_{31} = 0.625e^{j0.6435}.$$

Hence

$$F(s) = \frac{1}{s} + 0.625 \left(\frac{e^{j0.6435}}{s + 1.2 + j1.6} + \frac{e^{-j0.6435}}{s + 1.2 - j1.6} \right).$$

By use of Table A2.1 to find the inverse Laplace transforms of the terms on the right-hand side of the preceding equation, the function of time is obtained as

$$f(t) = 1 + 0.625[e^{j0.6435}e^{(-1.2-j1.6)t} + e^{-j0.6435}e^{(-1.2+j1.6)t}].$$

To simplify the form of $f(t)$, first group the exponents of the exponential terms:

$$f(t) = 1 + 0.625(e^{j0.6435-1.2t-j1.6t} + e^{-j0.6435-1.2t+1.6t}).$$

Moving the common factor outside the parentheses yields

$$f(t) = 1 + 0.625e^{-1.2t}[e^{-j(-0.6435+1.6t)} + e^{-j(0.6435-1.6t)}].$$

Now, substitute equivalent trigonometric expressions for the two exponential terms to obtain

$$f(t) = 1 + 0.625e^{-1.2t}[\cos(-0.6435 + 1.6t) - j \sin(-0.6435 + 1.6t) + \cos(0.6435 - 1.6t) - j \sin(0.6435 - 1.6t)].$$

By use of basic properties of sine and cosine functions, the final form of the solution is obtained:

$$f(t) = 1 + 1.25e^{-1.2t} \sin(0.9273 + 1.6t).$$

EXAMPLE A2.3

Consider the mass–spring–dashpot system shown in Fig. 11.7. Use Laplace transformation to find the system transfer function and obtain the step response $x(t)$ for a step change of force $F(t)$.

SOLUTION

The system differential equations of motion are

$$m\frac{dv}{dt} = F(t) - bv - kx,$$

$$\frac{dx}{dt} = v.$$

Laplace transformation of these two equations yields

$$m[s\mathbf{V}(s) - v(0^-)] = \mathbf{F}(s) - b\mathbf{V}(s) - k\mathbf{X}(s),$$

$$s\mathbf{X}(s) - x(0^-) = \mathbf{V}(s),$$

where $x(0^-)$ and $v(0^-)$ represent the initial conditions for displacement and velocity of mass m just before the input force is applied. Now, combining the Laplace-transformed equations yields

$$\mathbf{X}(s)(ms^2 + bs + k) = \mathbf{F}(s) + x(0^-)(ms + b) + mv(0^-).$$

Assuming zero initial conditions, $x(0^-) = 0$ and $v(0^-) = 0$, the system transfer function is found to be

$$\mathbf{T}(s) = \frac{\mathbf{X}(s)}{\mathbf{F}(s)} = \frac{1}{ms^2 + bs + k}.$$

Next, the system step response is to be found for the input given by

$$F(t) = \Delta F \cdot U_s(t).$$

The Laplace transform of the input signal is

$$\mathbf{F}(s) = \frac{\Delta F}{s}.$$

By use of the expression for the transfer function, the Laplace transform of the output signal is found to be

$$\mathbf{X}(s) = \mathbf{F}(s)\mathbf{T}(s) = \frac{\Delta F}{s(ms^2 + bs + k)}.$$

To obtain the solution in the time domain, two cases must be considered. First, assume that both roots of the quadratic term in the denominator of $\mathbf{X}(s)$ are real and distinct, that is,

$$ms^2 + bs + k = m(s + s_2)(s + s_1),$$

where s_1 and s_2 are real numbers and $s_1 \neq s_2$. The inverse transform of $\mathbf{X}(s)$ in this case is

$$x(t) = \mathcal{L}^{-1}\left\{\mathbf{X}(s)\right\} = \frac{\Delta F}{m}\left(\frac{1 + \dfrac{s_2 e^{-s_2 t} - s_1 e^{-s_1 t}}{s_1 - s_2}}{s_1 s_2}\right).$$

Now, assume that the roots of the quadratic term in the denominator of $\mathbf{X}(s)$ are complex conjugate, that is,

$$ms^2 + bs + k = m\left(s^2 + 2\zeta \omega_n s + \omega_n^2\right).$$

The inverse Laplace transform in this case is

$$x(t) = \mathcal{L}^{-1}\left\{\mathbf{X}(s)\right\} = \frac{\Delta F}{k}\left[1 + \left(\frac{1}{\sqrt{1 - \zeta^2}}\right)e^{-\zeta \omega_n t}\sin(\omega_n \sqrt{1 - \zeta^2}\, t + \phi)\right],$$

where

$$\phi = \tan^{-1}\left(\frac{\sqrt{1 - \zeta^2}}{\zeta}\right).$$

It should be noted that the initial conditions needed when using Laplace transforms, as well as for computer simulations, are at $t = 0^-$, whereas solutions using the classical methods of Chap. 4 need initial conditions at $t = 0^+$.

MATLAB Tutorial

MATLAB OVERVIEW

MATLAB,[1] an interactive and powerful engineering package from The MathWorks, Inc., is one of the most common software tools used in the analysis and design of dynamic systems and the control systems implemented on them. The basic package has many powerful numerical analysis functions as well as the ability to write scripts and functions to aid in repetitive analysis and to allow the user to augment the environment with custom capabilities.

Although it is likely that the reader will have a working knowledge of MATLAB, this tutorial is included to bring the uninitiated reader up to speed and to expose the casual user to several important features that are pertinent to this text. In this tutorial we cover the basic running environment of MATLAB (with its default windows), the fundamental data structure (the matrix), writing scripts and functions, and plotting data. We also explore some capabilities of the Control Systems Toolbox, which has many tools designed to aid in the analysis of dynamic systems. Appendix 4 contains a tutorial for Simulink, a MATLAB add-on, which allows for the rapid development of system simulations.

A3.1.1 Launching MATLAB

Find MATLAB from the START menu on your computer and launch the application. Figure A3.1 shows the default frame with the three standard windows. The "current directory" window has an alternative window that you can access by clicking on the tab. Figure A3.2 shows the alternative window and describes its function.

A3.1.2 The Command Window

The most important part of the user interface is the command window and the MATLAB command prompt. The "≫" character combination is the indication that the MATLAB command window is available for input. By default, the cursor is there. If you've clicked over on another window, you can click back to the prompt to enter a command. Among the activities you can perform at the command prompt are

- assigning numerical values to variables,
- performing mathematical operations on those values,

---

[1] MATLAB is a registered trademark of The MathWorks, Inc.

The Command window: This is where most of the work gets done.

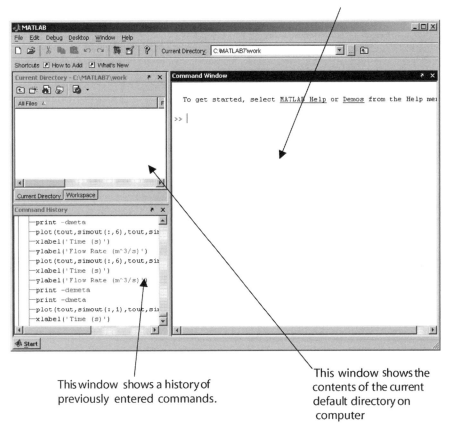

This window shows a history of previously entered commands.

This window shows the contents of the current default directory on computer

Figure A3.1. The MATLAB window.

- displaying those values graphically,
- storing and retrieving data.

Variables are storage locations in the computer that are associated with an alphanumeric name. For example, you can store a numerical value representing your age in years as a variable named "age." The syntax for this in MATLAB would be

```
>> age=22
```

Go ahead and enter this (or the correct equivalent) at your command prompt. The response should look like this:

```
age =
   22
```

Note that MATLAB echoes the results of any command to the screen, unless the command is terminated with the semicolon. When a semicolon is placed at the end of a command line, the output is suppressed. In this case, MATLAB confirms that a variable called "age" is being used and its numerical value is 22.

The Workspace Browser lets
you examine matrices that you
have stored in your workspace

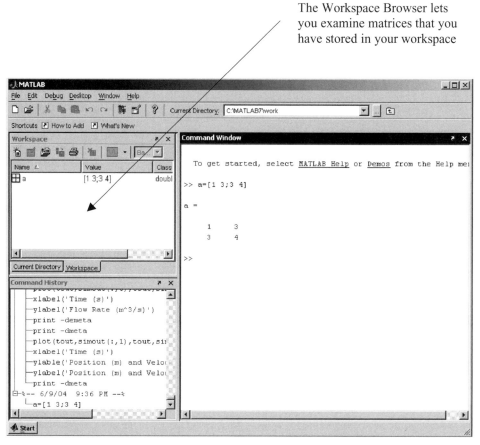

Figure A3.2. Alternative windows in the default frame.

Anything stored in a variable becomes part of the workspace and can be accessed at any time. To see the value of a variable, simply enter its name at the command prompt. You can also use the variable in a mathematical expression and MATLAB will evaluate it, as we will see shortly.

In MATLAB, the fundamental data unit is the "matrix," which is a regular array (a table of rows and columns) of numerical values. The smallest matrix is 1 row by 1 column (referred to as 1×1), which is a single numerical value known as a scalar.

EXERCISE 1: CONVERT TEMPERATURE FROM °F TO °C

Let's explore the data manipulation capabilities of MATLAB. We will store a number in a variable. That number is a value of a temperature in degrees Fahrenheit. We will convert this value to centigrade. Enter the following commands:

```
>> tf=45;
>> tc=(tf-32)*5/9
```

the results of which will be shown on the screen:

```
tc =
7.2222
```

Now we will store a series of numbers in a single variable. This is a matrix with only one row, but many columns. The subsequent commands illustrate the manner in which you can enter a series of numerical values in a vector:

```
≫tf=[45 46 34 38 55 60 65 70]
```

A very powerful feature of MATLAB is that you can perform operations on the entire matrix at the same time:

```
≫tc=(tf−32)*5/9
```

which results in the following matrix:

```
tc =
7.2222   7.7778   1.1111   3.3333   12.7778   15.5556   18.3333
21.1111
```

Finally, we will find the average of these temperatures by using MATLAB's mean function:

```
≫avgf = mean(tf)
≫avgc = mean(tc)
```

the results of which are left to the reader to discover.

A3.1.3 The WORKSPACE Window

Now that you've performed some operations, click over to the workspace window in the upper left-hand panel. You will see a list of all variable names you've used as well as some information regarding how much memory space they take up. Double-clicking on any of the variable names brings up a dialog box that allows you to see, and modify, all values of the matrix in a spreadsheetlike window.

The previous example introduced a matrix made up of a single row (a data structure sometimes called a vector). In the context of computer programming, we would call the two variables introduced in the preceding example (tf and tc) "1 by 8 matrices," meaning that each matrix has 1 row and 8 columns. If you carried out the example on your own version of MATLAB, this would be a good time to look at the workspace window where the matrices and their dimensions are listed.

A3.2 MULTIDEMENSIONAL MATRICES

More complex data structures are also easily included. Take for example, the following 3×3 matrix:

$$a = \begin{bmatrix} 1 & 2 & 4 \\ 6 & -3 & -1 \\ 9 & 4 & -10 \end{bmatrix}.$$

This matrix can be entered into the MATLAB environment with the following command:

```
» a=[1 2 4; 6 -3 -1; 9 4 -10]
```

Note that the different columns are separated by spaces, the rows by semicolons. Alternatively, the following command will also work:

```
» a= [1 2 4
6 -3 -1
9 4 -10]
```

In this case, the columns are still separated by spaces and the rows by carriage returns. Either method works and the choice is left up to the user's discretion.

Now we have a 3 × 3 matrix entered in the MATLAB environment. What if you made an error in typing in the data and needed to change just one value of the matrix? It would be great if you could alter just one value of the matrix without retyping the entire data set. MATLAB gives you an easy method of doing just that. If, for example the first element of the second row (second row, first column) was mistyped and you wanted to replace the erroneously entered data with the correct value of 6, you would issue the following command:

```
»a(2,1) = 6
```

It sometimes helps to put words to a command like this, to be clear what the MATLAB environment will do in this case. This command would be stated as "Take the element in the 2nd row and 1st column of the matrix designated as 'a' and store the value of '6' in that place (regardless of whatever is stored there previously)."

A3.2.1 Linear Algebra

If the reader is at all familiar with linear algebra, the power of the MATLAB environment may now be coming clear. Take, for example the following set of algebraic equations in x_1, x_2, and x_3:

$$x_1 + 2x_2 + 4x_3 = 15,$$
$$6x_1 - 3x_2 - x_3 = 8,$$
$$9x_1 + 4x_2 - 10x_3 = 0.$$

This set of three equations with three unknowns can be recast in matrix form:

$$\begin{bmatrix} 1 & 2 & 4 \\ 6 & -3 & -1 \\ 9 & 4 & -10 \end{bmatrix} \begin{bmatrix} x_1 \\ x_2 \\ x_3 \end{bmatrix} = \begin{bmatrix} 15 \\ 8 \\ 0 \end{bmatrix}.$$

It might aid our discussion to assign names to the matrices in this equation. Let's call the square matrix **A**, the vector containing the unknowns **x**, and the vector on the right-hand side **b**. The equation now becomes

$$\mathbf{Ax = b},$$

and we can find the solution by inverting the matrix **A**:

$$\mathbf{x} = \mathbf{A}^{-1}\mathbf{b},$$

where the inverse of the square matrix is shown as the exponent.

There's a variety of ways to solve this problem, and many numerical algorithms have been proposed to invert a matrix. No doubt you have a calculator that is capable of inverting a matrix of modest dimensions. MATLAB also has this capability. Note that the matrix **A** is the same matrix we have already entered into the MATLAB environment previously. We now wish to enter the **b** matrix:

```
≫ b = [15; 8; 0]
```

Note that **b** is a 3 × 1 matrix and each row is separated by a semicolon. Now we solve the equations:

```
≫ x = inv(a) * b
```

and the solution is

```
x =
2.3471
1.1676
2.5794
```

It is left as an exercise for the student to verify this solution.

A3.2.2 Manipulating Matrices

Sometimes it's important to manipulate entire rows or columns of a matrix (as we will see when we consider plotting command). In this situation, the colon (:) operator is invaluable. The colon acts as a placeholder or wild card when addressing the individual elements. Recall that, in the previous section, we were able to address a single element in the matrix by specifying the specific row and column of the element we wished. These numbers specifying the placement are called indices of the matrix. The following command would allow you to see (and hence, manipulate) the entire second row of the matrix:

```
≫ a(2,:)
```

Likewise, if you wished to see the entire first column:

```
≫ a(:,1)
```

Following the previous example, it's helpful to put this command in words. The last command says "take the matrix **a** and show me all rows in the first column." Similarly, the first command would say "take the matrix **a** and show me the second row, all columns".

This notation introduces yet another way of entering matrices into the workspace. Instead of a single command in which all elements are entered at once, you can enter the data one row or one column at a time. For example, the 3×3 matrix previously used could be entered this way:

```
>> a(1,:) = [1 2 4]
>> a(2,:) = [6 −3 −1]
>> a(3,:) = [9 4 −10]
```

A3.3 PLOTTING IN MATLAB

One of the strengths of the MATLAB environment is its data visualization capabilities. From simple $x - y$ plots of paired data to complex three-dimensional visualization and rendering techniques, MATLAB provides users with a complex array of flexible tools for plotting and graphing. In this section, we limit our discussion to those tools that are most often used to plot time-based sequences, like the time response of dynamic systems to various inputs or initial conditions.

First, we introduce another use of the " : " operator in the environment. At the command prompt, issue the following command:

```
>> t = [0: 0.01: 1.0]
```

This command creates a matrix that is 1 row by 101 columns. The elements are a regularly spaced sequence starting at 0 (the first element), spaced by 0.01 (the second element), and ending at 1.0 (the third element). This is a handy and easy method of generating an array of regularly spaced elements and works for both positive and negative increments.

In Chap. 4, we solved first- order differential equations that describe the response of systems with one energy-storing element. Let us look at one such response:

$$y = 5\left[1 - e^{-3t}\right].$$

This describes a step response with a time constant of $1/3$ s and a steady-state response of 5.0.

To visualize this response, first we must compute it over a range of values for time. The following command uses MATLAB's flexible matrix operations to perform this in one command:

```
>> y=5*(1−exp(−(1/0.333)*t));
```

which should result in a new matrix in the workspace, which is also 1 row by 101 columns. To see what this response looks like, simply use the plot command

```
>> plot(t,y)
```

which results in a plot like the one shown in Fig. A3.3.

Although space limitations demand that we skim only the surface of MATLAB's plotting capabilities, let's explore this a little further.

Now compute two different responses, both with a steady-state value of 5 but with time constants of 0.5 and 1, respectively:

```
>> y1=5*(1−exp(−(1/0.5)*t))
>> y2=5*(1−exp(−1*t))
```

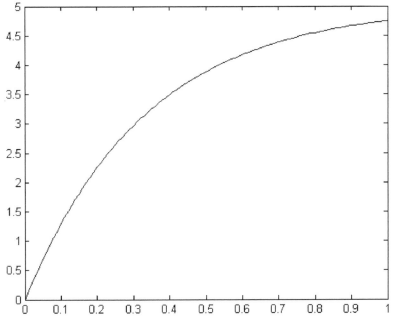

Figure A3.3. Response vs. time plot by use of the MATLAB plot command.

Each response can be plotted one at a time, as indicated in the preceding discussion, but it would be more instructive to plot them on the same plot:

```
≫ plot(t,y,t,y1,t,y2)
```

The results are shown in Fig. A3.4.

As an exercise, the reader is instructed to further explore the plot function using the following command,

```
≫ help plot
```

and to try out various options.

A3.4 PROGRAMMING IN MATLAB

Although the tutorial to this point describes the capabilities of the interactive environment, the true power and flexibility of MATLAB can be achieved only through user programming. In this section, we explore the two different methods available to add custom additions to the environment: scripts and functions.

A script is a series of MATLAB commands that are stored in a separate file on your computer hard drive. When that script is invoked (by typing the filename at the MATLAB prompt), the commands stored in the file are automatically executed. All variables created by the script are stored in the workspace. Likewise, any variable referenced (accessed) by commands in the script are assumed to exist in the workspace.

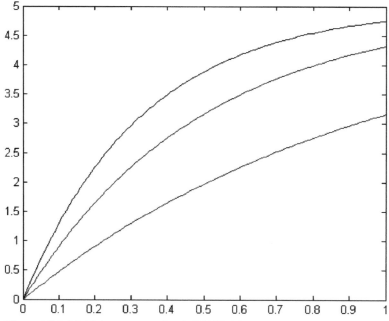

Figure A3.4. Multiple response plots in MATLAB.

To truly appreciate the flexibility of scripts, you should learn to use program flow control commands (`while`, `if`, `until`, `for`) and to use variables as indices to matrices. These topics are beyond the scope of this tutorial, but are similar to any discussion applied to programming languages such as C, C++, FORTRAN, and BASIC.

As an example, let us assume that the three step responses that we used to illustrate the plot command are to be used repeatedly in future sessions. It would be nice if we could gather up all the commands we used to generate those matrices and then we would have to issue only a single command every time we restart our work.

We begin by starting the m-file editor, an integral part of the MATLAB environment. We easily do this by clicking on the standard "new document" icon at the very far left end of the MATLAB toolbar. The m-file editor is a text editor that is "aware" of the MATLAB syntax and automatically flags errors as you type.

In the new document window, type in the commands that generated and plotted the three responses we previously saw. The m-file editor window should look something like Fig. A3.5.

Not that the ; operator was used at the end of each of the commands. The semicolon does not change the computation of the command but does suppress the output to the screen. For debugging purposes, removing the semicolons allows the user to see intermediate computational results. On the other hand, suppressing output greatly speeds performance.

Before you can run the script, you must store it as a file on your hard drive. Using the standard Windows "Save As" menu calls up the default subdirectory (usually one called `work` within the MATLAB directory structure.) Save this file as "myfile" (the Windows extension .m is automatically appended.) Note that, once the file has been

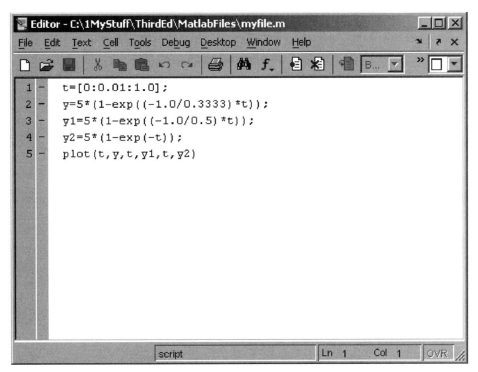

Figure A3.5. M-file editor window with a sample script.

stored, the title bar of the editor window shows the name, and directory path, of the file you're working on. Note also that an asterisk sometimes appears on the title bar. This indicates that changes made to the file have not yet been saved. The file must be saved before the changes will be in effect.

Before you run the file, you should clear out the workspace so that you can observe the results of your script. The clear command is simple and intuitive:

```
≫ clear
```

the results of which can be observed in the workspace window.

To run the file, click over to the command window and simply type the name of the file:

```
≫ myfile
```

The workspace window should immediately show the returned matrices and the figure window should show the plot as before. If you made a mistake in your typing, an error message would have been visible in the command window, indicating the line in your m-file where the execution was halted.

Script files are very handy for taking care of repetitive tasks and recalling frequently used groups of commands. However, the second kind of m-file, the function m-file, is far more powerful.

Keeping with our example, let us assume that we wish to compare the responses with systems with three different time constants, but the values of those time constants are likely to change. Further, let us assume that we wish not only to see the responses plotted, but we wish to store all the responses in one large matrix, with each response occupying a single column of the matrix. For the sake of completeness, and because this is a common data structure in MATLAB, we will also include the time sequence as the first column of this matrix. The resulting matrix is 4 columns (1 for time, 3 for the responses) and 101 rows.

Much of the code we've already written will form the basis of our new program. So we make another copy of this file by saving it as "myfunc" using the "Save As" function.

The main thing that distinguishes functions from scripts is the declaration command at the top of the file. The first line for our function is

```
function myfunc(tau1,tau2,tau3)
```

The first word is a reserved word in the environment and hence becomes blue in the editor window. The next is the name of the function (same as the name of the file) followed by a list of arguments. These are variables that represent the data that will be passed to the function when it is called from the command line or from another m-file. In this case the three variables represent three different time constants.

Now we edit the script file so that the time constants in the responses are replaced with our arguments. The editor window will now look something like Fig. A3.6.

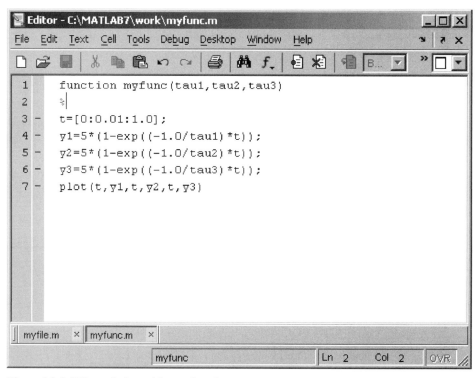

Figure A3.6. Editor window with new function.

Now go back to the command window, issue the clear command, and invoke this function with the same three time constants (0.333, 0.5, 1.0):

```
≫clear
≫myfunc(0.333,0.5,1.0)
```

Note the results in both the plot and workspace windows.

The plot window should have the same appearance as Fig. A3.4, but you will note that the workspace window is empty. What happened to the matrices that were formed within the function? This question points out another important difference between script files and functions. Script files operate just as if the commands were issued at the command prompt with access to all workspace variables and any new data generated stored in the workspace. Functions, on the other hand, do not interact directly with the workspace nor are the data they generate stored in the workspace. All variables generated within the functions are local to that function and cease to exist once function execution is completed. To get data back to the workspace, the function must *return* the data structure as part of the call.

Now let's modify the function to generate the 101 × 4 matrix containing all the responses and return the function to the calling workspace.

The following command will build up our larger matrix:

```
resp=[t' y' y1' y2'];
```

Note that the individual vectors now make up the columns of our new matrix resp. Note also that the apostrophe operator is used to indicate the transpose of the individual matrix is stored. The transpose of a matrix is one in which all rows are swapped to columns. Therefore the transpose of a 1 × 101 matrix (like all of our matrices to date) becomes a 101 × 1 matrix. Because we wish to store the responses in columns of our new matrix and because the original vectors are all row vectors, transposes become necessary.

The final step to complete our function is to modify the declaration statement to indicate that some data structure will be returned. The editor window, as seen in Fig. A3.7, shows the correct syntax.

Finally, we will demonstrate two different uses of the function from the command line.

First, type this command:

```
≫myfunc(0.333,0.5,1.0)
```

Once again, the plot window should show the expected results, but two different effects are visible in the command window. First, a new variable appears in the workspace, called ans, and second, the entire new 101 × 4 matrix is echoed to the screen. The variable name ans is used by MATLAB whenever it is asked to generate a result, but the user didn't provide a variable name. It is the default for results to be stored.

Now issue the following command:

```
≫ mymat = myfunc(0.3333,0.5,1.0);
```

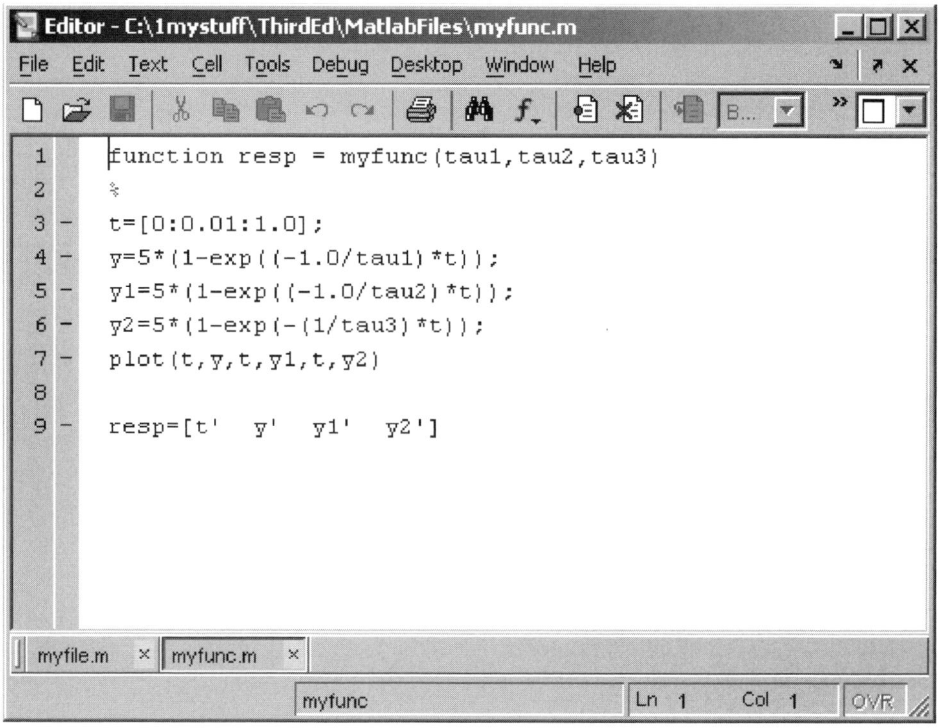

```
    Editor - C:\1mystuff\ThirdEd\MatlabFiles\myfunc.m                    _□×
 File   Edit   Text   Cell   Tools   Debug   Desktop   Window   Help         ×

 1        function resp = myfunc(tau1,tau2,tau3)
 2        %
 3 -      t=[0:0.01:1.0];
 4 -      y=5*(1-exp((-1.0/tau1)*t));
 5 -      y1=5*(1-exp((-1.0/tau2)*t));
 6 -      y2=5*(1-exp(-(1/tau3)*t));
 7 -      plot(t,y,t,y1,t,y2)
 8
 9 -      resp=[t'   y'   y1'   y2']
```
myfile.m × myfunc.m ×

                    myfunc   Ln 1   Col 1   OVR

Figure A3.7. MATLAB Function returning a matrix.

In this case, the semicolon will suppress output to the screen, but the new matrix, `mymat`, is now present in the workspace. It's important to note that the names you use for variables (either returned or as arguments) in the workspace are not related at all to the names you chose for variables within the function itself. Once the function is up and running, any user should be able to use it with no knowledge of its inner workings at all. This concept is a very important cornerstone in good programming practice.

One final point before we move on. The following command uses the MATLAB `plot` command along with the colon operator to plot just the second response from the matrix returned from the function call (recall that the first column of the matrix `mymat` contains time, the second the first response, and so on). The reader is encouraged to investigate this syntax and explore its use further:

>>`plot(mymat(:,1),mymat(:,3))`

A3.5 SOLVING ORDINARY DIFFERENTIAL EQUATIONS IN MATLAB

A3.5.1 Mathematical Overview

As discussed in the text, an important implication of state-space theory is that all lumped-parameter systems can be represented as a system of first-order differential equations. This makes it possible to develop a general numerical method for the solution of these systems. Chapter 5 describes the general nature and structure of these algorithms, and Chap. 6 and Appendix 4 deal with the use of higher-level simulation packages like Simulink to construct complex, yet well-structured, models of physical systems.

MATLAB offers yet another approach to the solution of ODEs by providing a set of algorithms that can be invoked directly from the command line, or from other files and functions. In this case, the model structure is provided in the form of an m-file function that takes the current state of the system as input and returns a vector of the values of the derivatives.

MATLAB has at least seven different ODE solver algorithms to choose from, and the plusses and minuses are discussed at some length in the MATLAB help documentation. For a large portion of electromechanical systems, `ode45` provides excellent results. This will form the basis of the examples to demonstrate the technique.

The simplest form of the solution command is

```
>> [T,Y] = ode45(odefun,tspan,y0)
```

with the components defined as follows:

T is a vector. On execution, the time steps for which the output (Y) was computed is stored there.

Y is a matrix. Each column corresponds to a state of the system; each row is the value of the states at the time step indicated in the corresponding row of T.

odefun is a pointer to the function that represents the system (evaluates the derivatives). Because it is a pointer, the form of this arguments is @myfunc.

tspan is a vector representing the time values over which the solution is to be carried out. At a minimum it should hold two values, the initial time (usually 0) and the final time.

y0 is a vector containing the initial conditions of the state variables.

EXAMPLE A3.1: FIRST-ORDER SYSTEM

Consider a simple system consisting of a large rotating mass (J) subject to a time-varying input torque (T) and a linear, velocity-dependent friction torque (B). The differential equation that describes this system can be written as follows:

$$J\dot{\Omega} + B\Omega = T.$$

Because the function describing the system is essentially the same as the state equations, this system is recast as a single state equation.

$$\dot{\Omega} = -\frac{B}{J}\Omega + T.$$

For the purposes of this example, we assume that the system is at rest with no input torque. After 1 s has passed, the torque suddenly changes to a value of 1.0. The m-file function that represents this would look like this:

```
function ydot = mysys(t,y0)
%
B =10;
J = 2;
%
if t > 1.0
   torque = 1.0;
else
   torque = 0;
end
%
ydot = -B/J*y0 + torque;
```

This file is stored in the working directory as mysys.m.
The solution is computed by the following command:

```
≫ [tout,yout]=ode45(@mysys,[0 3],[0])
```

Note the manner in which the m-file function is indicated in the function call. The @ sign is an operator that allows MATLAB to easily find your function.
After this command is successfully executed, the results can be plotted:

```
≫plot(tout,yout)
```

Figure A3.8 shows the results.

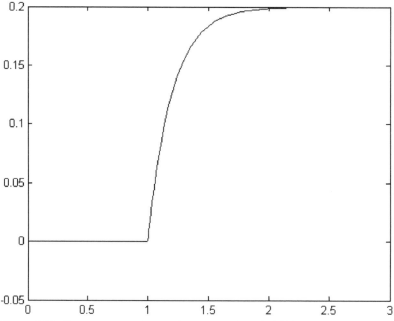

Figure A3.8. Step response of example system using ode45().

To further explore this technique, the reader is encouraged to implement this example and explore the response for various parameter values (B and J) and various step times and torque magnitudes.

EXAMPLE A3.2: NONLINEAR AND HIGHER-ORDER PROBLEMS

Although Example A3.1 clearly illustrates the method, the astute reader will recognize that the response can easily be found analytically and little was gained by using a computer-based solution. Such is not often the case, and, in this example, a more complex and nonlinear system is examined.

In the study of electronic systems, a class of systems called relaxation oscillators is often encountered. One of these is a well-studied nonlinear system called the Van Der Pol oscillator. The model of the Van Der Pol oscillator is a second-order nonlinear ODE,

$$\frac{d^2 y}{dt^2} - \mu \left(1 - y^2\right) \frac{dy}{dt} + y = 0,$$

where μ is a parameter of the system and related to the physical properties of the circuit components.

The equation must first be recast in state-space form. You can begin the process by solving for the highest derivative:

$$\frac{d^2 y}{dt^2} = \mu \left(1 - y^2\right) \frac{dy}{dt} - y.$$

The two states are defined as follows:

$$x_1 = \frac{dy}{dt},$$
$$x_2 = y.$$

The state equations are now easily written:

$$\dot{x}_1 = \mu \left(1 - x_2^2\right) x_1 - x_2,$$
$$\dot{x}_2 = x_1.$$

Translating these equations to an m-file is a straightforward procedure:

```
function xdot = vandp(t,x)
%
% x(1) — first state variable
% x(2) — second state variable
%
mu = 3;
%
xdot1 = mu*(1 — x(2)^2)*x(1)—x(2);
xdot2 = x(1);
xdot=[xdot1;xdot2];
```

Note how the added dimension of this example requires some additional details. First, the argument that passes the current value of the states (called x here) is a vector, the length of which is equal to the order of the system (2, in this case). Second, the value that is returned (xdot) is also a vector. Finally, note that the MATLAB solvers expect the derivatives to be returned as a column vector. The last line in the preceding function ensures that this requirement is met.

Now the equation is solved for time from 0 to 30 s and initial conditions of zero for the value of x_1 and 1.0 for the value of x_2:

```
>> [tout,yout]=ode45(@vandp,[0 30],[0 1]);
```

Plot the solution with this command

```
>>plot(tout,yout(:,1),tout,yout(:,2))
```

the results of which can be seen in Fig. A3.9.

For second-order nonlinear equations, it is often illuminating to plot the two states against each other, eliminating time as an explicit coordinate. The result is a graphic illustration of the actual state trajectory; a plot such as this is called a phase plot of the system. The follow command will generate a phase plot:

```
>> plot(yout(:,2),yout(:,1))
```

which is shown in Fig. A3.10.

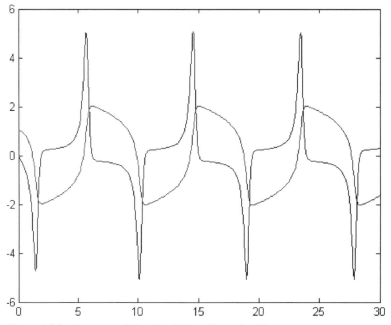

Figure A3.9. Response of Van Der Pol oscillator for 30 s.

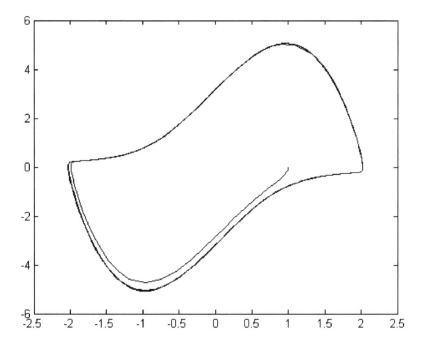

Figure A3.10. Phase plane plot of the system.

A3.6 SYSTEM ANALYSIS IN MATLAB

The main object of the tools and techniques presented in this text is the development of a model to describe the dynamic behavior of a physical system. The most useful model has the form of linear differential equations with constant coefficients. In the field of control systems, this subclass of system models is known as LTI, or linear time-invariant, models. As we've seen throughout the text, such models can take on various forms, the most common of which are state-space models and transfer function models. The Control Systems Toolbox, a very widely used extension of the MATLAB environment, provides a series of powerful algorithms and routines that allow the user to store, manipulate, and analyze LTI models.

There are four ways of specifying LTI models in MATLAB:

- Transfer function (TF) models: These are ratios of polynomials in s, as introduced in Chap. 11 of the text.
- Zero-pole-gain (ZPK) models: Similar to the transfer function models, except that the polynomials are factored and the zeros and poles are explicitly visible in the model representation.
- State-space (SS) models: These models comprise the four state-space matrices as discussed in Chap. 3.
- Frequency-response data (FRD) models: These are models based on empirical frequency-response data and are beyond the scope of this text.

The routines used to store the first three models in the MATLAB environment are briefly described here.

A3.6.1 Transfer Function (TF) Object

This object and the functions for storing and manipulating it are described in some detail in Section 11.7, but the essence is described here as well.

Consider the following transfer function:

$$\mathbf{T}(s) = \frac{10s + 50}{s^3 + 19s^2 + 104s + 140}.$$

MATLAB recognizes a vector of coefficients as representing a polynomial with those coefficients, with the highest-order term coming first. Therefore we can store the two polynomials as row vectors in MATLAB:

```
>> num=[10 50]
>> den=[1 19 104 140]
```

The transfer function object is generated with MATLAB's tf() function:

```
>> mytf=tf(num,den)
```

to which MATLAB responds by echoing the transfer function to the screen in readable format:

```
Transfer function:
    10 s + 50
- - - - - - - - - - - - - - - - - - - - - - - -
s^3 + 19 s^2 + 104 s + 140
```

A3.6.2 Zero-Pole-Gain (ZPK) Object

ZPK models are transfer functions that have been factored into the following form:

$$\mathbf{T}(s) = k\frac{(s - z_1)(s - z_2)\ldots(s - z_m)}{(s - p_1)(s - p_2)\ldots(s - p_n)},$$

where z_i is the ith zero of the transfer function and p_i is the ith pole.

Consider a transfer function that has a single zero at -5 and three poles, located at $-10, -7$, and -2. The steady-state gain (value of the transfer function as s approaches 0) is 0.357:

$$\mathbf{T}(s) = k\frac{(s + 5)}{(s + 10)(s + 7)(s + 2)}.$$

To find k, we must find the expression for the steady-state gain of the transfer function by taking its limit as s approaches zero and set that gain equal to the desired value of 0.357:

$$0.357 = k\frac{(5)}{(10)(7)(2)}.$$

Solving for k:

```
k=10.0
```

To create the ZPK object, first store the zeros and poles as vectors, then use the ZPK command:

```
≫ k=10;
≫ z=[-5];
≫ p=[-10 -7 -2];
≫ myzpk=zpk(z,p,k)
Zero/pole/gain:
 10 (s+5)
-------------------
(s+10) (s+7) (s+2)
```

Again, MATLAB provides direct feedback to verify that you've entered the data correctly.

A3.6.3 State-Space Object

Consider a third-order linear system represented by the following state-space model:

$$\dot{\mathbf{x}} = \mathbf{Ax} + \mathbf{Bu},$$
$$\mathbf{y} = \mathbf{Cx} + \mathbf{Du},$$

where

$$\mathbf{A} = \begin{bmatrix} -19 & -1.625 & -0.2734 \\ 64 & 0 & 0 \\ 0 & 8 & 0 \end{bmatrix},$$

$$\mathbf{B} = \begin{bmatrix} 0.5 \\ 0 \\ 0 \end{bmatrix},$$

$$\mathbf{C} = \begin{bmatrix} 0 & 0.3125 & 0.1953 \end{bmatrix},$$

$$\mathbf{D} = [0].$$

The ss command takes these matrices as arguments and creates the state-space object:

```
≫ mysys=ss(a,b,c,d)
```

```
a =
            x1          x2          x3
x1         -19        -1.625      -0.2734
x2          64          0           0
x3           0          8           0


b =
            u1
x1          0.5
x2          0
x3          0
```

```
c =
                x1              x2              x3
y1              0               0.3125          0.1953
d =
                u1
y1              0
```

A3.6.4 Converting Between Model Formats

The three functions previously described, tf, zpk, and ss, were shown as ways of creating the various forms of LTI system models in the MATLAB environment. They can also be used to convert between the various forms. To do this, simply use the LTI object as the argument in the function call:

```
≫newtf=tf(mysys)
≫newss=ss(myzpk)
≫newzpk=zpk(mytf)
```

It is left as an exercise for the reader to verify that the three models used in the preceding examples are all realizations of the same LTI model.

A3.6.5 Analysis and Manipulation of LTI Models in MATLAB

Once the objects are stored in the MATLAB workspace, they can be manipulated in a fairly intuitive manner. As described in Section 11.7, two systems in series (cascaded) can be combined by multiplying the objects in MATLAB. Similarly, two systems in parallel can be combined by adding their objects. In addition, several other functions are provided to compute unit step responses, frequency-response (Bode) plots, and root-locus diagrams. Tables A3.1–3.3 summarize these functions and indicate the chapters in the text that provide theoretical background for these functions.

Table 3.1. Summary of system object creation and conversion functions

MATLAB command	Description	Reference
≫newtf=tf(mysys)	Converts any other system object to a TF object	Chaps. 3 and 11
≫newzpk=zpk(mysys)	Coverts any other system object to a ZPK object	
≫newss=ss(mysys)	Converts any other system object to a SS object	Chaps. 3 and 11
≫mysys=zpk(z,p,k)	Creates a ZPK object from vectors of zeros, poles, and gain	
≫mysys=ss(A,B,C,D)	Creates a SS object from state-space matrices	Chap. 3
≫mysys=tf(num,den)	Creates a internal system representation TF object from transfer function polynomials	Chap. 11

Table 3.2. Summary of system time-domain functions

MATLAB command	Description	Reference
≫step(mysys)	Generates the time response of the system to a unit step input	Chap. 4
≫impulse(mysys)	Generates the time response to an impulse of unity strength	Chap. 4
≫initial(mysys,x0)	Generates the time response of the system to initial conditions	Chaps. 3, 4, 6
≫lsim(mysys,u,t,x0)	Generates the time response of a system to arbitrary inputs and initial conditions	Chaps. 3, 4, 5, 6

EXAMPLE A3.3

Start by entering the model used in the preceding discussion regarding the creation of system objects. In this case, we'll use the transfer function form, but any would work:

```
≫ num=[10 50]
≫ den=[1 19 104 140]
```

The transfer function object is generated with the tf function:

```
≫ mytf=tf(num,den)
Transfer function:
10 s + 50
- - - - - - - - - - - - - - - - - - - - - - - - -
s^3 + 19 s^2 + 104 s + 140
```

Table 3.3. Summary of system frequency-domain functions

MATLAB Command	Description	Reference
≫bode(mysys)	Plots the Bode plot (frequency response) for the system	Chap. 11
≫nyquist(mysys)	Plots the Nyquist diagram (polar frequency response) for the system	Chap. 12
≫rlocus(mysys)	Generates an interactive root-locus plot for control system design	Chap. 13
≫evalfr(mysys,f)	Evaluates the frequency response at a single (complex) frequency	Chap. 12
≫H=freqresp(mysys,w)	Evaluates the complex frequency response over a grid of frequencies	Chap. 12
≫nichols(mysys)	Plots the Nicols plot of the system (used in control system design)	
≫sigma(mysys)	Generates a plot of singular values for multi-input–multi-output systems	

Now, enter a new transfer function as shown:

```
>> num=10;
>> den=[1 2 100];
>> tf2=tf(num,den)
Transfer function:
      10
----------------
s^2 + 2 s + 100
```

Assuming the two transfer functions represent two systems that are cascaded (connected in series), create a new transfer function representing the overall system dynamics:

```
>> tfall=mytf*tf2
Transfer function:
      100 s + 500
---------------------------------------------------------
s^5 + 21 s^4 + 242 s^3 + 2248 s^2 + 10680 s + 14000
```

Although it's possible to use the analytical tools described in Chap. 4 to compute the step response of this system, MATLAB provides a very straightforward approach:

```
>>step(tfall)
```

which generates the plot shown in Fig. A3.11.

Equally straightforward, but of considerably more utility in control system design, is the Bode plot:

```
>>bode(tfall)
```

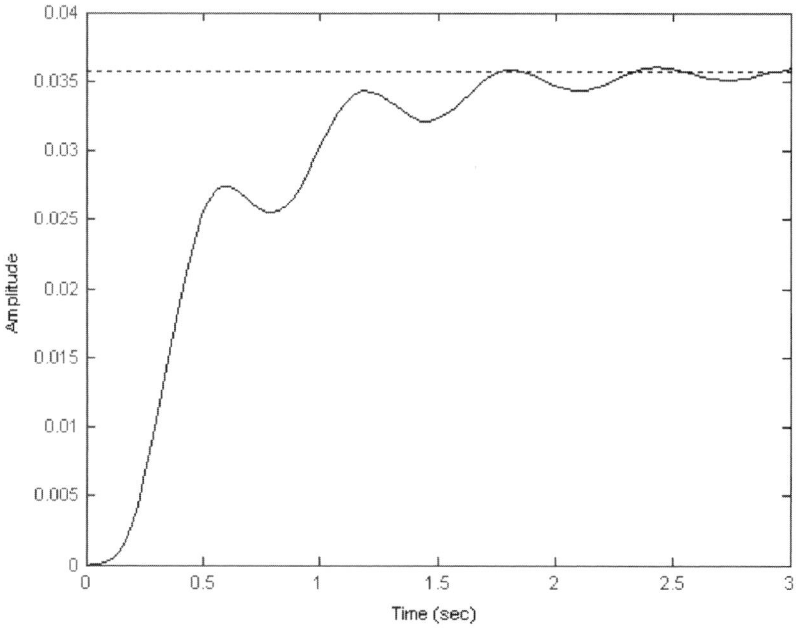

Figure A3.11. Step response obtained with the Control Systems Toolbox step function.

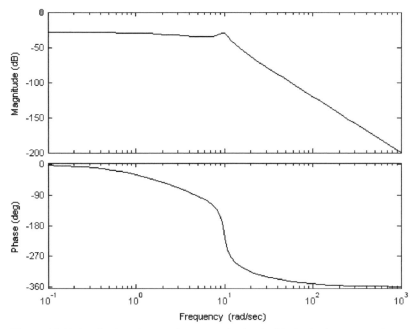

Figure A3.12. Bode plot of example system obtained with the bode command.

which results in the frequency-response plot shown in Fig. A3.12 for this system.

This result is best appreciated by those who have done the computations required for such a plot by hand or other more laborious means.

A3.7 GOING FURTHER: STORING DATA AND GETTING HELP

Often, it's useful to store the contents of the MATLAB workspace and restore the data at a later session. The MATLAB save and load functions are useful in this case. Follow the subsequent instructions to become familiar with these features.

(1) Use the "current directory" window to set the current directory to a suitable work directory such as a floppy disk that you've supplied or a directory on your installation's network drive, where you wish to store data.

(2) Use MATLAB's HELP command to find out more information about the save command. Type help save at the command prompt. Note that the command has many options, but you need concern yourself with only the simplest implementation (save filename).

(3) Save the contents of your workspace to a filename of your choosing.

(4) Verify that the file exists on your floppy or in the directory of the command drive.

(5) Use MATLAB's clear command to erase all variables in the workspace.

(6) Verify that there is nothing stored in the workspace (try help who to learn about a useful command here).

(7) Use the load command to restore your data (help load to find more) and verify that your workspace has been restored.

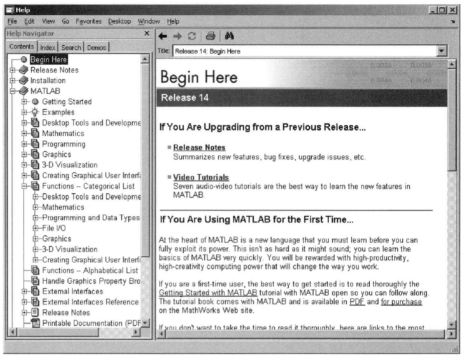

Figure A3.13. MATLAB's powerful Helpdesk feature.

A3.7.1 The HELP Functions

In the previous exercise, you've used the `help` command to get details about a specific function. As helpful as that is, you will find that the `help` command is not very useful for finding information when you don't know the specific function name. For that, use MATLAB's `helpdesk` feature. Enter this command at the prompt:

```
≫ helpdesk
```

A new window is opened featuring a weblike browser interface on the right and a typical help window on the left, as seen in Fig. A3.13. Together, they make a powerful tool for discovering more about MATLAB's features and capabilities. Take some time now to discover two or three features about MATLAB that you think might be useful.

Simulink Tutorial

OVERVIEW OF SIMULINK

This tutorial is intended to acquaint the reader with the rudiments of Simulink,[1] the MATLAB add-on that allows engineers and researchers to rapidly develop and run computer simulations of dynamic systems by using a block-diagram-oriented graphical environment. As with the MATLAB environment itself, the Simulink package is a very rich and versatile tool that is constantly developing. The best we can hope to accomplish with this tutorial is to inspire interest and start you on the path to developing a strong competence in the area of computer simulation of dynamic systems.

LAUNCHING THE SIMULINK LIBRARY BROWSER

Figure A4.1 shows the default configuration when MATLAB launches in a Windows environment. If Simulink is installed as part of the package, the multicolored icon will appear in the tool bar. The first step in building models is launching the Simulink Library Browser by clicking on that icon.

Figure A4.2 shows a typical library browser window. Your installation may look different because the installation of different MATLAB and Simulink add-ons such as the Signal Processing, Artificial Neural Networks, or Fuzzy Logic toolboxes.

A4.2.1 Starting a New Model

Start modeling in Simulink by clicking on the standard "new document" icon on the Library Browser toolbar. That opens a new, blank window on your desktop similar to the one shown in Fig. A4.3.

Building a simulation requires two operations; dragging and dropping blocks from the library and connecting the ports on the blocks by using mouse clicks and drags (like any drawing program). Let's proceed with a very simple example to demonstrate the operation of a simulation. Drag three blocks into the new model window, an integrator block (from the continuous block library), a step input block (from the sources library), and a scope block (from the sinks library). Connect the output of the step to the input of the integrator and the output of the integrator to the input of the scope. Your model window should now look something like the one shown in Fig. A4.4.

---

[1] Simulink is a registered trademark of The MathWorks, Inc.

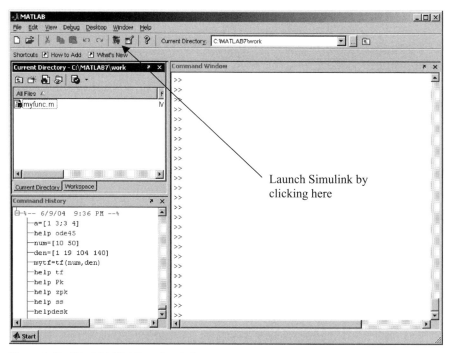

Launch Simulink by clicking here

Figure A4.1. The MATLAB main window.

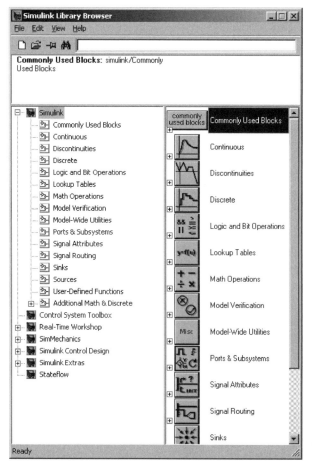

464   **Figure A4.2.** The Simulink library browser.

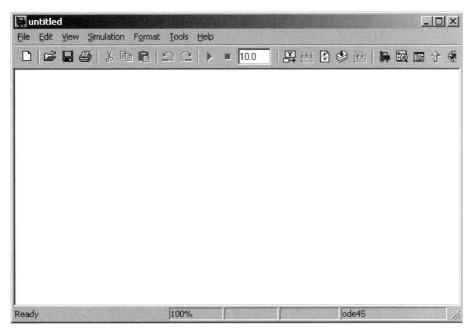

Figure A4.3. New model window in Simulink.

To run the simulation with default settings, double-click on the scope block so you can see its display and click on the run button (the right-facing triangle on the window toolbar) and observe the results. The scope display should be similar to the one in Fig. A4.5.

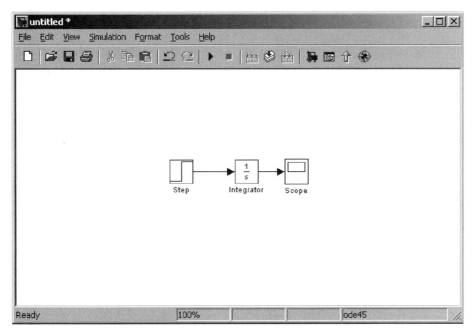

Figure A4.4. Simple model with input, integrator, and output scope.

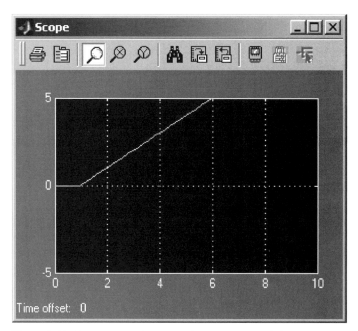

Figure A4.5. Scope output from simple integrator model.

Some discussion of this result is in order. By default, the step-input block starts at 0 and abruptly changes to 1.0 at $t = 1.0$ s. Thereafter it remains at 1.0. Note that the output of the integrator starts at zero (the initial condition of the integrator) until the step time (1 s) is reached, then rises in a straight line, the slope of which is 1.0 (value of the step) for the duration of the simulation. By default, Simulink simulations run for 10 s.

All of the features discussed in the preceding paragraph can be adjusted to suit the simulation: step time, step magnitude, initial conditions of the integrator, and simulation time. By double-clicking on the blocks, you bring up a dialog box that allows the user to change the settings of each box. Make the following changes to your simulation:

- Double-click on the step input and change the step time to 2.0 s and the final value to 0.5.
- Change the initial condition of the integrator to –1.0
- Using the "Simulation|Configuration Parameters" menu selection, set the stop time to 5.0 s. (*Note*: In Version 6.0 and later of Simulink, this can also be accomplished directly on the toolbar of the model window.)

The simulation window looks the same as it did before, but this time we would expect the output to begin at –1 (the initial condition of the integrator), to change at 2.0 s (the step time), and the slope of the ramp to be about half as steep (the magnitude of the step becomes the slope of the ramp). Run the simulation and observe the results. If all goes well, the output looks something like Fig. A4.6.

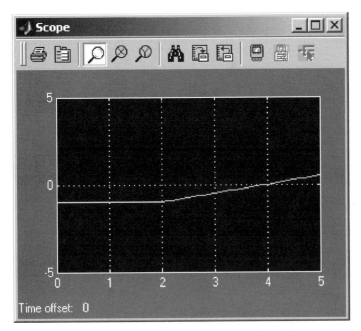

Figure A4.6. Scope output for simple integrator with modified parameters.

EXAMPLE A4.1: SIMULATING A FIRST-ORDER SYSTEM

Consider the following differential equation in $y(t)$ with input $u(t)$:

$$a_1 \frac{dy}{dt} + a_0 y = b_0 u. \tag{A4.1}$$

The first step in simulating the system represented by this equation is solving for the highest derivative:

$$\frac{dy}{dt} = \frac{b_0}{a_1} u - \frac{a_0}{a_1} y. \tag{A4.2}$$

Once the equation is in this form, it suggests the form of the simulation block diagram. The simulation will have one integrator block (indicated by the fact that it's a first-order system). The output of the integrator is the dependent variable, $y(t)$. Therefore the input to the integrator block is the derivative of y, which is defined by Eq. (A4.2). So now we can expect to see a simulation block diagram in which the input to the integrator is the difference between two signals, one proportional to the input that drives the system and one proportional to the variable in question, y. We can use the simple integrator from the previous example and modify it to build this simulation. We will need two new kinds of blocks, the gain block that multiplies a signal by a constant gain (the same as a constant coefficient in an equation) and a summation block that adds or subtract signals. Both can be found in the math operations library. Drag a copy of each of these blocks into your simulation window, as shown in Fig. A4.7.

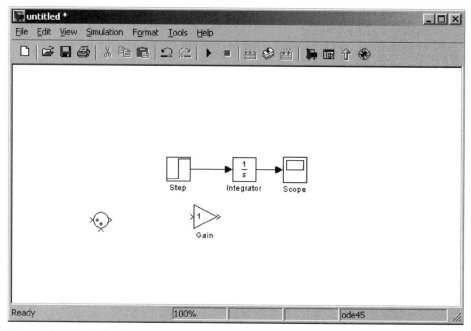

Figure A4.7. Beginning the modifications for the new simulation.

On inspection, you can see that we really need two gain blocks in this simulation because there are two constant coefficients. Instead of dragging a new gain block from the library, right-click on the gain block in this window and drag to create a copy. This is a very useful feature of Simulink that the reader is well advised to learn. Now break the existing connections (select, then press the delete key) and form a new block diagram similar to the one shown in Fig. A4.8.

We're not done yet, but let's take a minute to point out two new operations that were required for getting this far. First, the second gain block (below the integrator) is pointing the other way. This is easily accomplished by selecting the block and choosing "flip block" from the format menu. Second, we had to "tap off" from a signal line to have the output of the integrator serve as input to both the scope block and the second gain block. Tapping a line is done by positioning the cursor over the line to be tapped and right-clicking, then dragging.

Now we need to make sure the settings are correct. The observant reader will notice that the summation block is adding the two signals together, not subtracting them as demanded by the equations. Fix this by double-clicking on the summing block. In the slot labeled "list of signs" change the string "|++" to "|+−" to get the desired effect. Adding signs to the string creates more ports on the block, allowing you to add and subtract an arbitrary number of signals for more complicated simulations.

This would be a good time to change the labels on the blocks to create a more readable block diagram. Click on the block labels to edit them. Figure A4.9 shows a possible set of labels. Note that the convention is to label the integrator blocks to indicate the output of the blocks.

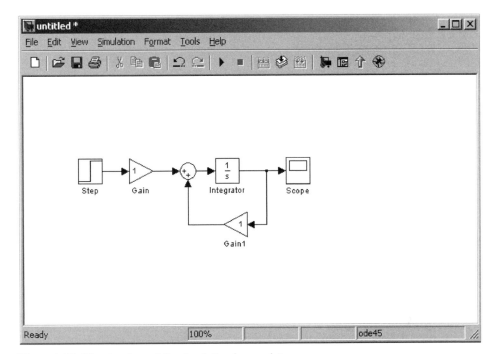

Figure A4.8. The structure of the simulation is complete.

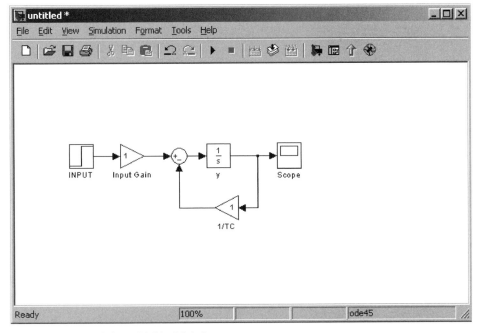

Figure A4.9. Simulation with block labels.

Before we proceed, we must assign numerical values to the coefficients, decide on an input, and verify the initial condition for the simulation. Let's choose the following values:

$$a_0 = 2.0,$$
$$a_1 = 4.0,$$
$$b_0 = 8.0,$$

which leads to the following ratios:

$$\frac{b_0}{a_1} = 2.0,$$

$$\frac{a_0}{a_1} = 0.5.$$

The first ratio is the value of the gain labeled "Input Gain" in Fig. A4.9, and the second ratio is the value for the other gain. Double-click on the gain blocks and set them accordingly. As you can see in Fig. A4.10, the gain blocks now show the values that you've set.

Finally, set the Step Input to transition from 0 to 1 at 1 s, and set the initial condition of the integrator to 0.0 and the simulation time to 10 s (all defaults). Open the scope block and run the simulation. Figure A4.11 shows the output from this simulation on the scope.

It is left to the reader to verify that this is the expected unit step response for this differential equation.

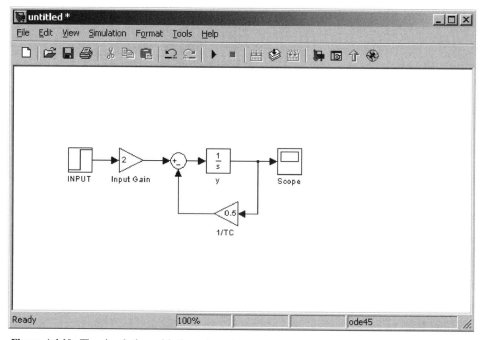

Figure A4.10. The simulation with the gains set.

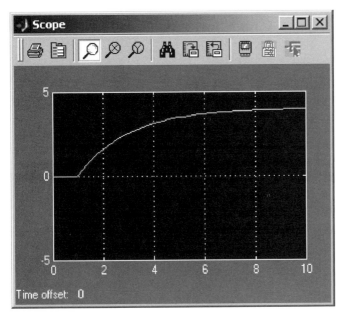

Figure A4.11. Scope output showing a step response of a first-order system.

EXAMPLE A4.2: NONLINEAR SIMULATIONS AND USER-DEFINED FUNCTIONS

The previous sections described the use of Simulink to simulate a linear first-order system, which is easily solvable with traditional analytical techniques. The true power of computer simulations lies in their ability to solve nonlinear problems and those of arbitrary order. In this section, we explore methods needed for the solution of nonlinear equations.

As described in Appendix 3, a very common nonlinear differential equation is the Van Der Pol equation, which describes the oscillations of circuits by use of vacuum tubes. A common form the the Van Der Pol equation is

$$\frac{d^2y}{dt^2} - \mu(1 - y^2)\frac{dy}{dt} + y = 0. \tag{A4.3}$$

This equation is a second-order equation; hence we expect a simulation with two integrators. Building the simulation is an exercise in manipulating the Simulink blocks to carry out the operations expressed in the equation. However, for equations of this, and greater, complexity, a more direct and elegant approach is called for. Simulink provides a number of ways to introduce user-defined functions, the most versatile of which is the MATLAB FCN block. We will use this block to compute the highest derivative of our system, given the values of the states. Figure A4.12 shows the beginning of the model structure.

The MATLAB function block allows you to call a user-defined m-file function (see the description of m-files in Appendix 3) from within a Simulink model. Before we get into the details of the function, it's important to complete the model definition.

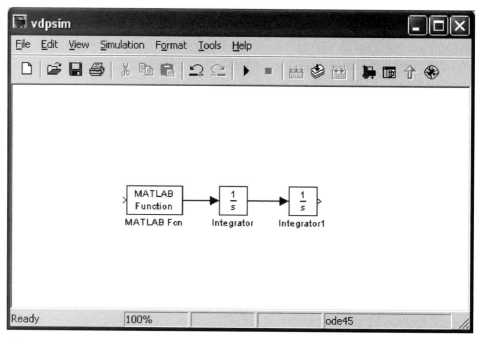

Figure A4.12. Introducing the MATLB FCN block.

From the preceding model, it should be clear that the function will compute the value of the second derivative. Mathematically, it should carry out this evaluation:

$$\frac{d^2y}{dt^2} = \mu\left(1 - y_2^2\right)\frac{dy}{dt} - y. \tag{A4.4}$$

The function that computes this must therefore have access to the current values of both y and its derivative. From Fig. A4.12, y is the output of the right-hand integrator whereas dy/dt is clearly the value between the two integrators. Figure A4.13 shows the next step in building the simulation.

Note that we have two values that need to attach to the input of the function block, but only one input port. We solve this difficulty by using yet another Simulink block, the Multiplexor or MUX block. The MUX block takes its name from an electronic component that allows two or more signals to share a common communication channel. The MUX block therefore has one output and an arbitrary number of inputs. You will find the MUX block in the "Signal Routing" library in Simulink. By default, the MUX block has two inputs, but you can easily change that by double-clicking on the block. Figure A4.14 shows the model with the MUX in place.

Finally, let's insert some blocks that allow us to monitor the simulation results. In this case, we'll use two different scope blocks. One will allow us to monitor both integrator outputs on the same scope (again using a MUX block), the other will allow us to plot one state against the other. The latter is called the XY Graph block and, unlike the Scope block, it generates a standard MATLAB graph in a figure window. Figure A4.15 shows the final configuration of the simulation model.

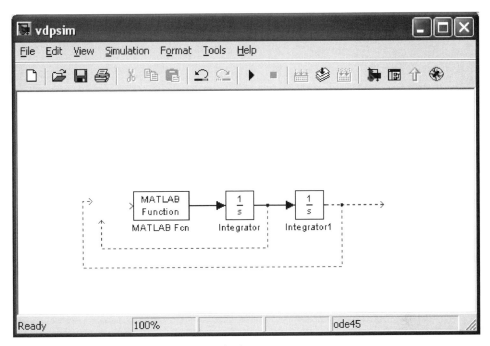

Figure A4.13. Van Der Pol simulation, continuing.

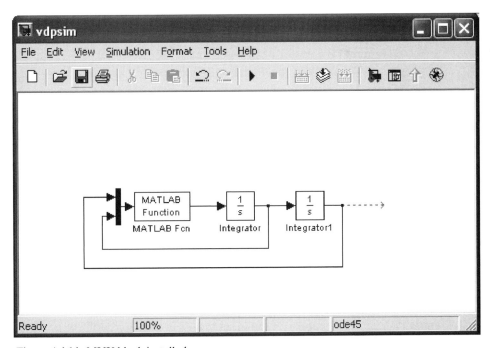

Figure A4.14. MUX block installed.

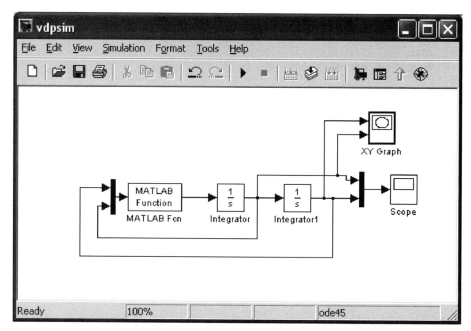

Figure A4.15. The model structure complete.

Figure A4.15 shows the structure of the model, but the actual equation has not yet been implemented. For that we turn to the MATLAB m-file editor (see Appendix 3) and write a function that has the following characteristics:

- It takes as an argument a vector containing the current values of *y* and *dy/dt*,
- the argument is a vector of two elements,
- it returns the value of the highest derivative,
- it is stored in the same subdirectory as the Simulink file.

One possible implementation of the function is shown in the subsequent program listing:

```
function y2d = vdpfun(y)
%
% y(1) = y
% y(2) = dy/dt
%
mu=3;
%
y2d = mu*(1-y(1)^2)*y(2)-y(1);
```

Note that the order in which the two state values are passed to the function are not arbitrary. They appear in the input vector in the same order they appear on the MUX block in the Simulink model (top to bottom).

Store this file as `vdpfun.m` and return to the Simulink model. Double-click on the `Fcn` block and provide the name of the function you just defined and indicate the dimension of the information being returned to Simulink. In this case, we are

Block Parameters: MATLAB Fcn

—MATLAB Fcn—

Pass the input values to a MATLAB function for evaluation. The function must return a single value having the dimensions specified by 'Output dimensions' and 'Collapse 2-D results to 1-D'.
Examples: sin, sin(u), foo(u(1), u(2))

—Parameters—

MATLAB function:

`vdpfun`

Output dimensions:

`1`

Output signal type: `auto` ▼

☑ Collapse 2-D results to 1-D

Sample time (-1 for inherited):

`-1`

OK	Cancel	Help	Apply

Figure A4.16. Dialog box for MATLAB Fcn.

returning a scalar, dimension 1. Figure A4.16 shows the dialog box appropriately filled out.

One final step, and we're ready to run the simulation. We want to set the initial value of y to be 1.0 and the initial value of dy/dt as 0.0. As before, double-click on the integrator blocks to accomplish this and run the simulation.

The scope block shows the two states during the first cycle of oscillation, as seen in Fig. A4.17.

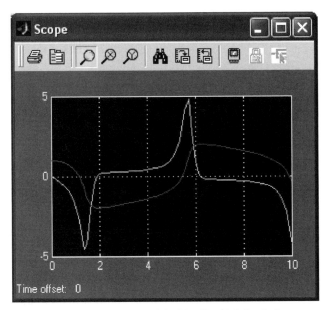

Figure A4.17. Scope output of the Van Der Pol simulation.

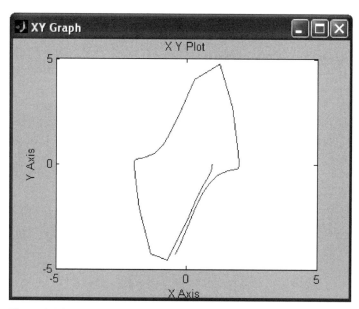

Figure A4.18. Output of XY graph after the scale has been properly set.

The XY Graph, however (you may have to find the figure window; it won't necessarily pop to the top of your desktop), is probably not very illuminating because most of the state trajectory is outside the plotting window set by the default settings of the bock. Referring to Fig. A4.18, we can see that all of the activity of the oscillator falls between −5 and 5. Go back to the Simulink model and double-click on the XY Graph block to reset the scale, and then rerun the simulation. The XY Graph produces a plot similar to that seen in Fig. A4.18.

Two features might be apparent from this graph. One is that it doesn't seem to have settled down to a steady oscillation (which would be indicated by the plot retracing itself around the plane several times.) The second is more subtle. Note the sharp "cusps" in the plot at the very top and the very bottom of the trajectory. This occurs because the variable-time-step algorithm in Simulink has taken fairly large steps at those points (while still maintaining specified accuracy) but the plotting algorithm has simply drawn straight lines between the computed points. We can improve the appearance of the plot by changing the accuracy requirements of the Simulink solver.

From the Simulation menu, choose "Configuration Parameters." In the dialog box, change the "Relative Tolerance" to 1E-6. Also, change the "Stop Time" to 30.0 and rerun the simulation. Figure A4.19 shows the improved plot that addresses both concerns.

A4.3   ADDITIONAL PLOTTING AND STORAGE OPTIONS

The Simulink "Sinks" library has several blocks that can be used to implement a variety of visualizations. However, the maximum flexibility is achieved by storing the results of the simulation as a MATLAB workspace variable (matrix) that can then

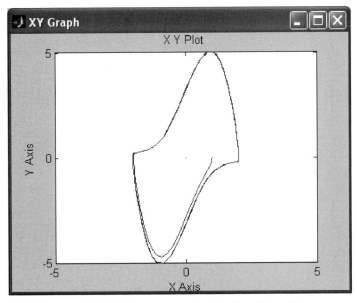

Figure A4.19. Typical state trajectory of a Van Der Pol oscillator.

be saved as a file or manipulated like a MATLAB variable. The "To Workspace" block in the "Sinks" library accomplishes this.

Keeping with the Van Der Pol simulation, drag a "To Workspace" block to the simulation window, and connect it to the simulation by tapping the signal that connects to the scope block. This is shown in Fig. A4.20.

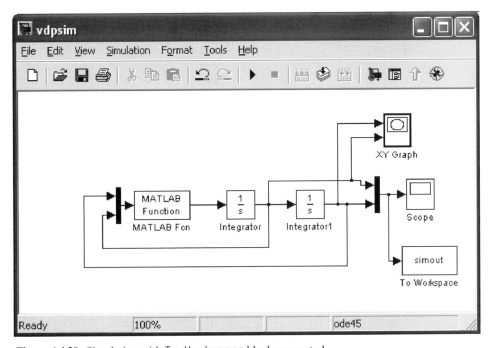

Figure A4.20. Simulation with To Workspace block connected.

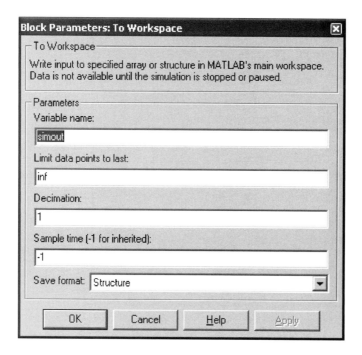

Figure A4.21. Dialog box for the To Workspace block.

Before we rerun the simulation, double-click on the new block and examine the settings, as shown in Fig. A4.21.

The five fields are described briefly:

Variable name: This is self-explanatory, the name of the MATLAB workspace variable that is created on completion of the simulation. The default is simout.

Limit data points to last: For very large simulations spanning long time periods, the amount of data that is stored may be problematic, depending on the amount of RAM in your computer. Therefore there is a capability to limit the storage to the last *n* data points. By default, the block stores all computed points.

Decimation: Related to the problem of storage, you have the option of storing every *n*th data point instead of every point computed, which is the default.

Sample time: By default, the workspace variable will receive data at the points in time when Simulink computes them. Because most Simulink algorithms are variable-step-time algorithms, the time increments are rather varied for most simulations. If a constant time increment between output data is desired, that can be accomplished here.

Save format: In Release 12 and later versions of MATLAB/Simulink, several new data structures (more complicated than matrices) were introduced. By default, one of these data structures is used to store the simulation results to the workspace. The drop-down menu allows two additional choices, the simplest of which is "Array." If you are following along with the tutorial, select "Array" now.

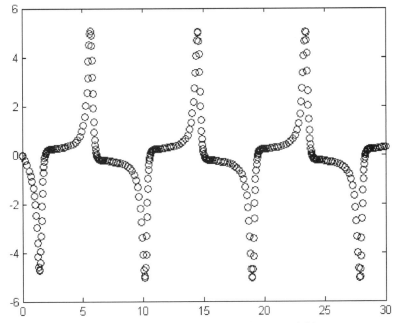

Figure A4.22. Plot of simulation output from workspace variable.

Run the simulation and click over to the MATLAB window. Inspect the "Workspace" subwindow in MATLAB (see discussion in Appendix 3). You should see two new variables in the workspace, tout and simout. The first variable, tout, is a one-dimensional vector containing the values of time at the points where the outputs are computed. The second, simout, contains the values of dy/dt (in the first column of the variable) and y (in the second column). Note that the order in which they are stored in simout is the same as the order in which they are connected to the MUX block in the simulation window.

As a final illustration, use the MATLAB plot command to plot the derivative of the output against time, but using a small circle to indicate each computation as a separate point without connecting lines,

```
>>plot(tout,simout(:,1),'o')
```

the results of which are seen in Fig. A4.22.

A4.4 CONCLUSION

In this tutorial, we've attempted to give you enough of an orientation to implement moderately complicated models in Simulink. Like any comprehensive engineering tool, the full potential is achieved through a combination of use, repetition, and exploration. The help facility that accompanies Simulink, like the one for MATLAB, is extensive and educational. The interested reader is well advised to explore.

Index

A-type elements, 4–7, 18, 31, 168, 198, 202, 219, 220, 222, 225
A-type variables. *See* Across variable
ac current, 172, 251
ac induction motor, 251
Acceleration
  rotational, 30
  translational, 16
Accumulator, 219
Across variable, 4–7, 31, 65, 170, 198, 213, 220, 222, 250, 254
Aliasing, 399
Amplifier
  operational, 179–186
  pneumatic, 239–243
Amplitude
  complex, 303
  Real, 303
Analytical solution, 81–111
Angle, shaft, 30
Antialiasing filter, 399
Asymptotic Bode diagrams, 309–311
Asymptotic frequency characteristics, 309–311
Asymptotic stability, 332
Automobile
  motion, 17–18
  steering, 330
  suspension system, 48

Backward-difference approximation, 391, 422
Bandwidth, 310
Bilinear transformation, 417
Biot number, 203
Block diagrams
  simulation, 143–147, 463
  system, 2
  transfer function, 286–293
Bode diagrams, 307–314
  asymptotic, 309–311
Break frequency, 310

Capacitance
  electrical, 170–171, 310
  fluid, 219–222
  pneumatic, 236
  thermal, 202

Capacitor, 4, 168, 170
  time-varying, 186
Capillary resistor, 223
Centrifugal pump, 253
Characteristic equation and roots, 83, 84, 93–98, 106–109
Charge. *See* Electrical capacitor
Classical solutions. *See* Analytical solution
Closed-loop control system, 356–357
Coefficient block, 143
Compatibility law, 23, 225
Compensation, 378–382
  feedback, 379–382
  feedforward, 379
  series, 379–382
Compensator
  lag, 380–382
  lag–lead, 381–382
  lead, 380–382
Complex number, 93
Complex plane, 274
Complex roots, 93–98
Complex variable
Conduction, 199–200
Continuity law, 225
Control laws
  on-off control, 366–367
  proportional (P) control, 367
  proportional–derivative (PD) control, 368
  proportional–integral (PI) control, 367
  proportional–integral–derivative (PID) control, 273, 368, 421
Controllers
  analog, 365–382
  digital, 421–423
Convection, 200–201
Convective heat transfer coefficient, 200
Corner frequency, 170–171, 310
Coulomb friction, 131, 154
Coupling coefficient, 249, 253
Critical damping, 96, 100
Current source, 169

D-type elements, 4, 7, 22, 33, 168, 173, 198, 219, 223
Damped natural frequency, 96, 415

Damper, 4
  rotational, 32, 39
  translational, 22
Damping ratio, 94–104, 415
dc motor, 251, 257
Decay ratio, 103
  logarithmic, 104
Decibel, 307
Delay time, 103, 369
Density
  mass, 222
  weight, 236
Differential equations
  linear, 54–71
  nonlinear, 34–44, 71–76
  solutions, analytical, 81–111
  solutions, numerical, 120–138
Digital control system, 410–424
Digital signal, 390
Digital simulation. *See* Simulation
Dirac's delta function, 87, 398
Direct-transmission matrix, 65
Discrete-time system, 389–407
Discretized model of continuous system, 389–407
Distributed digital control system, 411
Disturbance, 361–364
  pulse transfer function, 420
  steady-state sensitivity, 361–364
  transfer function, 362
Divider, 144
Dominant roots, 106–109

Electric-field energy, 168, 171
Electrical capacitor. *See* Capacitor
Electrical circuits, 168–189
Electrical compensator. *See* Compensator
Electrical inductor. *See* Inductor
Electrical resistor. *See* Resistor
Electrical systems, 168–189
Energy
  converters, 249–254
    electromechanical, 250, 257
    fluid mechanical, 252
  dissipation, 4, 7
    elements, 7
electric field, 168, 171
  kinetic, 6, 18
  magnetic field, 4, 168, 172
  potential, 4, 19
  storage, 4–7
Equation(s)
  characteristic, 82–84, 93, 106
  differential, 54–76, 81–111
  input–output, 55–60, 68–71
  Laplace transformed, 276–277, 437
  state variable, 61–71
  s-transformed, 273–276
Error
  in numerical integration, 123–132
  steady-state, 357–361, 418–421

Error pulse transfer function, 418
Error transfer function, 358
Euler's equation, 274, 428
Euler's method, 121–123
  improved, 125–126
Expansion
  partial fraction, 403, 437
  Taylor series, 34–44
Exponential input function, 274–276

Feedback
  compensation, 378–382
  intentional or manmade, 329–331
  natural, 7, 146, 273
  negative, 329–331
  path, 146
Feedforward compensation. *See* Compensation
Field
  electric. *See* Electric field
  magnetic. *See* Magnetic filed
Filter
  antialiasing, 399
  Low-pass, 310
Final-value theorem, 358–364, 433
First law of thermodynamics, 202
First-order models, 84–92
Flow rate
  charge (electric), 168–169
  heat, 198–201
  mass, 235
  volume, 220–228
  weight, 235
Fluid
  capacitance, 219–222
  capacitor, 220–222
  coupling, 33, 39, 236–237
  inertance, 4, 222, 237
  inertor, 222, 237
  pump, 253
  resistance, 4, 223, 237–238
  resistor, 223, 237–238
  sources, 224
Flux linkage, 171, 187
Force, 16
Forcing function
  impulse, 87
  ramp, 360
  sinusoidal
  step, 85, 359
Fourier series, 307, 427–431
Fourier transform, 427–431
Free-body diagram, 19
Free response, 82, 84, 93
Frequency analysis, 302–323
Frequency response, 302–323
Frequency, natural. *See* Natural frequency
Friction
  Coulomb, 131, 154
  nonlinear, 39, 131, 153
  viscous, 22, 32

Gain, 341, 467
Gain margin, 341
Gating function, 398
Gear
    ratio, 115
    train, 115
Guard filter. *See* Antialiasing filter

Huen's method, 125
Higher-order models, 106–109
Holding device, 396–400
Homogeneous solution, 82, 84, 93, 100, 109
Hooke's law, 19
Hurwitz determinants, 335
Hurwitz stability criterion, 334
Hydraulic orifice, 224, 227, 230
Hydraulic pump, 253
Hydraulic resistance. *See* Resistance
Hydraulic turbine, 253

Impulse function, 87
Impulse response, 90, 94
Incremental resistance, 175
Incremental spring constant, 20, 39
Incremental variables, 37–44
Inductance, 4, 168
Inductor, 168, 171–172
    time-varying, 187
Inertance. *See* Fluid inertance
Inertia
    rotational, 30
    translational. *See* Mass
Initial conditions, 82, 142, 144, 148, 276,
        437
    at $t = 0-$, 437
    at $t = 0+$, 82
Initial-value theorem, 433
input, 1, 55
    exponential, 274–276
    matrix, 65
    simulation, 142
    variables, 1, 55
Input–output models
    of continuous models, 55–60, 71–76
    of discrete-time systems, 390–394
Instability
    numerical computation, 133–138
    system. *See* Stability
Integration
    numerical, 120–138
    role of, 4, 18
    step-size, 129–132
Integrator, 144, 186
Inverse Laplace transform, 432
Inverse z transform, 402

Kinetic energy, 6, 18
Kirchhoff's current law, 170
Kirchhoff's voltage law, 170

Lag. *See* Time lag
Laminar flow, 224
Laplace transform, 276–277, 437
    basic properties, 433
    definition, 432
    equation, 199
    inverse, 432
    table of basic transforms, 435
Linearization, 34–44
    error, 43
Logarithmic decay ratio, 104
Loop method, 177
Loop variables, 177
Low-pass filter, 310
Lumped-parameter models. *See* Model
Lyapunov's definition of stability, 332

Magnetic-field energy. *See* Energy
Mapping between the s and z planes,
        412–418
Marginal stability, 94, 339
Mass, 16–19
Mathematical model. *See* Model
Matrix
    direct transmission, 65
    input, 65
    output, 65
    state, 65
Maximum overshoot, 103
Mechanical systems, 14–45
    rotational, 16–30
    translational, 30–34
Mixed systems, 249–261
Model, 2
    continuous, 4, 54–76
    discrete, 4, 390
    discretized, 4, 390
    distributed, 4
    Input–output, 55–60, 71–76
        of discrete-time systems, 390–396
    Linear, 4, 54–76
    Linearized, 34–44
    Lumped, 4
    Nonlinear, 4, 71–76
    Second-order, 92–105
    State
        of continuous systems, 61–68
        of discrete-time systems, 394–396
    Stationary, 251–252
    time-varying, 4
Motion
    rotational, 30–34
    translational, 16–30
Motor
    ac induction, 251
    dc, 251, 257
    electrical, 251–252, 257
    hydraulic, 252, 259
Multi-input, multi-output system, 60
Multiplier, 144

Natural frequency, 60, 94–100, 311
Newton's laws, 16
Node method, 176
Node variables, 176
Nonlinear system. *See* Model
Normal operating point, 37
Numerical integration, 120–138
Nyquist frequency, 398
Nyquist stability criterion, 338–341

Octave, 311
On–off control. *See* Two-position control
Open-loop control system, 356
Operational amplifier, 179–186
Ordinary differential equations, 54–76
    analytical solution, 81–111
    numerical solution, 120–138
Orifice
    hydraulic, 224, 227, 230
    pneumatic, 237
Oscillations, 94, 101, 370
Output
    matrix, 65
    variables, 2, 55, 142
Overdamped system, 101
Overshoot, 103

Partial differential equations, 4, 15, 199
Partial fraction expansion method, 403, 437
Particular integral, 82
Passive systems, 329
Peak time, 102
Period of oscillations, 101
Periodic functions, 307, 427
Phase, 303, 307
Phase margin, 342
Pneumatic amplifier, 239–243
Pneumatic capacitance, 236
Pneumatic inertance, 237
Pneumatic orifice, 237
Pneumatic resistance, 237–238
Pneumatic systems, 235
Polar plots, 317–319
Poles, 344–347, 412–418
Porous plug resistor, 237
Power, 249–254
Pressure
    absolute, 220
    gauge, 220
    reference, 220
Proportional (P) control, 367
Proportional–derivative (PD) control, 368
Proportional–integral (PI) control, 367
Proportional–integral–derivative (PID) control,
        273, 368, 421
    position form, 422
    velocity form, 422
Pulse function, 89
Pulse transfer function, 405–407

Pump
    centrifugal, 253
    hydraulic, 253

Quantization, 390

Rack and pinion, 115
Radiation thermal, 201
Ramp input, 360, 419
Reference
    frame of, 16
    nonaccelerating, 16
    pressure, 220
    velocity, 16
    voltage, 169
Resistance
    electrical, 169, 173
    hydraulic, 220, 223
    incremental, 175
    pneumatic, 237–238
    thermal, 202
        conductive, 203
        convective, 203
        radiative, 203
Resistor
    capillary, 223
    electrical, 169, 173, 186
    fluid, 223, 237–238
    nonlinear, 174
    time-varying, 186
Rise time, 103
Root-locus method, 344–347
Rotational system, 16–30
Round-off error. *See* Error, numerical integration
Routh stability criterion, 336–337
Runge–Kutta method, 126–129

s domain, 274–276, 413
s plane, 274–276, 413
    mapping to z plane, 413
Sampling device, 396–400
Sampling frequency. *See* Sampling time
Sampling time, 392, 398
Second-order models, 92–105
Self-tuning, 370
Sensitivity to disturbances, 361–364, 420
Series compensation, 379–382
Sevoactuator, electrohydraulic, 228–235
Settling time, 103
Shannon's theorem, 398
Signal
    digital, 390
    periodical, 427
    sampled-data, 390
    sinusoidal, 275
Simplification (dominant roots), 106–109
Simulation, 141–164
Simulation block diagrams, 143–147, 463
Single-input–single-output system, 55

Sinusoidal excitation, 275
Small-perturbation analysis, 37–44, 203, 240
Source
    current, 169
    flow, 224
    pressure, 224
    voltage, 169
Specific heat, 199, 202
Spring
    constant, 4
    real, 14
    rotational, 31
    stiffness, 19
    translational, 19–22
Stability
    asymptotic, 332
    of discrete-time systems, 415
    gain margin, 341
    Hurwitz criterion, 334
    Lyapunov's definition, 332
    marginal, 94, 339
    necessary condition for, 334
    numerical computation, 133–138
    Nyquist criterion, 338–341
    phase margin, 342
    Routh criterion, 336–337
State, 61, 62
    matrix, 65
    model equations, 63, 71–76
    model of discrete-time systems, 394–396
    output equations, 64
    space, 61, 62
    trajectory, 62
    variables, 61
        auxiliary, 69
        selection, 65
    vector, 61, 62
Static position error coefficient, 359, 419
Static velocity error coefficient, 360, 420
Stationary system, 251–252
Steady sinusoidal excitation, 302–323
Steady–state control error, 357–361, 418–421
    in response to ramp input, 360–361, 419–421
    in response to step input, 359–360, 419
Steady-state disturbance sensitivity, 361–364, 420
Stefan–Boltzmann law, 201
Step function, 85
Step response, 86, 98
Stiff system, 133–138
Superposition, 89
System
    continuous, 4, 54–76
    discrete-time, 4
    dynamic, 1–4
    electrical, 168–189
    fluid, 219–244
    linear, 4
    lumped-parameter, 4

mass–spring–damper, 25–26, 104–105, 145
mechanical, 14–45
mixed, 249–261
model, 1–4, 54–76
nonlinear, 4
pneumatic, 235
spring–damper, 23
stable, 329–352
stationary, 251–252
thermal, 198–213
type, 4, 359–361, 419–420

Table of Laplace and z transforms, 435
Taylor series, 34–44
Thermal capacitance, 203
Thermal conductivity, 199
Thermal resistance, 198, 202–204
Thermal systems, 198–213
Third-order models, 106–109
Through variable, 4–7, 22, 32, 65, 170, 172, 213, 222, 254
Time constant, 85
Time delay, 103, 156, 369
Time step size, 129–132, 142, 150, 390
    See also Sampling time
Torque, 30–34
Transducer
    electromechanical, 250, 257
    energy converting, 249–254
    fluid mechanical, 252
    mechanical translation to mechanical rotation, 249
    signal converting, 254
Transfer function, 273–299
    block diagrams, 286–293
    closed-loop, 289, 334
    frequency-response, 302–307
Transformer, 172
Transient response, 84–111, 364–365, 412–418
Translational systems, 16–30
Truncation error. See Error, numerical integration
T-type element, 4–7, 22, 32, 168, 219, 222
T-type variable. See Through variable
Tuning rules of Ziegler and Nichols, 368–370
Turbine, 253
Turbulent flow, 224
Two-position control, 366–367
Type of system, 4, 359–361, 419–420

Underdamped system, 99–100
Unit impulse function. See Impulse function
Unit impulse response. See Impulse response
Unit step function. See Step function
Unit step response. See Step response
Unstable system. See Stability

Valve, 228
Variable-displacement hydraulic motor, 259

Velocity
  rotational, 30
  translational, 16–19
Vibration absorber, 28
Viscous friction, 22, 32
Voltage source, 169
Volume chamber, 221

Weighting sequence, 406
Wind-up of a control error, 422

z domain, 400–407
z transform, 400–407
  inverse, 402
  one-sided, 400
  table of basic transforms, 435
  theorems, 402
Zero-order hold (ZOH), 396–400
Zeros, 347
Ziegler–Nichols tuning rules,
  368–370